PERSPECTIVES FOR PEAS AND LUPINS AS PROTEIN CROPS

WORLD CROPS:
PRODUCTION, UTILIZATION, AND DESCRIPTION

volume 8

Other volumes in this series:

1. Stanton WR, Flach M, eds: SAGO The equatorial swamp as a natural resource. 1980. ISBN 90-247-2470-8.
2. Pollmer WG, Phipps RH, eds: Improvement of quality traits of maize for grain and silage use. 1980. ISBN 90-247-2289-6.
3. Bond DA, ed: Vicia faba: Feeding value, processing and viruses. 1980. ISBN 90-247-2362-0.
4. Thompson R, ed: Vicia faba: Physiology and breeding. 1981. ISBN 90-247-2496-1.
5. Bunting ES, ed: Production and utilization of protein in oilseed crops. 1981. ISBN 90-247-2532-1.
6. Hawtin G, Webb C, eds: Faba Bean improvement. 1982. ISBN 90-247-2593-3.
7. Margaris N, Koedam A, Vokou D, eds: Aromatic plants: Basic and applied aspects. 1982. ISBN 90-247-2720-0.

Perspectives for Peas and Lupins as Protein Crops

Proceedings of an International Symposium on Protein Production from Legumes in Europe, organized by University of Naples, held in Sorrento, Italy, 19–22 October 1981

Sponsored by the Commission of the European Communities, Directorate-General for Agriculture, Coordination of Agricultural Research

edited by

R. THOMPSON
Scottish Crop Research Institute, Dundee, Scotland

R. CASEY
John Innes Institute, Norwich, England

1983

MARTINUS NIJHOFF PUBLISHERS
THE HAGUE / BOSTON / LONDON
for the
COMMISSION OF THE EUROPEAN COMMUNITIES

Distributors:

for the United States and Canada
Kluwer Boston, Inc.
190, Old Derby Street
Hingham, MA 02043
USA

for all other countries
Kluwer Academic Publishers Group
Distribution Center
P.O.Box 322
3300 AH Dordrecht
The Netherlands

Library of Congress Cataloging in Publication Data CIP

Main entry under title:

Perspectives for peas and lupins as protein crops.

(World crops ; v. 8)
1. Peas--Congresses. 2. Lupine--Congresses.
2. Lupine--Congresses. 3. Plant proteins--Congresses.
I. Thompson, R. II. Casey, R. III. EEC Programme of Coordination of Research on the Improvement of the Production of Plant Proteins. IV. Università di Napoli. Plant Breeding Institute. V. Commission of the European Communities. Coordination of Agricultural Research. VI. Series: World crops (Hague, Netherlands) ; v. 8.
SB343.P47 1983 635'.656 82-23207
ISBN 90-247-2792-8

ISBN 90-247-2792-8 (this volume)
ISBN 90-247-2263-2 (series)

Publication arranged by:
Commission of the European Communities
Directorate-General Information Market and Innovation,
Luxembourg

EUR 8295 EN

Proceedings prepared by
Janssen Services, 33a High Street, Chislehurst, Kent BR7 5AE, UK

PRINTED IN THE NETHERLANDS

CONTENTS

PREFACE

Previous seminars in this series, sponsored by the EC Common Research Programme on Plant Protein Improvement, have been concerned mainly with *Vicia faba*. A broader approach to the question of protein production from legumes in Europe has for some time been considered prudent. The seminar reported here was therefore timely and was held in Sorrento, Italy on 19th - 22nd October 1981. Organisation of the meeting, including selection of the topics and speakers, was expertly done by Professor Monti and his colleagues in the University of Naples, to whom much gratitude is due.

The main aim of the seminar was to explore the potential of peas and lupins as major components in the protein budget of the CEE. Contributions on both peas and lupins are drawn from disciplines spanning basic physiology to field crop production and biochemistry to nutritional value. Current activity in plant breeding is covered and additional breeding objectives are identified.

SESSION I

Chairman: G.T. Scarascia Mugnozza

OPENING

L.M. Monti *(Italy)*

It is a great pleasure for me to welcome all of you here to this seminar, which has been organised within the framework of the EC Plant Protein Programme. In 1979 the EC Plant Protein Programme Expert Committee decided to organise this meeting in order to discuss perspectives for peas and lupins as protein crops in Europe.

Highest priority has been given to *Vicia faba*, as a continuation of the research carried out on this species during the previous five year programme. Peas and lupins were considered as species which needed less research because successful breeding had already been achieved in some European countries. But it was decided that it was important to hold a meeting on peas and lupins to assess present progress and future potential.

In recent years there has been increasing interest in peas as a protein crop, and so the purpose of this seminar is to discuss the potential of the crop in Europe and to decide on priorities among research objectives. The conclusions of this seminar will be presented to the Plant Protein Expert Group, of which M. Calet is Chairman, and then to the Permanent Committee for Agricultural Research of the EC. Professor Scarascia-Mugnozza is a member of that Committee.

Director Magnificus is unable to be with us as he had to attend an important meeting and, for the same reason, the presence of the Dean of the Faculty of Agriculture, Professor Mariello, will be delayed.

I would like to take this opportunity to acknowledge the financial support of the EC for the organisation of this seminar. We have also received financial support from the University of Naples, from the Campagna Region and from the Province of Naples.

Once again, I welcome you here, and hope that the next few days will be both fruitful and enjoyable. This part of Italy is famous, but not as yet for peas and lupins.

I now call on Professor Scarascia-Mugnozza to address the seminar. He is President of the Committee for Agricultural

Research of the Italian National Research Council, and a member of the EC Permanent Committee for Agricultural Research.

G.T. Scarascia-Mugnozza *(Italy)*

It is also my privilege and pleasure to welcome you to this meeting in this lovely part of southern Italy. I welcome you on behalf of the Agricultural Research Committee of the National Research Council of Italy, and also as a member of the Italian delegation on the EC Steering Committee for Agricultural Research. I wish you a very fruitful and successful meeting.

We must recognise that great scientific effort over recent decades has been devoted to the improvement of crops such as cereals and the success of this work is well known, whilst yield increases for leguminous crops (except for soya) have been limited and inconsistent (Table 1).

TABLE 1

WORLD AVERAGE YIELD (t/ha)

Crops	1948 - 52	1975
Glycine max	0.99	1.47
Pisum sativum	0.85	1.08
Vicia faba	0.96	1.17
Lupinus spp.	0.54	0.67
Lens culinaris	0.53	0.64
Cicer arietinum	0.53	0.59
Phaseolus vulgaris	0.44	0.54
Zea mais	1.63	2.82
Oriza sativa	1.63	2.44
Triticum spp.	1.11	1.56

The Italian acreage and production of leguminous crops for 1955/56 and 1975/76 are given in Table 2. There has been a clear reduction in production of all these species, but particularly in lupins and *Pisum*. The reduction of both acreage and production is due to a shortage of agronomic, agrotechnical, genetic and breeding research on these species.

TABLE 2

HECTARAGE AND PRODUCTION DURING 1955 - 56 AND 1975 - 76 IN ITALY

Crops	Mean 1955 - 56		Mean 1975 - 76	
	Area 000 ha	Prod. 000 q	Area 000 ha	Prod. 000 q
Pisum sativum	18	105	4	47
Vicia faba	547	3 742	203	2 454
Phaseolus vulgaris	423	1 461	71	1 079
Cicer arietinum	34	488	16	174
Lens culinaris	25	140	2.5	25
Lupinus spp.	48	409	6	78

The continued reduction and eventual disappearance of these crops would have far reaching consequences. The residual soil fertility of a legume crop is between 40 - 50 kg N/ha. The substitution of a cereal crop for a legume crop requires increased levels of chemical fertilisers which involve the use of valuable fossil fuels and lead to increased dangers of pollution. A valuable topic for a future meeting would be to examine the consequences of such changes on the present balanced agro-ecosystems, especially in semi-arid environments such as this Mediterranean region. It is, therefore, a matter of some urgency to promote research on these crops. As you know, the CEE has already organised technical meetings on *Vicia faba* and I am happy to see here colleagues who took part in those meetings.

The EC interest in these crops is also motivated by the need to reduce European dependence for plant protein on external markets, for animal feed as well as for human consumption.

These are only some of the reasons which make this meeting important and topical. The potential of two species will be discussed. Peas are already well known, and grown mainly for canning and for industrial purposes. Research into this species is already being carried out in all European countries. In Italy, for instance, for some 20 years the National Research Council has promoted and supported research on peas, mainly for manufacturing purposes but recently for protein production. Our host, the organiser of this meeting, Professor L. Monti, has worked very intensively on this species, and three cultivars resulting from this work are now under registration: Proteo, VP, Paisa.

Lupins have practically disappeared from farms in the European Mediterranean countries, although some species are of Mediterranean origin, such as *L. luteus*, *L. albus* and *L. angustifolius* and these are probably the most important.

I would like to take this opportunity of expressing thanks and appreciation on your behalf to Professor Monti for having organised this meeting, and once again I wish you all a fruitful meeting and a pleasant stay here in Sorrento.

C. Calet *(France)*

First of all, I would like to congratulate Professor Monti on the excellent organisation of this meeting and on the choice of such a lovely place in which to hold it.

It is well known that human consumption of meat and other animal products in developed countries increases each year, and it is necessary to produce more cows, pigs and poultry to satisfy this need. Greater efficiency of production is required to avoid too great an increase in the price of these products. In order to achieve this, animals must be fed well balanced diets containing a large amount of protein.

Proteins for this purpose are obtained from oil cakes, mainly soyabean oil meal, which has traditionally been imported from the United States, but now principally from Brazil. The cost of this importation is a heavy burden to the Community, as I believe Dr. Schiratti will tell you later. The potential within the EC for the production of plant protein is also very great, and this is why the Community decided, five years ago, to establish a Plant Protein Improvement Programme in order to encourage agricultural production and utilisation of protein rich seeds and plants. Two ways of doing this were proposed:

- financial advantages for farmers and feed manufacturers who produce and utilise European proteins
- a greater effort to encourage research in this field.

In connection with the second of these, a Working Group was set up to study research programmes and co-ordinate the work of the different European laboratories involved in protein studies. Therefore, separate funds are involved: one is used for experimental studies on plant breeding and culture, including genetic, physiological, pathological and microbiological aspects of crops such as cereals, legumes and oilseeds. Secondly, funds are devoted to co-ordination activities such as meetings, workshops and seminars for the dissemination and discussion of research. These meetings are most important, acquainting participants with up to date results and also identifying gaps in our knowledge. Priorities may then be worked out for updating research programmes, which can then be presented to the EC Agricultural Research Committee.

Such meetings have already been held on a number of different plant and seed proteins. The present meeting on peas and lupins is particularly important. On the one hand, peas are increasingly being cultivated in Europe, principally in France and, on the other hand, the development of lupins could follow as an important source of seed protein given EC backing. We expect consideration of this last point in the near future and results are required for this purpose.

I am sure that this will be a very fruitful seminar, and that we shall obtain interesting and useful results which will allow increased economic independence by the EC in relation to protein.

ECONOMIC PERSPECTIVES FOR PEAS AND LUPINS IN THE EC

M.G. Schiratti
Head, Division for Oil and Protein Crops in the Commission of the European Communities
Brussels, Belgium.

On my way here from Brussels, travelling in the pouring rain, I was thinking to myself, "What will be my role here in Sorrento, as I am an economist, and you are all specialists in a research field?". We live in two different worlds, with different problems. I realised, however, that I could give you some information which might be helpful to you in your daily work. Professor Scarascia-Mugnozza has said that research is urgently needed on these crops. However, research cannot be an end in itself. It is only useful if there is an economic base from which the results of research can be exploited to allow farmers to choose crops and to expand production, to expand their farms and to utilise better the available land. From that point of view it is quite clear that there is a link between the economists and the research workers. We must tell you what kind of guidelines we intend to follow in the future and you can help us to reach the results for which we are aiming.

Over the next three days you will be discussing peas and lupins to see what can be done to improve productivity. I think it will be interesting for you to see the future perspectives for peas and lupins within the Community, and the guidelines which you should keep in mind when you carry out your very important work.

We must look at the Common Agricultural Policy. Everybody talks about this, mainly to say that they don't like it and that it should be changed. Even if it is not perfect, I think that the Common Agricultural Policy has had a very important role in the past, and that it will remain as a part of our life in future. We have to take it into consideration, therefore, and should not forget the guidelines of this Policy.

As you know, the Policy started in 1962 with the Cereals Common Market Organisation, followed by other Common Market Organisations such as Sugar Beet, Milk, Meat and Fruit and

Vegetables, which in turn were followed much later by Fats and Oils, and Wine. In fact, there has never been any discussion of a protein policy in the EC. There are a certain number of reasons which justify this choice on the part of the Agriculture Ministers of the Community.

Initially, for historical, political, social, economic and trade reasons, attention was concentrated on the great crops: cereals, sugar beet, meat and milk. After the destruction by the Second World War, we needed to put agriculture back in shape and it was inevitable that we should think of traditional crops, in particular cereals and meat. Politically, member countries had support for all these products, so that it was easy to get agreement on transforming national policy into Community policy, because all that was involved was moving an existing policy at national scale to a different scale. A political and social reason was the number of people involved in the production of these crops. Two million families are involved in the production of milk. That is at least six million votes, and politicians are very sensitive to votes. Finally, there were trading reasons. We could not produce all we needed, so we had to choose what we should support and what we should import. Taking into account our trade engagements we decided to depend on imports to cover our needs of proteins.

In the protein field, then, the only products which were supported from the beginning were rapeseed and sunflowers. The reason why the EC decided to support them was that they were already supported in certain member states. From the beginning, however, the intention of the Community was quite clear: no increase in production. Production was protected but that protection was limited.

What was the reason for this policy? It was very simple. We were dependent for our needs on imports from other countries. We decided that since we had limited land, we would choose to utilise that land for the big crops I have listed, and to import the protein we needed to feed our animals. Furthermore, we did not realise at the time that there would be such a development of the need for those proteins within the EC. We did not realise that there would be an incredible increase in demand for soya cake and cake in general.

Another reason was that in the 1960s the supply of cake was plentiful and the prices were stable. Why bother producing something that we could buy on the world market in unlimited quantities and at stable prices?

This situation prevailed until 1970, when something happened which made us wake up. In 1973 production in the United States was below forecast and the US government became scared. They stopped the export of soya from the United States, and so suddenly we discovered that we were very vulnerable. It was enough for the Dakotas and the other states where soya is grown to have one bout of bad weather, reducing the crop, to create a problem in the EC. Soya went from $200/ton to $800/ton in a few weeks.

For the first time, the Council for the European Community decided that it was time to look into the matter, and asked the Commission to prepare a report on the supply of protein for animal feed in the Community, the present situation and the perspectives for reducing our dependence on other countries. I had the opportunity of helping to prepare that report in 1973, and what we said in that report was firstly that there was no possibility of reducing in any great degree our dependence on imported products. For that reason, we should try to diversify imports and at the same time try to obtain guarantees from the exporting countries. At the same time, we said that although the result would not be a dramatic reduction in our import needs, we should try to increase our production of protein in order to create a minimum of production which could help to soften blows when new crises arose. The Council accepted this point of view.

Nothing came of the first part of the report. We are now importing from Brazil, but this is not a result of our own policy, it was because Brazil decided to grow more soya. We did, however, do something to increase our own production of protein and protein rich products.

First of all, we began to increase support for rapeseed and sunflower, giving them higher prices than products such as cereals, for instance, which compete with them. The result was that the production of rapeseed which was only 400 000 tons when we began has increased to 2 million tons. This is still very little in comparison with our requirements, but it is a step on the way. We still have some problems with varieties of sunflower seeds, and results which are sometimes misleading, but nevertheless, this year we will have a crop of about 0.5 million tons, and the prospects for the future are encouraging. We did start supporting the production of dry fodder, and we tried to support soya, but with very unhappy results. We also tried to support linseed, but the results there were not very good either. And last but not least we decided to support the production of peas and beans for animal consumption. We were convinced that the consumer could pay a reasonable price for peas and beans for human consumption, but there was no hope of persuading the feed manufacturers to use our peas and beans in feed if the price of those peas and beans was not competitive with soya. We did realise that we had to pay the farmers the difference between a reasonable price which would account for the cost of production and the price which the feed manufacturer would pay, taking into consideration the price at which soya bean is sold on the market.

As a result of the 1978/79 campaign, we decided to support the consumption of peas and beans for animal feed. We had to put some limits on that scheme, because as with every scheme into which money is paid, there has to be a control to avoid fraud. So we had to limit the aid to peas and beans used by feed manufacturers, which is where control is possible, and the Italians were not happy because that meant that a large proportion of production in Italy was excluded from that scheme. We are now trying to find ways of making the quantities which are consumed on the farm to benefit from this scheme, but we must be sure that the controls are possible and are feasible, taking into consideration the realities of the Italian situation.

Each year, we fix the price of the peas and beans at a level at which the producer can obtain a reasonable price which takes into consideration the costs of production. We compare that price with the price which can be paid by the feed manufacturers, taking into consideration the price of the competing proteins, especially soya cake. If the price of the feed manufacturer is below that of the producer, we give the feed manufacturer aid which allows him to pay the farmer a price which we consider fair. This may sound simple, but I can assure you that it is complicated.

When a farmer grows peas and beans, he very often does not know whether they will be used for human or animal consumption. The Council decided that we shouldn't try to make the import of peas and beans more difficult so as not to upset our friends who export peas and beans. This system should allow us to pay the feed manufacturer for peas and beans which have been grown in the EC, whilst we are unable to control what happens to the imported products. As a result we have to create a complicated system involving a lot of paper, and I am not sure that all the papers which we ask the interested party to fill in give us a sufficient guarantee of the quantity which the aid benefits. Since we are an evolving community, I hope that we will be able to improve this system in the not too distant future and create a system which is easy to administer and as effective as the system we have now.

We are looking into the problem of the quantities which are consumed at producer level to see whether it is possible to find a system which will allow those quantities to benefit from the aid without too much money being spent. We are also looking into the problem of imports in order to avoid paying aid for peas and beans which have been imported from other countries. In addition, we are looking into the problem of peas and beans which go for human consumption because we realise that very often, at producer level, problems are created and our system is a factor which makes the utilisation of peas and beans for human consumption more difficult than it was before our system was created.

What is the result of the system which was introduced in 1978? I think some figures are necessary to explain this. In 1978/79, we paid aid for the following quantities: 76 000 tons of peas, 89 000 tons of beans, a total of 165 000 tons. 1979/80 saw the first important increase in peas: 154 000 tons of peas and 116 000 tons of beans, a total of 270 000 tons. We saw an even bigger success in 1980/81. We paid aid for 229 000 tons of beans and for 102 000 tons of peas, a total of 331 000 tons. We expect an even greater quantity in the future.

The two most important producing countries are France and the UK. Most of the EC production comes from those two countries, and whilst production in the UK has remained more or less stable, the production in France has seen an important development. The Netherlands and Denmark have a certain production, but I would say that three quarters of the production is concentrated in France and the UK.

Although the production concentration is here, the utilisation is not at all concentrated in those two countries. The country where there is the highest utilisation of peas and beans produced for human consumption is the Netherlands. Dutch feed manufacturers saw the interest in utilising those crops immediately, and with the flair typical of all Dutch traders, they asked for the produce. French and British feed manufacturers, who were in a better position to take advantage of this production, took a couple of years to realise that the product was usable and could be included in the formula for compound feed. As a result, we are now seeing an increase in the utilisation of these products, especially in France which is still catching up with the Dutch. The Belgian feed industry has also been using a large quantity of peas and beans, although only half what the Dutch are using, and interest is now being shown by the UK feed manufacturers.

I was in East Anglia last June and I had a meeting with all the most important feed manufacturers in the UK. I got the impression that they were more and more interested in using the products, but they were complaining that the paper work was still excessive. I accept that complaint, because I know that to someone who is not an administrator like myself, paperwork is just a nuisance.

Without accepting all the farming organisations' programmes for the future (in France, the farming organisation is talking of 1 million tons of peas and beans) we still think that the door is open for a development in production of peas and beans, but especially of peas. Beans seem to have encountered a resistance on the part of the feed manufacturers, and the possibilities seem to be less encouraging. I would say that the greater our production, the greater our chances of interesting industry. In the past, industry has said to us, "The quantities are too small".

In order to understand the problem, we have to look at some figures. I am not talking about the 50 million tons of cereals used by the feed manufacturers, but remaining in the protein sector, I would like to remind you that in 1981 we will be using around 23 million tons of oil cakes, around 2 million tons of dry fodder, almost 4 million tons of corn gluten feed and not far off 2 million tons of fishmeal and meat meal. We are talking about millions of tons, and when you compare that with a production of 150 000 tons of peas, that production can be seen to be too small to interest our feed manufacturers. Production needs to become greater in order to attract the industry, and we are now approaching the size which will interest industry. As we produce more, we will have more and more feed manufacturers who are ready to introduce peas and beans into the compounds, and to buy the products if the price is right.

There is a drawback in this rosy picture, and that is the cost. The greater the production, the greater the cost. We started with a very small budget and whilst we are not big spenders, attention is being focussed more and more on our products. We are far from being the big spenders of the milk sector, and we are far from being the big spenders of the cereal sector, or of the fats and oil sectors. The cost is increasing, and I expect that we will spend about 50 - 60 million units of account this year. As you know, the cost of agricultural policies is now under criticism. We are spending a lot of money, sometimes not as efficiently as we should, and as a result, last year the Heads of State gave the Commission a mandate to revise all the mechanisms of the agricultural policy and to find a way to try to obtain a maximum and a minimum cost. We must now become thrifty people.

The Commission has already given a paper, earlier in the year, in which a policy is outlined which should reduce the cost whilst still maintaining an efficient agricultural policy, concentrating on the sectors where cost efficiency is the highest. This paper will shortly be followed by a second paper in which the Commission analyses, product by product, the problems and what should be done to reduce unnecessary cost. For instance, we should reduce production where we are over producing, and push production where the cost per hectare or per ton is lower, or where we need to increase the production for economic, social, or political reasons.

The question is, what role can the products in which you are interested, peas and beans, play in the new context? That question is difficult to answer, because the Commission paper has not yet been published and I am not authorised to tell you what will be said in it. There are, however, a certain number of arguments which plead in favour of maintaining, and even perhaps developing slightly, the support for peas and beans for animal consumption. One of those reasons was mentioned by Professor Scarascia-Mugnozza; the importance that those crops have in the rotation practised by farmers. I would add that from a cost point of view, they are not more expensive, they are less expensive than cereals, and oilseeds, to mention but two of the crops which are competing for funds and the farmer's favour. We must add that we are still dependent on other countries for our requirements in proteins. We produce only 6% of our requirement of oil cakes, the most important source of protein for animal feed in the EC. In that case, it is quite clear that everything which we can do to alleviate that situation at a reasonable cost and within reason, cannot be criticised.

The Commission is the first step. Under the Treaty of Rome, the Commission has the power to propose. Nothing can be decided if the Commission has not first mentioned it. We have the power to put forward ideas to the Council. The members of the Council cannot say that they want to discuss a subject which has not been proposed by the Commission. The Commission can decide what the Council can and will discuss, but the power of decision remains in the hands of the Council. If the Council does not like an idea, it will remain an idea and will never become law. I do not know whether the Council will approve the approach of the Commission to changes in the agricultural policy.

As you know, there are some member states which are not very much in favour of continuing the support scheme for peas and beans. As recently as last year, one of the member states demanded the end of the support for peas and beans. The discussion will be difficult, but we must be optimistic, because the arguments are there and because we can also hope that the cost will diminish and not increase. If you can find the way to increase productivity so that you can get more peas and beans per ha than you did before, then you can assure a fair revenue for the producer on one hand, and on the other an equitable price for the feed manufacturer. If, instead, productivity is low and we do not increase it, then we are in trouble because we will be forced to increase the price.

Professor Scarascia-Mugnozza showed us that the increase in peas, beans and legumes productivity was much lower than productivity in the cereals sector. This is one of the causes of difficulties which we have to foresee. It could become a major reason for abandoning the system. In other words, we need your support. Without it, we will never be able to convince the finance ministers that what we propose should be maintained and improved.

You will have noticed that so far I have not mentioned the word 'lupin' and you are probably wondering why. I will tell you, as part of my conclusion. It is simple: we have no support scheme for lupins. When we prepared our report in 1973, we looked at the crops which, at that time, seemed to us the most promising, whose development was most probable and at those crops which already had a certain development in the Community. At that time we thought that peas and beans were the crops we should support, because they met all the requirements and because we thought that they had a possibility of increased productivity. Lupins did not seem very interesting to us at that time. However, the Spanish government has officially asked the EC to consider the possibility of extending the support system which exists for peas and beans to lupins. We will be bound to reconsider this problem in the near future in the context of the Spanish accession and the attitude that the ministers will take around the table will depend on the possibility of development of these crops. I think that this seminar might have something to say which could add a little to the arguments which we are having about accepting or refusing the Spanish demand.

That brings me to the end of my speech. I hope that what I have said has given you an idea of the framework within which you work. I think it is encouraging to know that you are not working in a vacuum, that what you do will be useful, and that what you are doing may not necessarily be useful to science but will be useful for agriculture in general and the agriculture policy in particular.

DISCUSSION

W. Williams *(UK)* You have shown us the complexities of the price support system which is most complicated and involves economic interactions between various crops such as cereals, peas, beans and rapeseed. A major problem seems to be that whilst rapeseed and sunflower produce protein which is in deficit in the EC, they also increase the vegetable oil surplus problem. Support for peas, field beans and lupins at the expense of that for rape and sunflower would perhaps provide a better balance of protein to oil in the EC.

M.G. Schiratti *(EC)* Oil seed rape and sunflower have been successfully exploited by farmers in Europe in response to encouragement from EC. About 5 m tonnes/annum of vegetable oil are consumed in the EC of which 1 m is home-produced from rape and sunflower. Of the 1.5 m tonnes of sunflower oil used, ½ m tonnes is home-produced. Currently 200 000 tonnes of rapeseed oil is exported due to the lack of a home-market. It may, therefore, be prudent to ensure that any further increases in rapeseed oil production can be absorbed by the market. On balance perhaps peas and beans should be favoured more than at present, but a fuller appraisal of the exact role and performance of legumes in this respect is needed.

SESSION II

GENETIC, BREEDING AND AGRONOMIC ASPECTS

Chairmen: G.T. Scarascia Mugnozza, J. Picard
W. Williams and D.R. Davies

NATURAL AND INDUCED VARIABILITY IN PEAS FOR PROTEIN PRODUCTION

L.M. Monti

Plant Breeding Institute, University of Naples,
Portici-Naples, Italy.

ABSTRACT

The traits characteristic of pea as a protein crop are reported, for both grain yield and protein quality. All the variability existing in the genus Pisum can be utilised, due to the absence of incompatibility between species.

Efficient methods of carrying out experimental mutagenesis are available for this species, and this permits the variability to be widened in the genetic background chosen for improvement; pollen irradiation has been shown to be the most efficient technique.

Correlations between the main traits involved in seed yield and protein quality have been worked out. These indicate that improvement for protein yield can be obtained either by increasing the seed yield whilst the protein percentages remain unchanged or by selecting for higher protein content in lines with a grain yield at least equivalent to that of the control.

The results of breeding work carried out in Portici with traditional and new types of pea are reported.

INTRODUCTION

Among the legume species in the EC peas offer the greatest potential as an alternative to soybean for protein production. Pea seeds have a high protein content with up to 40% and an average of about 23% (Table 1). Furthermore the quality of their protein is good when compared with other legume species, owing to the higher lysine content (Table 2). In addition to the above advantages, the crop is highly adaptable and its culture is fully mechanised.

In Italy the average dry seed yield is about 25 - 30 q/ha in the north with spring sowings, and about 30 - 35 q/ha in the south from November - December sowings. It is estimated that to make the crop more competitive in relation to other crops, a 10 - 15% increase in the yield is required. Peas have been grown mainly for domestic use (canning and freezing), a single harvest being made whilst the seeds are immature. The main breeding objectives were, therefore, aimed at obtaining synchronous fruiting to achieve large numbers of seeds all of

TABLE 1

RANGES OF SEED PROTEIN CONTENT (% d.m.) IN LEGUME SPECIES

Species	Protein content Minimum	Protein content Maximum
Cicer arietinum	14.9	29.6
Glycine max	34.8	40.5
Lens esculenta	25.0	29.3
Lupinus albus	17.0	38.7
Phaseolus vulgaris	17.0	39.4
Pisum sativum	15.5	39.7
Vicia faba	22.0	37.0

TABLE 2

RANGE OF SULPHUR AMINO ACID, TRYPTOPHAN AND LYSINE CONTENT (g/100 g PROTEIN) IN A RANGE OF LEGUMES

	Methionine	Cystine	Tryptophan	Lysine
Cicer arietinum	0.5 - 1.7	-	0.9	6.9 - 7.2
Glycine max	1.4 - 1.6	1.4 - 1.6	1.3 - 1.4	6.3 - 7.0
Lens esculenta	1.6	1.9	0.9 - 1.0	7.2 - 7.3
Lupinus albus	0.5 - 1.0	1.7 - 2.5	0.7 - 1.6	4.3 - 5.8
Phaseolus vulgaris	1.1 - 1.3	1.0	1.0	6.7 - 7.2
Pisum sativum	0.6 - 1.0	0.8 - 1.7	0.7 - 1.9	6.9 - 8.2
Vicia faba	0.6 - 1.0	0.6 - 1.7	0.8 - 1.0	6.2 - 6.5
FAO standard	3.5		1.0	5.4

the same age and maturity. Most canning and freezing cultivars have green and wrinkled cotyledons, a uniform seed size, and quality traits suitable for processing for culinary use. Varieties are available, covering a wide range of maturities to ensure continuity of supply to the processing factories over a long period of time.

As a protein source, it is the yield of dry seed and its protein content and quality that are important. Consideration must be given therefore to the yield components such as the number of branches, the number of flowering nodes, of flowers and of pods per plant, as well as the flower fertility. Resistance to lodging, drought, pod shattering and to pests and diseases at maturity *(Oidium* and *Pernospora)*, are other aims required to achieve high yield stability.

It is evident that the ideotype for culinary use and the one for protein production are quite different from each other. The only traits common to the breeding of both types are disease resistance at the seedling stage (to *Fusarium oxysporum* and *F. solani*, *Ascochyta* spp and other agents) and the resistance to low temperatures for autumn sown crops.

It follows that much potentially useful material for protein production has been discarded from the culinary pea breeding programmes for reasons such as lack of synchronous flowering or poor fresh seed quality. Such material could well be worth re-examining with the seed protein crop in mind.

The basis of the genetic variability of the most important traits also requires study in order to set up *ad hoc* breeding programmes for pea protein production.

This paper will analyse the genetic basis of characters in relation to protein production and explore correlations among them. Methods of increasing variability will be examined and an outline of the work carried out in Portici presented.

NATURAL VARIABILITY

Peas are Mediterranean in origin and the Near East and Anatolia are considered secondary habitats. Several species (all with $2n = 14$ chromosomes) have been ascribed to the genus *Pisum*, but Lamprecht (1966) recognised only one species *(P. arvense)* and five ecotypes of it *(Abyssinicum*, *Arvense* or *Elatius*, *Fulvum*, *Humile* or *Syriacum* and *Sativum)*.

Ben-Ze'ev and Zohary (1973) recognised *P. sativum* and the wild species *P. elatius*, *P. humile* or *P. syriacum* and *P. fulvum*: the first three species, being closely related on the basis of botanic, genetic and cytological data, are grouped together into a single biological species. *P. fulvum*, on the other hand, represents a divergent species differing from *P. elatius* - *P. humile* - *P. sativum* groups in the length ratios of several chromosome pairs, for the presence of an additional satellited pair on chromosome V and for the occurrence of a large satellite on the chromosome VII.

Despite the presence of one or two translocations (Saccardo, 1971) and some differences in the cytoplasm (Monti, 1971), crosses are possible among all the species, at least in one direction (Table 3), so that all the genetic variability which is present in the genus can be used in pea breeding. This is particularly important in the breeding for dry seeds, because the adaptive value of the traits related to dry seed production is greater than that related to the traits associated with the production of fresh seeds which ripen at the same time on the plant. The use of wild types should be of great interest for breeding purposes, because some characters of agronomic interest appear under systems of genetic control different from those found in the cultivated varieties (Snoad and Arthur, 1973).

On the basis of the linkage map studied by Lamprecht (1964), several duplicated segments are present in pea chromosomes, which can explain the ability of this species to withstand large chromosomes aberrations.

TABLE 3

CROSSES AMONG *Pisum* SPECIES (Based on Ben-Ze'ev et al., 1973)

♀ \ ♂	*P. sativum*	*P. elatius*	*P. humile*	*P. fulvum*
P. sativum		semi-sterile F_1 plants	fertile or semi-sterile F_1 plants	semi-sterile F_1 plants
P. elatius	semi-sterile F_1 plants		fertile or semi-sterile F_1 plants	semi-sterile F_1 plants
P. humile	fertile or semi-sterile F_1 plants	fertile or semi-sterile F_1 plants		semi-sterile F_1 plants
P. fulvum	no F_1 seeds	lethal F_1 seedlings	lethal F_1 seedlings	

Large pea germplasm collections are present in Europe (Leningrad, Gatersleben, Landskrona, Bari) and the wide knowledge on the genetics of this species has allowed an analysis of the effects due to the presence of the dominant or recessive alleles of some genes on dry seed production (Blixt, 1978). Table 4 reports the average yield for lines being dominant and lines being recessive for a marker gene. Though the genetic background was different, data are useful for estimating the effects of the following six genes: *A/a* (presence/absence of anthocianin), *I/i* (yellow/green cotyledons), *R/r* (smooth/ wrinkled cotyledons), *St/st* (normal/reduced stipules) *Fa/fa* (normal/fasciated stem), *Le/le* (long/short internodes).

TABLE 4

YIELD (g/PLANT) OF LINES WITH DIFFERENT ALLELES OF SOME GENES (from Blixt, 1978)

Gene	Character	Yield	
		X	x
A	Anthocianin present/absent	18.8	25.6
I	Cotyledons yellow/green	22.3	17.9
St	Stipules normal/reduced	21.9	13.3
Le	Internodes long/short	22.6	16.1
Fa	Stem normal/fasciated	21.5	18.1
R	Cotyledons smooth/wrinkled	22.2	16.8

Larger seed yields are obtained when the alleles *a*, *I*, *R*, *St*, *Fa*, and *Le* are present, and indeed most protein pea varieties possess all these characters, except the last one, as short internodes are required to avoid lodging.

Genetic control of leaf morphology has also been intensively studied in relation to the yield components. The three genes controlling leaf morphology included in the study were *Af/af* (leaflets normal/converted to tendrils), *St/st* (normal/ reduced stipules), *Tl/tl* (tendrils normal/converted to leaflets). Our study included eight isogenic lines in which the three genes were present either alone or in different combinations (Monti and Frusciante, 1978). The results are reported in Table 5 and

TABLE 5

WEIGHT OF 100 SEEDS IN ISOGENIC LINES OF PEA DIFFERING FOR ONE, TWO OR THREE OF THE GENES *af*, *st* AND *tl*.

Genotype *Af*	*St*	*Tl*	100 seeds weight (g)	
+			23.13 ± 0.96	NS
af			20.72 ± 0.70	
	+		23.48 ± 0.95	X
	st		20.43 ± 0.73	
		+	21.52 ± 0.66	NS
		tl	22.26 ± 0.90	
+	+		24.73 ± 1.40	XX
af	*st*		19.82 ± 0.80	
+		+	22.89 ± 0.95	NS
af		*tl*	21.35 ± 0.96	
	+	+	23.95 ± 0.89	NS
	st	*tl*	21.57 ± 0.99	
+	+	+	24.90 ± 1.17	X
af	*st*	*tl*	20.86 ± 1.05	

Student's test

NS = not significant; X = significant at 5% XX = significant at 1%.

the negative pleiotropic effect of the *st* allele, mainly on the seed weight, was confirmed. The *af* and *tl* genes alone or together did not adversely affect yield. These data indicate considerable promise for these two factors in breeding programmes The lines *af* and *af-tl* are particularly interesting to us and elsewhere, the *af-tl* being characterised by a large number of minute leaflets which confer good standing ability. Our studies at Portici have been concentrated on the *af* and *af-tl* types for which we have obtained good harvest indices.

The negative pleiotropic effect of the *r* gene (wrinkled cotyledons) on seed yield described by Shia and Slinkard (1977) was also confirmed by our results (Table 6). Lines with smooth seeds *(R-)* yield more than lines with wrinkled seeds *(rr)*, because of a greater average seed weight. Another important

TABLE 6

MEAN YIELD OF LINES WITH SMOOTH AND WRINKLED SEEDS

Seeds	Lines (number)	Grain yield (g/plot)	Protein (%)	Protein yield (g/plot)	100 seed weight (g)	Protein/seed (mg)
Smooth	13	1 811	25.85	466.98	20.03	52
Wrinkled	11	1 611	29.23	470.41	18.59	54

effect of the *r* gene is its greater protein level. Yields of protein per hectare, however, are similar for smooth and wrinkled isogenic lines since the lower protein percentage of the former is balanced by its higher seed yield. Consequently, it is sensible only to breed lines with wrinkled cotyledons where seed quality is of prime concern.

INDUCED VARIABILITY

Experimental mutagenesis on seed or other plant part in peas has resulted in many mutants presently included in germplasm collections.

Extensive mutagenic research has been carried out in Europe and the problems relating to the chimeric structure of the M_1 plants after seed treatments have been dealt with in several papers, demonstrating improved methods for isolating the induced mutants. Methods have also been devised for early discrimination of M_1 plants damaged by the direct effect of the mutagenic agents on the leaf primordia present in the embryos of the treated seeds (Blixt and Gottshalk, 1975; Monti, 1968).

Irradiation of gametes and particularly of pollen seems to be the most efficient technique for inducing mutations useful for pea breeding (Monti and Saccardo, 1979). When compared with seed treatment, pollen irradiation does not result in the chimeric condition in M_1 plants, but gives a greater amount of point mutations due to the sieve action of haplontic selection at zygote formation of the M_1.

Table 7 gives the frequencies of mutations induced in two pea varieties (the fodder cv. Parvus and the canning cv. Sprinter) following X-irradiation of pollen and seed, and seed treatment

TABLE 7

FREQUENCIES OF MUTATIONS INDUCED IN PEA AFTER TREATMENTS OF POLLEN AND OF SEED WITH X-RAYS AND WITH DIETHYLSULPHATE (DS).

Variety	Irradiation on	Mutagenic agent	M_1 plants analysed (no.)	Mutations/ M_1 plant (%)	Relative frequency of mutations Chloro-phyll (%)	Morpho-logical (%)
Parvus	Pollen	X-rays	227	3.3	22.2	77.8
	Seed	X-rays	349	41.5	65.5	34.5
		DS	247	155.9	68.3	31.7
Sprinter	Pollen	X-rays	271	14.8	17.5	82.5
	Seed	X-rays	1 107	16.0	80.3	19.7
		DS	372	22.3	63.8	36.2

with diethylsulphate (DS). DS was more effective as a seed treatment than was X-irradiation. Although irradiation of pollen caused fewer mutations than irradiation of seed, a greater proportion of the pollen mutations involved morphological characters rather than chlorophyll apparatus. Indeed, pollen irradiation produced more high yielding lines than seed irradiation (Monti et al., 1980).

Several mutants for seed size and shape and for a modified plant structure more resistant to lodging were induced and many of them have been included in our protein pea breeding programme.

Pollen treatment has also proved to be useful for producing duplication of chromosome segments (Saccardo et al., 1980). Through the induction of dicentric chromosomes, variegated mutants were obtained, the progenies of which produced mutants with duplicated chromosome segments, due to the breakage - fusion - bridge cycles of the dicentric chromosomes (Grillo et al., 1982). The importance of chromosome duplication is well known as a source of new genetic material. The lines obtained up to now have been analysed. Differences between a line with the chromosome aberration and the mother line in protein content and protein quality indicate good potential for pea breeding, as it is possible to obtain duplications in different chromosome segments in different generations.

Clearly, a major advantage of such mutagenesis is the production of wide variation in some traits against a genetic background closely approximating to the chosen ideotype.

BREEDING FOR PROTEIN PRODUCTION

As already shown (Table 1) the amount of variability for protein percentage is considerable but insufficient (Slinkard, 1980) for improving this trait, because part of the genetic variability is due to the pleiotropic effect of the gene for wrinkled seeds. Also only moderate heritability was found for this trait. Slinkard also found negative correlations between protein content and seed yield. Blixt (1978) and Pandey and Gritton (1975) on the other hand found no significant correlations between the two traits. There was no negative correlation between the protein content and the 1000-seed weight. Therefore, the negative correlations between protein content and seed yield must have been due to fewer seeds per plant, as found in other legume species (Monti and Grillo, 1982).

Screening for improvement in protein content is difficult because of sampling problems; the above correlation between yield and protein content indicates that adverse environmental conditions causing reduced fertility will correspond with an increase in protein content. Within plant variation for seed protein content has been established by Cousin (1979) and ourselves, with a lower protein content for seeds on the upper part of the plant.

It is concluded that because of the predominantly negative relationship between yield and protein content, it is prudent to select for protein content among reasonably high yielding lines.

Another possibility is to increase protein production by improving seed yield without changing protein percentage. Thus the contribution of the main traits to yield are of great importance. The correlations obtained by Blixt (1978) (Figure 1) among the main characters of agronomic interest, in the germplasm available in Landskrona showed that 100-seed weight, the number of seeds per pod and the number of flowers per plant were most directly linked to yield. Numbers of flowers per plant were correlated with the number of branches and the number of

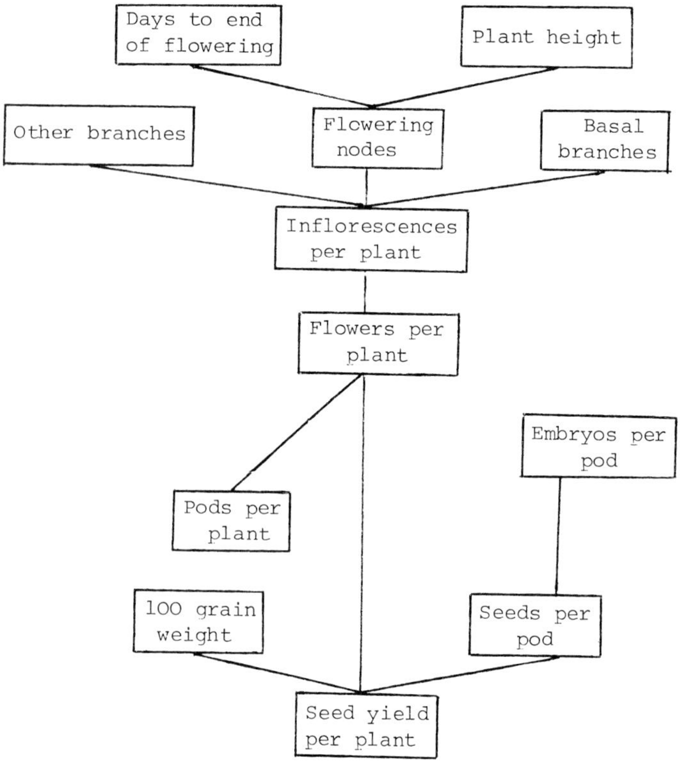

Fig. 1. Correlations found (at p 0.001) among some characters in peas (Blixt, 1978).

flowering nodes, which was related to plant height and the duration of flowering. Variability for these traits with the minimum, mean and maximum values are given in Table 8. Such variability seems to be sufficient to allow progress in selection for seed yield, and this has been a main aim at Portici in recent years.

Drought stress during ripening is a serious yield limiting factor in Italy; autumn sowing can be used to avoid drought stress in many areas, and for this purpose genetic resistance to low temperatures at the seedling stage is necessary. In many areas in southern Italy, for instance, sowings are made in November - December and the crop is at the seedling stage in January - February, and minimum temperatures of -7°C are uncommon. Screening in controlled environmental

TABLE 8

AVERAGE VALUES, MAXIMUM AND MINIMUM VALUES FOR SOME CHARACTERS IN THE NON-INDUCED MUTANT MATERIAL OF THE WELBULLSHOLM COLLECTION (from Blixt, 1978)

Character	Average	Value	
		Maximum	Minimum
Plant height (cm)	115.96	201.2	20.8
Days to end of flowering (number)	98.07	134.2	16.0
Flowering nodes on main stem (number)	8.48	17.5	1.9
Basal branches (number)	2.76	17.4	1.0
Other branches (number)	2.51	16.5	0.0
Inflorescences per plant (number)	16.44	71.3	2.6
Flowers per plant (number)	31.87	106.9	2.6
Pods per plant (number)	19.00	136.9	2.4
Embryos per pod	7.65	15.1	4.4
Seeds per pod (number)	5.93	12.4	1.6
100 grain weight (g)	19.54	48.0	3.4
Seed yield per plant (g)	21.35	113.2	0.4
Protein content (%)	23.75	32.1	15.8

conditions at this temperature is therefore carried out in Portici and, as already found by Cousin (1976), the more advanced the plant development, the lower is their ability to survive (Figure 2). The fodder cv. Imposant was the most resistant of the lines screened. In crosses involving this variety, resistance to low temperature was monogenic and dominant, with a linkage of about 27% between this trait and the gene *le* localised on chromosome IV which determines internode length. In 1980, 24 lines from previous crossing and mutagenic programmes were assessed for cold tolerance. The best lines plus other new lines were also tested in the following year at the same place, that was in the Lucania Region, near Potenza. Significant differences were found among the lines for protein yield in both years (Table 9) and were due to differences both in the grain yield and in the protein content of seed. The yield and the protein data for the highest yielding lines in the two years are reported in Table 10 and a good performance was obtained by the cv. Proteo, a new protein pea variety which has been produced by ourselves.

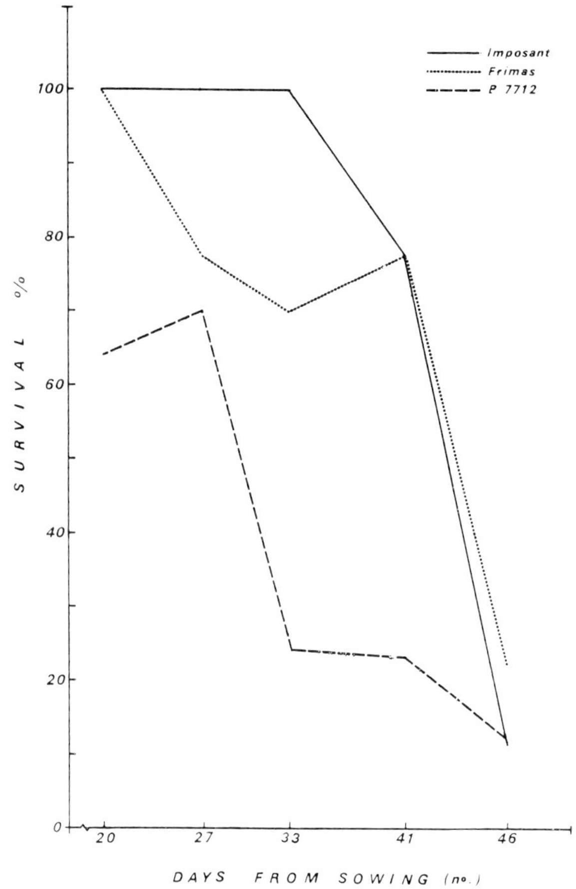

Fig. 2. Relationship between seedling age and survival following -7°C low temperature treatment.

TABLE 9

ANALYSIS OF VARIANCE FOR DRY SEED YIELD, PROTEIN CONTENT AND PROTEIN YIELD IN PEAS IN 1980 AND 1981. RANDOMISED BLOCKS WITH 24 AND 16 LINES RESPECTIVELY

	1980			1981		
Item	Dry seed yield	Protein content	Protein yield	Dry seed yield	Protein content	Protein yield
Block	**	NS	NS	NS	NS	*
Lines	*	**	*	**	**	**

NS not significant
* significant at the 0.05 level
** significant at the 0.01 level

TABLE 10

PROTEIN YIELDS AND MAIN TRAITS OF THE HIGHEST YIELDING LINES IN TWO YEAR TRIALS (PLOT SIZE OF 4 m^2)

Line	Dry seed yield (g)	Protein content (%)	Protein yield (g)	Main traits Cotyledon	Plant
Proteo	2 140	27.67	592.14	Green, smooth	Deep green
Finale	2 213	26.27	581.36	Green, smooth	Deep green
P 77185	1 910	28.89	551.80	Yellow, wrinkled	Afila
P 7838	1 763	31.29	551.64	Green, wrinkled	
Imposant	1 821	26.46	481.84	Yellow, wrinkled	Anthocianic
P 77180	1 828	25.74	470.53	Yellow, smooth	Afila-acacia
P 77191	1 841	25.02	460.62	Yellow, smooth	Afila-acacia

Proteo has smooth green cotyledons, good standing ability and more than 27% protein content. Noteworthy among the other lines were one 'afila' line (P77185) and two 'afila-acacia' lines (P77180; P77191). The P77185 has yellow, wrinkled cotyledons and a protein yield similar to Finale, one of the control varieties. Both types, 'afila' and 'afila-acacia' show an interesting canopy structure with good standing ability and work is in progress with these types to improve harvest index and stability.

Experiments have evaluated mixtures of pea lines with differing leaf form, that is mixtures with normal leaves and a line with 'afila' or 'afila-acacia' leaves. The aim was to improve the standing ability of a high protein line with normal leaves, by the unique structure of the other component.

Several mixtures among different genotypes have been tested and Table 11 reports the data referring to one of them, line P7833, which is a normal-leaf type and line P77191, which is an 'afila-acacia' type, both with yellow and smooth cotyledons. These lines were blended in different proportions in a trial with 3 replicates. All the mixtures of the two lines produced higher protein yields than either of the two lines alone. Confirmatory experiments are continuing.

There is sufficient variation in protein quality, in relation to the major limiting amino acids (methionine, cystine

TABLE 11

PROTEIN YIELD IN DIFFERENT MIXTURES OF TWO GENOTYPES (PLOT SIZE OF 4 m^2)

Genotype	Mixture (%)				
P 77191 *af af - tl tl*	100	75	50	25	0
P 7833 *AF AF - TL TL*	0	25	50	75	100
Dry seed yield (g)	1 670	2 000	2 110	2 110	1 380
Protein content (%)	25.02	26.20	26.05	26.00	26.80
Protein yield (g)	417	524	549	548	369

tryptophan and lysine), to provide optimism for improvement through breeding. It has been found that the variability for these four amino acids was greater than that found for any of the others (Holt and Sosulsky, 1979). Nevertheless, the selection for this objective can be difficult. For instance, Slinkard (1980) found little genetic variation in methionine content (mg/g protein), because of a large environmental effect. Also it must be remembered that an increase in one amino acid can result in lower values of another as observed by Lafiandra et al. (1979) in *Vicia faba*.

Genetic improvement in sulphur amino acids and in lysine is likely to result in decreased total protein contents, due to the negative correlation between sulphur amino acid and protein content (Holt and Sosulsky, 1979). According to Boulter (1977), this is because the selection of lines with higher protein content results in higher content of storage proteins which are characterised by a lower sulphur amino acid content. Perhaps a better way to increase the sulphur amino acid contents in peas would be by improvement of the legumin fraction, which has a larger amount of sulphur amino acids than the vicilin fraction (Monti and Grillo, in press).

Another approach is to improve the albumin/globulin ratio, because of the greater amount of sulphur amino acids in the albumin fraction (Bayay, 1971; Muller, 1973; Gottshalk et al., 1976).

As emphasised in the introduction, no antinutritional, or toxic factors are present in pea seeds, which can therefore be used also in human nutrition, either directly or through protein concentrates. Oligosaccharides, particularly stachyose, which are believed to be responsible for flatulence, are present in only small amounts in pea seeds, and more in wrinkled than in smooth seeds. Selection programmes to decrease the content of flatulence factors have a good chance of success, probably similar to those obtained for soybeans by Hymoniwitz et al. (1974).

CONCLUSIONS

Pea is one of the genetically best known species, which provides a sound base for improvement. It possesses high levels of natural and induced variability, represented in several gene banks. So far this variability has been used mainly to improve the synchrony of the fruiting and to improve quality of fresh peas.

The potential of peas for dried seed production is good. Recent breeding work with this aim has produced some good varieties, through improvement of the dried seed yield. Co-ordination of work in relation to detailed analysis of gene bank material within the framework of the EC plant protein programme, should lead to further progress. Clear improved understanding of the biochemistry and inheritance of protein levels, and of genetic engineering techniques, should enable improvements in protein yield to be achieved.

REFERENCES

Bayay, S., Mickelsen, O., Lillevik, H.A., Baker, L.R. and Bergen, W.G., 1971. Prediction of protein efficiency ratio of peas from their albumin content. Crop Science 11, 813-815.

Ben-ze' ev N. and Zahary, D., 1973. Species relationship in the genus *Pisum* L. Israel J. of Botany 22, 73-91.

Blixt, S. and Gottshalk, W., 1975. Mutation in the Leguminosae. Agr. Hortique Genetica 33, 38-85.

Blixt, S., 1978. Problems relating to pea-breeding. Agr. Hortique Genetica 36, 56-87.

Blixt, S., 1978. Natural and induced variability for seed protein in temperate legumes. FAO/IAEA Int. Symp. Seed Protein Improvement in Cereals and Grain Legumes. Neuherberg Sept. 4-8.

Boulter, D., 1977. Quality problems in protein plants with special attention paid to the proteins of legumes. EC Seminar on protein quality from leguminosae crops, Dijon (France), November 3-5, 1976.

Cousin, R., 1976. Amélioration Génétique du Pois d'hiver Ann. Amélior. Plantes 26, 235-263.

Cousin, R., 1979. Breeding for protein and yield in peas. CNR meeting on Prospettive delle Proteaginose in Italia-Perugia (Italy), May 30 - June 2.
Gottshalk, W., Muller, H.P. and Wolff, G., 1976. Further investigations on the genetic control of seed production in *Pisum* mutants. Proc. Symp. Vienna IAEA, 157-177.
Grillo, S., Saccardo, F. and Addeo, F., 1982. Induced duplications in chromosome segments in peas and their significance for seed protein studies. In: Perspectives for peas and lupins as protein crops, (Eds. R. Thompson and R. Casey). This publication.
Holt, M.W. and Sosulski, F.W., 1979. Aminoacid composition and protein quality of field peas. Can. J. Sci., 59, 653-660.
Hymoniwitz, T., Collins, F.I., 1974. Variability of sugar content in seed of *Glycine max* (L.) Merril and *G. Soya* Sied and Zucc. Agronomy J., 66, 239-240.
Lafiandra, D., Polignano, G.B. and Colaprico, G., 1979. Protein content and aminoacid composition of seed in varieties of *Vicia faba* L.Z. Pflanzenzuchtg, 83, 308-314.
Lamprecht, H., 1964. Partielle Sterilitat und Chromosomenstruktur bei *Pisum* Agri. Hort. Genet. 22, 56-148.
Lamprecht, H., 1966. Die Entstehung der Arten und hoheren Kategorien-Spingher-Verlag-Wien.
Monti, L.M., 1968. Mutations in pea induced by diethylsulphate and X-rays. Mutations Res., 5, 187-191.
Monti, L.M., 1971. Genome cytoplasmic interactions in *Pisum* species. *Pisum* Newsletter, 3, 28.
Monti, L.M. and Frusciante, L., 1978. Pea breeding by using genes drastically influencing the leaf morphology. Genet. Agr., 32, 365-373.
Monti, L.M. and Saccardo, F., 1979. The gamete irradiation technique in pea genetics and breeding. Monografia IV di Genetica Agr., 141-152.
Monti, L.M., Saccardo, F. and Vitale, P., 1980. Pea mutations of practical value. *Pisum* Newsletter, 12, 54-55.
Monti, L.M. and Grillo, S., 1982. Legume seed improvement for protein content and quality. European Congress on Plant Proteins - Nantes (France), 5-7 Oct., 1981, (in press).
Muller, H.P., 1973. Content in globulins, albumins and essential amino-acids in seeds of pea mutants and recombinants. *Pisum* Newsletter, 5, 37-38.
Pandey, S. and Gritton, E.T., 1975. Genotypic and phenotypic variances and correlations in peas. Crop Science, 15, 353-356.
Saccardo, F., 1971. Crosses among *Pisum* species. *Pisum* Newsletter 3, 38.
Saccardo, F., La Gioia, N. and Monti, L.M., 1980. Induction of chromosome duplications in peas. *Pisum* Newsletter, 12, 62-63.
Shia, G. and Slinkard, A.E., 1977. Relationship of seed shape and cotyledon colour to percent protein of peas. Crop Sci., 17, 183-184.
Slinkard, A.E., 1980. Breeding field peas for increased yield and protein. In: Production, utilization and marketing of field peas. Final Report (I) (Sub projects I-VI) Section 1. Breeding peas. Crop Development Centre, University of Saskatchewan, Saskatoon, Canada. 1-11.
Snoad, B. and Arthur, A.E., 1973. Genetical studies of quantitative characters in peas. A six parent diallel cross of cultivars and primitive forms. Euphytica, 22, 510-519.

DISCUSSION

C. Calet *(France)* Do you think it is more important to select for energy, protein or amino acid content?

L.M. Monti *(Italy)* Although negative correlations exist between seed yield and protein content, and between protein content and some amino acids, they are not invariable and it is possible to select for both seed yield and quality. The first priority in Italy is to increase seed yield and improved protein quality is the next priority. For instance, it is important to increase the tryptophan content because this could be a limiting amino acid in some diets with grain legumes.

R. Davies *(UK)* You mentioned that in the Nantes meeting there had been a recommendation that wrinkled peas should not be used because of the difficulties of extraction. Do the difficulties arise in milling the seed or in extraction of the protein? If the latter is the case, is there any known difference in the lipid content of round and wrinkled seeds?

L.M. Monti The case I referred to related to difficulties in separating the seed coats from the cotyledons. Some differences between the lipid contents of the wrinkled and smooth seeds have been found.

M. Frauen *(FRG)* You seemed to favour the *Aphila acacia* type because of its high lodging resistance. Is the lodging resistance of a canopy better than that for a plant with many tendrils?

L.M. Monti We find that the *Aphila acacia* types resist lodging better than normal leaf types.

B. Snoad *(UK)* Our evidence suggests that increased amounts of tendrils markedly improve standing ability. However, the combination of *af* and *tl*, which gives the small leaflets, hardly improves standing ability, and indeed, it may reduce it.

COLLECTION AND ASSESSMENT OF THE GENE POOL OF *LUPINUS ALBUS* IN SOUTHERN EUROPE

M.J.A. Simpson

Department of Agricultural Botany,
Plant Science Laboratories, University of Reading,
Whiteknights, Reading RG6 2AS, UK.

INTRODUCTION

Our *Lupinus albus* genetic resources are now being enriched by collection of land races from countries bordering the Mediterranean. The variation pattern among populations of *L. albus* is not well understood and it is important to study methods that will characterise and quantify variation in such a way as to help identify useful genotypes as well as regions of maximum diversity.

Such a study requires methods which can use a range of separate measurements to create an objective representation of the similarities and differences between accessions. For this reason, some techniques used in numerical taxonomy are particularly useful.

This study aims to explore the range of morphological and physiological variation contained within a collection of *L. albus* cultivars and land races, using the numerical taxonomic method called principal component analysis. Emphasis is given to new land race material acquired during a seed collecting expedition to Iberia in 1980.

MATERIAL AND METHODS

The seed samples used were taken from the Reading University *Lupinus* collection and detailed information on each accession is available.

Two sets of accessions, one from the original collection (Experiment 1) and one from the Iberian collection (Experiment 2), were sown at Reading in March 1980 and March 1981 respectively. In both experiments, two well known accessions were sown for comparison with the other accessions (Table 1).

TABLE 1

THE STANDARD ACCESSIONS

Experiment	Standard	Provenance	Description
1	Lupini Bean	Italy	Tall, late flowering, winter type
1 and 2	Kievskij Mutant	USSR	Short, early flowering, spring type
2	Lucky	France	Intermediate between spring and winter types

Principal component analysis

The numerical taxonomic method called principal component analysis (PCA) was used to create an objective diagrammatic representation of the variation contained within the two sets of accessions.

Using PCA, the degree of similarity between accessions can be estimated by projecting their distribution in multi-dimensional space (with respect to all the characters, each representing an axis) on to two or three axes, chosen in such a way that discrimination of accessions is maximised. By plotting the first axis (principal component) against the second followed by the first axis against the third, two views (scatter diagrams) of the same distribution are obtained, in which proximity of accessions indicates similarity. The axes of the scatter diagrams represent all the characters in the study but in varying proportions. Those characters that contribute most (the largest variable loadings) to each axis can be discerned in the analysis. The principal components can contain positive or negative variable loadings meaning that a constituent character may increase in either direction with respect to the axis. This direction can be indicated by an arrow.

Experiment 1, 1980: Original Collection

Sixteen plants each of 61 accessions representing named

cultivars and land races were sown in four randomised blocks in the field. Twenty plants each of the two standard accessions, Lupini Bean and Kievskij Mutant, were sown in each block.

TABLE 2

CHARACTERS AND CHARACTER STATES IN EXPERIMENT 1

1	Presence or absence of alkaloids
2	Length of secondary flower colour (1 - 3)
3	Secondary flower colour (1 - 5)
4	Pattern of secondary flower colour (1 - 3)
5	Flower length
6	Flower breadth
7	Presence or absence of keel tip colouration
8	Length of petiole
9	Breadth of petiole
10	Leaflet length
11	Leaflet breadth
12	Leaflet number
13	Stipule length
14	Plant height to apex
15	Primary pod number
16	Plant height to first pod
17	Pod length
18	Pod breadth
19	Seed number per pod
20	Seed length
21	Seed breadth
22	Number of days until plant height exceeds 10 cm
23	Number of days to flowering
24	Flower length/breadth ratio
25	Petiole length/breadth ratio
26	Leaflet length/breadth ratio
27	Stem length/plant height ratio
28	Seed length/breadth ratio
29	Pod length/breadth ratio

A list of 23 primary characters and six ratios (Table 2) was scored on each plant. Characters were chosen from all parts of the plant and all stages of the life cycle. Averages from the data matrix were then subjected to PCA.

Experiment 2, 1981: Iberian Collection

Variation between populations

Twenty plants each of 59 accessions representing the full range of ecological conditions in which Iberian *L. albus* exists were sown in four randomised blocks in the field. Twenty five plants each of the two standard accessions, Lucky and Kievskij Mutant, were sown in each block. An improved list of 25 characters and eight ratios (Table 3) was scored on each plant. The list was the result, firstly of a greenhouse study of the level of discrimination of accessions facilitated by different characters and secondly, of the practicability of scoring these characters on large numbers of field grown plants. PCA was carried out on averages from the data matrix.

Variation within populations

A comparison of variation levels within accessions was obtained by using the figures for each character in the data matrix to calculate the coefficient of variation (i.e. the standard deviation as a proportion of the mean) for each accession in turn. By scanning the resultant table it was possible to pick out those populations that were more variable than others with respect to the 25 characters. Comparison of variation levels within the standard cultivars with that of the land races was thus possible.

RESULTS AND DISCUSSION

Experiment 1, 1980: Original Collection

The scatter diagram from the first two principal components is shown in Figure 1. The largest variable loadings for each principal component are indicated. It can be seen that separation

TABLE 3

CHARACTERS AND CHARACTER-STATES IN EXPERIMENT 2

1	Number of days to germination
2	Cotyledon length
3	Cotyledon breadth
4	Number of days from germination until mainstem is twice the length of the first leaf petiole
5	Number of days to first flowering
6	Number of leaves at first flowering
7	Leaflet length
8	Leaflet breadth
9	Stipule length
10	Stipule breadth
11	Petiole length
12	Petiole breadth
13	Standard petal length
14	Standard petal breadth
15	Sepal length
16	Sepal breadth
17	Secondary flower colour (1 - 5)
18	Presence or absence of alkaloids
19	Number of primary pods
20	Pod length
21	Pod breadth
22	Pod wall thickness
23	Number of seeds per pod
24	Plant height
25	Height of first pod
26	Cotyledon length/breadth ratio
27	Leaflet length/breadth ratio
28	Stipule length/breadth ratio
29	Petiole length/breadth ratio
30	Standard petal length/breadth ratio
31	Sepal length/breadth ratio
32	Pod length/breadth ratio
33	Height of first pod/plant height ratio

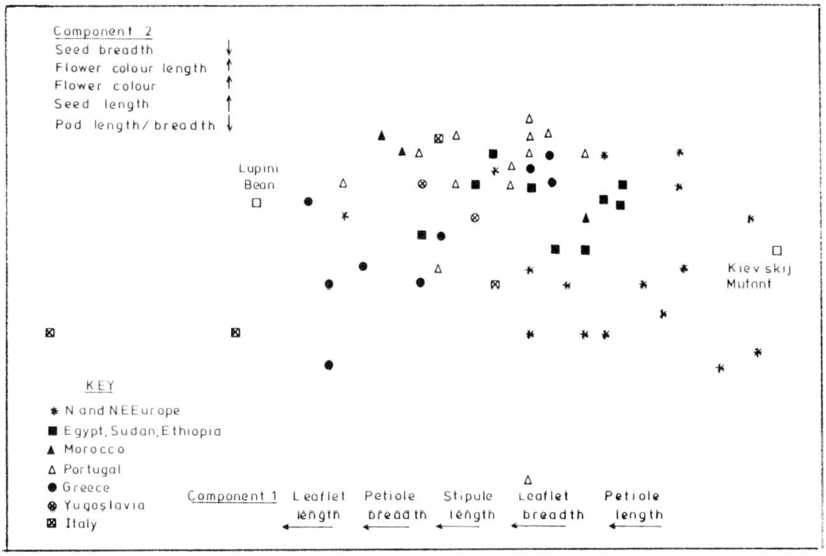

Fig. 1. Original collection: component 1 (accounting for 24.26% of total variation) against component 2 (accounting for 9.60% of total variation).

along the first principal component is largely due to metric characters such as leaflet length and stipule breadth with smaller types tending more towards the right of the diagram. Separation along the second principal component involves more multistate characters and ratios such as flower colouration and pod length/breadth ratio.

The two standards, Kievskij Mutant and Lupini Bean are widely separated as might be expected and the majority of the other accessions are placed between them. The named cultivars and other accessions from north and north-east Europe form a scattered, though well defined, group tending towards Kievskij Mutant. All groups overlap in the centre of the diagram although the accessions from Sudan, Egypt and Ethiopia appear to be very similar to each other and more similar to northern European accessions than to other groups. Two out of the three Moroccan accessions cluster nearer to the Portuguese accessions than to the others from south of the Mediterranean. Most of the Portuguese accessions form a fairly discrete cluster near the centre of the diagram and is seen more clearly in Figure 2. The Greek accessions are more widely scattered from the centre towards

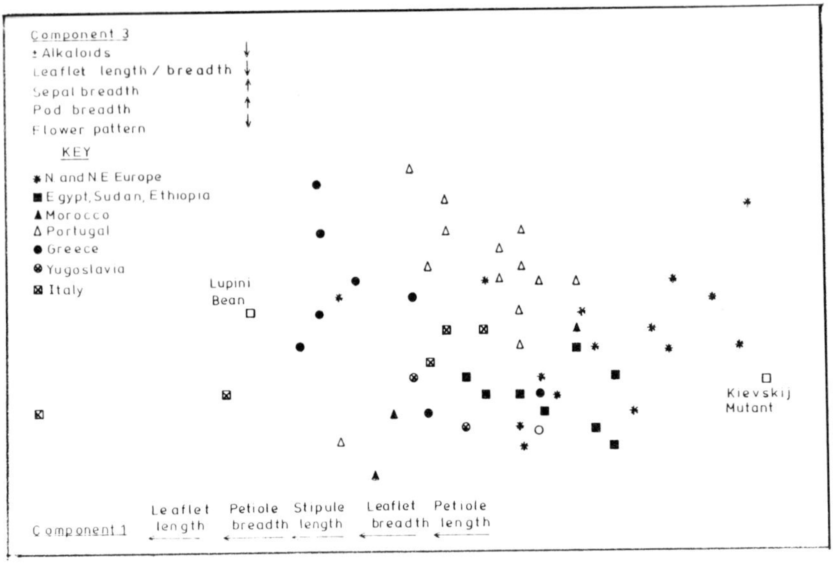

Fig. 2. Original collection: component 1 (accounting for 24.26% of total variation) against component 3 (accounting for 9.00% of total variation).

Lupini Bean. The wider range of variation that this indicates might be expected from Vavilovian theory if Greece is the country of origin of the crop. The few Italian accessions are, however, also widely spaced and tending more towards the left of the diagram, demonstrating that these two groups consist of larger, later flowering types compared to those from northern Europe.

Experiment 2, 1981: Iberian Collection

Separation along the first principal component (Figures 4 and 5) is again largely due to metric characters with the duration of developmental stages, multistate characters and ratios playing parts in the second and third principal components.

Variation between areas

The majority of the accessions are clustered around the centre much nearer to Lucky than to Kievskij Mutant. In spite of extensive overlap between groups from different areas (Table 3), some tendencies are evident. For example, the accessions from north Portugal tend to be bigger plants that those from just

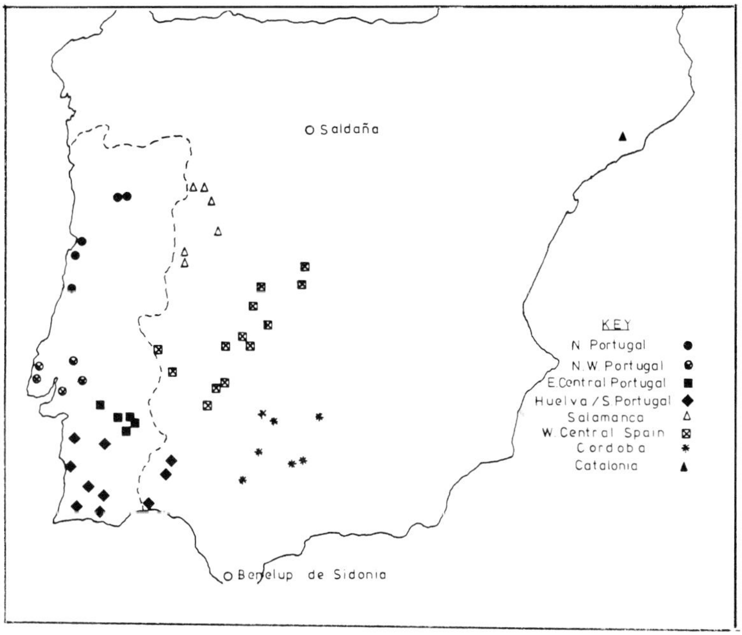

Fig. 3. Accession provenances

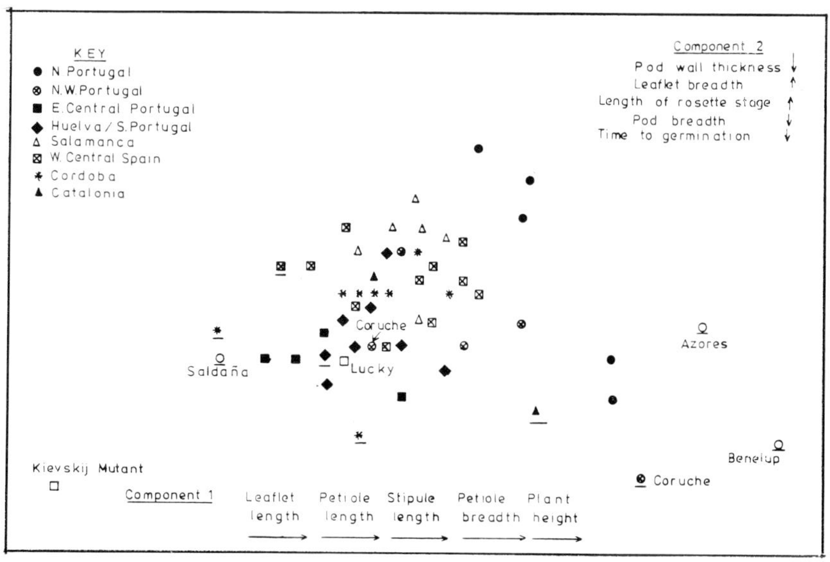

Fig. 4. Iberian collection: component 1 (accounting for 33.16% of total variation) against component 2 (accounting for 14.94% of total variation).

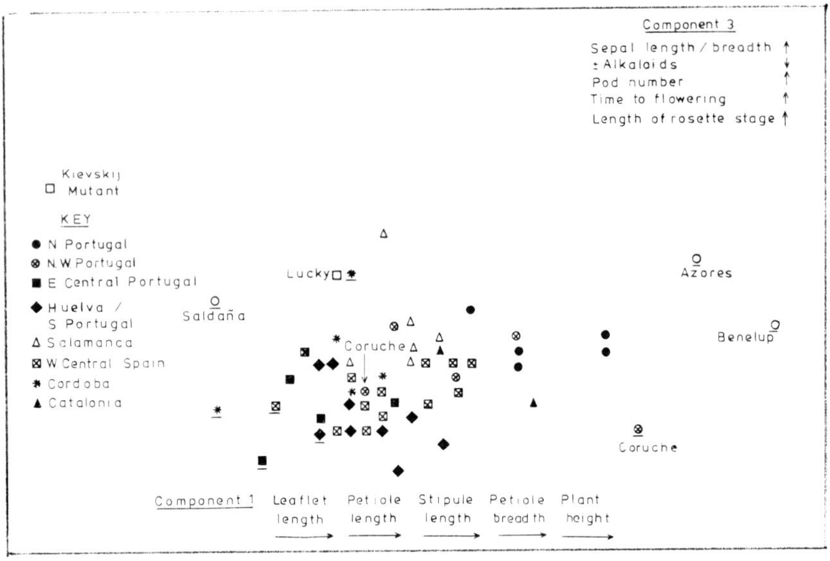

Fig. 5. Iberian collection: component 1 (accounting for 33.16% of total variation) against component 3 (accounting for 9.97% of total variation).

north of Lisbon and they are even more widely separated on the second principal component. The three groups from south of Portugal and Huelva, Cordoba and west central Spain tend to overlap while accessions from Salamanca are more similar to those from Huelva/south Portugal than to northern Portuguese types.

The single accessions from Saldaña in Palencia and Benelup de Sidonia near Cadiz are at opposite ends of the spectrum. The former is more similar to Kievskij Mutant than is any other accession but it is much more similar to accessions from Cordoba and central Portugal. It is also more similar to those southern Iberian types than to Salamanca accessions. This probably reflects the fact that *L. albus* around Saldaña is traditionally spring sown, unlike crops in the rest of Iberia which are almost invariably autumn sown. At the opposite side of the diagram, the accession from Benelup de Sidonia is isolated from the others as is a single accession from the Azores, although the two are more similar to each other (being tall, late flowering, almost winter types) than to any others.

Variation within areas

It is clear that in Portugal as a whole, the level of variation in *L. albus* is greater than it is in Spain. Populations from central, north and north west Portugal are much more scattered than those from any other group.

Differentiation of populations might be expected in areas where habitats are more diverse, however, two examples from the scatter diagrams show that local farming practices can also be important in maintaining crop diversity. The two Catalonian accessions were collected from neighbouring gardens and yet were noticeably different in stature and seed size. This difference is also evident from the scatter diagrams and might be associated with the fact that immigrants from west central and south Spain have introduced their own seed stocks which they now grow alongside native types. A similar separation of plant types was seen near Coruche in central Portugal where two crops producing seeds of markedly different size were growing in neighbouring plots. The scatter diagrams confirm that the two accessions are quite different. (The two accessions are indicated by arrows.)

In Spain, the most variable group is from Cordoba, although most of these are close to the central cluster composed mainly of the less variable central Spanish and Huelva/southern Portuguese groups.

Variation within populations

The coefficient of variation for each character for each accession indicates that variation levels within the majority of the land races studied here are probably not significantly higher than those within the cultivars Lucky and Kievskij Mutant. This might be expected if, as is often the case, farmers retain seed for resowing in successive years and do not use seed from other areas. Particular characters, namely flower colour, pod number and stipule breadth are generally more variable than the others, even within the standards.

Nine accessions (underlined in Figures 4 and 5) appear to be more variable than the others in terms of these characters as well as plant height and leaf size. All nine are placed towards the periphery of the main cluster and are usually separated from other members of their groups. It can be seen that one of each pair of the Catalonian and Portuguese accessions mentioned earlier appear to be more variable than the others perhaps indicating some intercrossing of different types. These observations await verification by multivariate statistical methods.

Implications for *L. albus* collection strategy

If the aim of seed collection is to obtain the maximum amount of genetic variation within a given time, the results of this study suggest that this would best be achieved by covering as many areas as possible, minimising time spent at each collection site. It has been seen that much variation can be found within fairly limited areas (e.g. north and north east Portugal) where there is a diversity of micro-climatic conditions, while even in neighbouring sites, crop differentiation might be evident, depending on local farming practices. In the absence of detailed information on these environmental factors, more genetic resources will usually be made available to the collector by longer journeys to different areas.

CONCLUSION

The grouping of plant types in Experiment 1 displays their similarities and differences and provides a summary of the range of variation available to the breeder. The diversity of types indicated among the Greek accessions indicates that Greece may be a rich source of genetic variation and study of material collected there this year should prove interesting.

The similarity of accessions from Sudan, Egypt and Ethiopia with some of those from north and north-east Europe suggests that collection in countries south-east of the Mediterranean may be valuable in the search for early-maturing *L. albus*.

The range of Iberian types examined in Experiment 2 can best be described as intermediate between spring and winter forms. The spring-sown type from Palencia in northern Spain was placed nearer to Kievskij Mutant than was any other accession, suggesting that this area is worthy of further exploration.

The distribution of variation among Iberian populations and its association with environmental factors will be studied in future analyses.

The useful picture created by PCA of the variation pattern in *L. albus*, especially in the less variable Iberian collection, suggests that extended studies using this and other numerical taxonomic methods could be rewarding.

DISCUSSION

R. Hardwick *(UK)* What non-metric characters did you use and did they contribute much to the analysis?

M.J.A. Simpson *(UK)* By non-metric characters I mean, for example, ratios which are measures of shape rather than size and also multistate characters such as flower colour. Although the characters contributing to the first principal component were mainly metric ones, ratios and multistate characters were usually more important in the second and third principal components.

R. Hardwick Have you thought of using methods such as minimum spanning trees.

M.J.A. Simpson I have not used this method although I have applied group average clustering to the data and obtained broadly similar groupings. It is important to compare results of several different techniques. An important feature of principal component analysis is its identification of characters of primary importance in group separation.

W. Williams *(UK)* You showed greater variation in the Portuguese population. Would you have been able to detect this increased variation by any other means than some form of numerical taxonomic analysis?

M.J.A. Simpson I would say that this is the only means.

DESIGNING AN IMPROVED PEA CROP PLANT

C.L. Hedley, M.J. Ambrose, and K.A. Pyke
John Innes Institute,
Colney Lane, Norwich, Norfolk, UK.

PROBLEMS OF THE CONVENTIONAL DRIED PEA CROP

The dried pea, in common with the field bean (*Vicia faba*) the other major grain legume grown in Britain, suffers from relatively low yields and poor stability of yield compared with cereals. A major contributing factor to the low performance of the dried pea crop is its poor standing ability. The lodged pea crop creates a favourable environment for fungal attack on pods which cause discoloration of the seeds. There is also an increased chance of pod shattering and seed loss. During harvesting the proportion of the lodged crop which may be lost has been shown to be between 13 and 47% (Proctor, 1963) and may be as high as 80% (Davies, 1977).

Poor standing ability is one symptom of a wider problem which is that existing dried pea varieties are not suited to the crop environment. This is perhaps not surprising since such varieties have, in general, been derived from existing horticultural or garden types, selected for their yielding ability as supported spaced plants, rather than for their yield when grown as a crop. Selection criteria would therefore have taken little account of the effects of plant interactions which occur between plants within populations. The main effect of these interactions is greatly to reduce the average size of the crop plant compared with individuals given ample space (Table 1).

One consequence of growing vigorous plants as a crop is that the distribution of yield between individuals within the population becomes skewed. It has been known for many years, mainly through the work of Kira and colleagues (Kira et al., 1953; Hozumi et al., 1955; Koyama and Kira, 1956) in Japan, that plant populations which have an initial normal distribution for dry weight become progressively more skewed as growth continues. The extent to which populations become skewed is dependent on the competitiveness or agressiveness of the individuals, the planting density and the environment, all of which can be

TABLE 1

COMPARISON OF CONVENTIONAL (JI 1198, JI 516 AND JI 1068) AND SMALL STIPULED 'LEAFLESS' (JI 1194, BS 151 AND BS 4) GENOTYPES GROWN AT HIGH (100 PLANTS/m^2) AND LOW (16 PLANTS/m^2) PLANTING DENSITIES.

	100 plants/m^2		16 plants/m^2	
	Plant biomass (g)	Seed weight/ plant (g)	Plant biomass (g)	Seed weight/ plant (g)
JI 1198	4.20	1.88	29.9 ± 3.6*	12.6 ± 2.70
JI 516	4.40 ± 0.72	1.71 ± 0.70	45.4 ± 6.00	23.1 ± 3.10
JI 1068	6.15 ± 0.78	3.30 ± 0.25	39.8 ± 1.33	24.0 ± 1.07
JI 1194	6.97 ± 1.05	2.80 ± 0.55	17.9 ± 4.60	8.26 ± 2.50
BS 151	5.96 ± 0.42	2.80 ± 0.05	23.6 ± 2.01	10.5 ± 1.78
BS 4	6.81 ± 1.10	3.37 ± 0.63	22.3 ± 2.80	12.5 ± 1.88

* One standard deviation from the mean.

manipulated to a greater or lesser extent by the breeder, agronomist or farmer.

It may be assumed that crops in which the yields of individual plants are distributed normally are to be preferred to those where the distribution is skewed. In other words a crop in which all of the individuals yield something is to be preferred to a crop where a few individuals yield much while others yield little or nothing.

Our task at the John Innes Institute is to identify for the breeder those plant characteristics which determine a good dried pea crop plant.

THE SMALL STIPULED 'LEAFLESS' CROP PLANT - THE FIRST IMPROVED MODEL

The first major step in producing a plant model suited to the crop environment was made by Snoad at the John Innes Institute (1972; 1974). Snoad introduced the *st* gene for reduced stipule size and the *af* gene which substitutes tendrils for leaflets into genetic backgrounds derived from commercial varieties. A detailed report of the advantages and disadvantages of this novel phenotype has been published elsewhere (Hedley and Ambrose, 1982). The most obvious advantage of this small stipuled 'leafless' pea is its improved standing ability.

This enables the crop to be more easily harvested and reduces the instances of seed spoilage due to those fungal diseases associated with lodged crops.

A major difference between the conventional and the small stipuled 'leafless' phenotype is the reduced effect on the whole plant and seed dry weights of average individuals when grown spaced or in crop environments (Table 1). The decreased vigour of the small stipuled 'leafless' plants is reflected in the extent of plant interactions within the crop. The degree of skewing from a normal distribution is less than that found for conventional varieties of similar seed size when grown at the same planting density. A greater proportion of the plants within the population will therefore contribute to the yield of the crop.

Although the small stipuled 'leafless' pea has definite advantages as a crop plant over conventional leafed varieties there are also inherent problems with these phenotypes.

Comparisons between conventional varieties and small-stipuled 'leafless' types with similar seed sizes have revealed major differences in crop growth rates between the two phenotypes when grown at the same planting density (Figure 1). This difference in growth rate results in the small stipuled 'leafless' crop being on average seven days behind the conventional in the accumulation of dry matter. A difference of seven days during the exponential phase of growth results on average in a two-fold difference in biological yield at any point in time between the two types of crop. The consequence of this difference in growth rate may be to reduce the total dry matter of the small stipuled 'leafless' crop at the end of the growing season. The seed yield of the small stipuled 'leafless' crop will also be reduced if there is a similar efficiency of partitioning into economic yield between the two phenotypes. If assimilate partitioning within the two crops begins at the same time from sowing then the lower vegetative growth of the small stipuled 'leafless' crop may result in insufficient photoassimilate to satisfy the developing reproductive sinks and to maintain plant growth. Premature senescence of the crop will then occur resulting in a relatively low yield. If partitioning is delayed in the

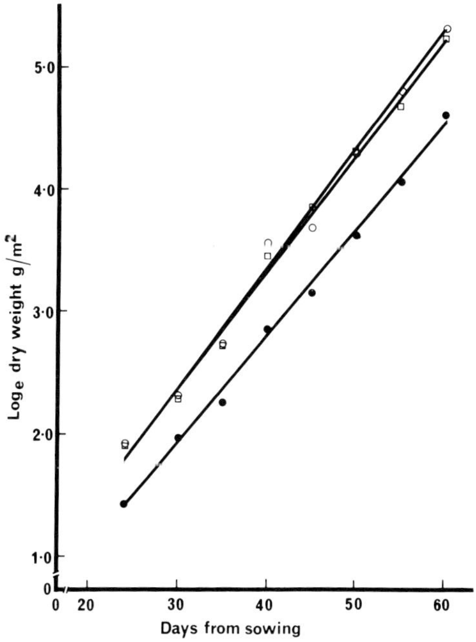

Fig. 1. Regressions of $\log_e$ dry weight g/m^2 on time for conventional (○), small-stipuled 'leafless' (●) and large-stipuled (□) peas.

small stipuled 'leafless' crop until a biological yield equivalent to the conventional crop has been attained, then seed yields may be reduced because of insufficient time in the growing season.

The low crop growth rate of the small stipuled 'leafless' type is due to a reduction in individual plant growth rates compared with conventional plants. Comparisons between the development of conventional and small stipuled 'leafless' genotypes have shown that the growth-rate differences are due mainly to a lower relative growth rate (RGR) for the small stipuled 'leafless' type during the first week after emergence of the seedlings (Figure 2b). No significant differences in RGR were observed prior to emergence (Figure 2a) and only small differences after about seven days from emergence.

There are two ways of overcoming the low crop growth rate in the small stipuled 'leafless' crop. The planting density may be increased, but this will also increase the seed cost compared with conventional varieties, with no guarantee of increased yields. An increase in crop growth rate can also be

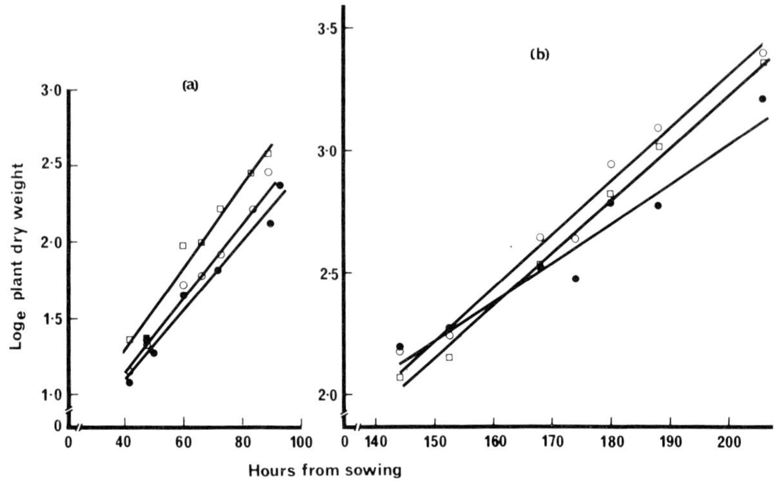

Fig. 2. Regressions of $\log_e$ plant dry weight on time for (a) pre-emergent and (b) post-emergent growth of conventional (○), small-stipuled 'leafless' (●) and large-stipuled 'leafless' (□) peas.

achieved by using genotypes with larger seeds. Large seeded genotypes however must still be sown at relatively high planting densities and intense competition for assimilates will occur within such plants, resulting in seed abortion and a low crop harvest index.

A third alternative for increasing crop growth rates in the 'leafless' phenotype is to change the crop plant model such that the plant growth rate is equivalent to that of the conventional crop, while maintaining the advantages of the small-stipuled 'leafless' plant. We have shown that 'leafless' plants with normal stipules (*ST*) have similar growth rates to conventional peas (Tables 1, 2a and 2b). We are at present studying the underlying basis of this similarity to conventional peas and apparent difference to the small stipuled 'leafless' type. 'Leafless' peas with excessively large stipules do not appear to have the required standing ability. We are determining therefore the optimum size and shape of stipules required to increase growth rate while maintaining the advantages of the small stipuled crop. This larger stipuled 'leafless' pea must still retain those features required for a good pea crop plant (Hedley and Ambrose, 1982) and these may be summarised as follows:-

1. The plant must be tolerant of neighbours within the crop and a normal distribution for plant weight within the population must be maintained throughout development.

2. To reduce competition during early crop development, the formation of non-productive branches must be inhibited by selecting plants which are genetically non-branching or which branch late and which will then be inhibited by competition within the sward.

3. The plant should be early flowering and have an indeterminate habit, to increase the duration of assimilate partitioning and decrease the level of competition between reproductive sinks for assimilates.

4. Each reproductive node will contain one and no more than two fruits selected for maximum efficiency of carbon refixation and recycling.

5. To reduce competition between seeds, seed size must not be large and the rate of seed development should be low.

REFERENCES

Davies, D.R., 1977. Restructuring of the pea plant. Sci. Prog. (Oxf.) 64, 201-214.

Hedley, C.L., and Ambrose, M.J., 1982. Designing 'leafless' plants for improving yields of the dried pea crop. Adv. Agron. (In press) Vol. 34, 225-275.

Hozumi, K., Koyama, H. and Kira, T., 1955. Intraspecific competition among higher plants. IV. A preliminary account on the interaction between adjacent individuals. J. Inst. Polytech. Osaka City Univ. Ser. D. Vol. 6, 121-130.

Kira, T., Ogawa, H. and Sakazaki, N., 1953. Intraspecific competition among higher plants. I. Competition-yield density interrelationship in regularly dispersed populations. J. Inst. Polytech. Osaka City Univ. Ser. D. Vol. 4, No. 1, 1-16.

Koyama, H. and Kira, T., 1956. Intraspecific competition among higher plants. VIII.Frequency distribution of individual plant weight as affected by the interaction between plants. J. Inst. Polytech. Osaka City Univ. Ser. D. Vol. 7, 73-94.

Proctor, J.M., 1963. An experiment to determine the effects of date of sowing on the yield and quality of harvesting threshed peas. J. Agric. Sci. 61, 281-289.

Snoad, B., 1972. Ann. Rep. John Innes Inst. Norfolk, England. No. 63, 29-33.

Snoad, B., 1974. A preliminary assessment of 'leafless' peas. Euphytica Vol. 23, 257-265.

DISCUSSION

R. Hardwick *(UK)* I found your resolution of the difference between your two genotypes to within the first 200 hours of growth very intriguing. Is this through differences in mobilisation of cotyledonary reserves, or of slower deployment of leaf area in the 'leafless' type?

C.L. Hedley *(UK)* At the moment we are studying photosynthesis during the early stages of development. Leaves and stipules act as both sources and sinks. It may be that at about emergence the 'leafless' type may rely on storage reserves for longer because of its smaller leaf area.

R. Hardwick The rate of decrease of the cotyledonary reserves may in part answer the question.

R. Thompson *(UK)* Can the yield of leafless peas be expected to increase much for plant densities exceeding 100 plants/m^2?

C.L. Hedley I don't think yield from 'leafless' varieties ever significantly exceeds that from conventional types, irrespective of plant density. The yield from 100 plants/m^2 is not necessarily the ceiling yield; it depends on the 'leafless' genotype used.

ENVIRONMENTAL FACTORS AFFECTING PLANT DEVELOPMENT IN *LUPINUS ALBUS*: THE EFFECT OF CHILLING AND PHOTOPERIOD DURING SEEDLING DEVELOPMENT ON FLOWERING

W. Williams and S.K. Brocklehurst
Department of Agricultural Botany, University of Reading, Whiteknights, Reading RG6 2AS, UK.

THE EFFECT OF CHILLING AND PHOTOPERIOD

The establishment of *Lupinus albus* as a major grain legume depends on the success that can be achieved in modifying its pattern of growth and maturation so as to increase the rate and synchronisation of stem and seed development. It has been established that growth rate in the species responds according to genotype, to temperature treatments (vernalisation effects) and to photoperiod during the seedling and pre-flowering stages. (Rahman and Gladstones, 1972; Almeida, 1966). These responses are, however, highly variable from season to season, making it impossible to predict the optimal sowing date for the promotion of a balanced development of the vegetative and reproductive phases and also ensure early maturity with a maximum harvest index.

The historical range of cultivation of the species covers countries bordering the north coast of the Mediterranean from the Iberian Peninsula to the coastal regions of Turkey and the Near East, as well as along the Nile Valley as far south as Sudan. Establishment of the crop has almost invariably been practised in the autumn and early winter when the seasonal rains promote establishment and development before the onset of periods of water stress during spring and summer. Most of the locally adapted populations cultivated in the regions of historical distribution tend to be leafy types possessing a relatively slow rate of stem extension during early plant development, especially when photoperiods are short, a large number of nodes on the mainstem, significantly greater leaf area per plant and markedly higher seed production potential. They also develop into tall plants with indeterminate growth of the lateral branches which therefore mature late.

Many of these growth characteristics are far from ideal in the shorter growing seasons that prevail after spring sowing in cooler latitudes. Improvements in agronomic adaptability over a wider geographical range have been achieved by selecting cultivars, e.g. Blanca (W. Germany), Kali (Poland) and Kievskij Mutant (USSR) which have a reduced response to low temperature after germination and a more rapid rate of early development. Even such cultivars, however, show varying growth patterns in different seasons which indicates that they retain the same degree of sensitivity to temperature and photoperiod effects at certain stages in development.

The experiments described here were designed to investigate the extent and timing of the sensitivity of cultivars of *L. albus* to temperature and to photoperiod with a view to identifying more accurately stages of development which might yield the most effective selection response in plant breeding programmes.

MATERIALS AND METHODS

The cultivars used, together with a brief description of their seasonal adaptation, chief agronomic characteristics and country of origin are given below:

a) Kali - white flower, Poland; Kievskij Mutant - USSR; Blanca - white flower, W. Germany; Pflug's Ultra - W. Germany. These cultivars are spring types, varying in rate of growth and earliness - Kali is the earliest and Pflug's Ultra the latest when grown at Reading.

b) Tapioszele - Hungary; Nahrquell - E. Germany; Multolupa - W. Germany. These are intermediate in growth between the most extreme spring cultivars listed in (a) and the typical winter types in group (c) below.

c) Hope - USA; VLS - USA. Extreme winter types with a dense rosette form of growth in the seedling stage, large number of nodes on the mainstem, very tall and late flowering.

Except for the field grown experiment, plants were grown under glass and provided with supplementary heat to provide a temperature of 65°F. The photoperiod treatments were provided by natural daylight, supplemented by daylight fluorescent lamps and, where stated, by incandescent tungsten bulbs. Low temperature treatments comprised three weeks at 2°C or 5°C at stages of development of the apical meristem which were determined by dissection of spare plants grown under identical conditions. Seeds of *Lupinus albus* possess four fully developed vegetative primordia at the apical meristem at the stage when they become fully imbibed with water. A further four primordia are differentiated by the stage when the cotyledons begin to emerge above the soil surface, and when the first true leaves become visible, the total number of primordia at the apex is ten.

Development was recorded according to three phases: Phase I from seed imbibition to the differentiation of first flower primordium, Phase II from end of stage I to the appearance of the first flower, and Phase III from the end of stage II to the development of the first pod 1 cm long.

RESULTS

The effect of photoperiod

The four cultivars, Kievskij Mutant, Tapioszele, VLS and Hope were used to study the effect of photoperiod. Two photoperiods, 14 h and 8 h were applied using 'daylight' fluorescent lamps to supplement the natural daylight which varied between 11¼ and 7½ h during the course of the treatments. The results are set out in Tables 1 and 2.

Development during Phase I was accelerated in all cultivars at 14 h day length; the reduction in development time ranging from 6 - 10 days. Although there was a two to three-fold difference in the duration of Phase I in the different cultivars, the effect of day length on development rate was remarkably constant suggesting that the response is quantitatively related to the genetical control of development and that no switching mechanism with major effect is involved.

There was also acceleration of development of Phase II, and here the later cultivars showed a much greater response relative to Kievskij Mutant and Tapioszele; the latter averaging only a seven day reduction with increased photoperiod while in VLS and Hope the equivalent difference was 14 days. It is not

TABLE 1

EFFECT OF PHOTOPERIOD ON DURATION (DAYS) OF PHASES OF DEVELOPMENT IN 4 CULTIVARS OF *L. albus*

Cultivar	Photoperiod							
	14 h Phase				8 h Phase			
	I	II	III	Total	I	II	III	Total
Kievskij Mutant	15	67	8	90	21	74	8	103
Tapioszele	25	73	7	105	35	80	8	123
VLS	35	63	11	109	37	79	13	129
Hope	47	57	12	116	52	69	19	140

Phase I - Germination - flower initiation

Phase II - Flower initiation - first open flower

Phase III - First open flower - 1 cm pod

TABLE 2

EFFECT OF PHOTOPERIOD ON MAINSTEM HEIGHT (cm), LEAF NUMBER AND LEAF AREA (cm^2) IN CULTIVARS OF *L. albus*

Cultivar	Photoperiod					
	14 h			8 h		
	Height	Leaf No.	Leaf area	Height	Leaf No.	Leaf area
Kievskij Mutant	26	12.5	74	22	12.8	82
Tapioszele	36	16.7	246	24	17.0	213
VLS	36	16.4	286	22	18.4	288
Hope	35	20.4	297	25	19.9	248

possible, however, confidently to relate this difference exclusively to photoperiod since the later cultivars were undergoing Phase II 10 - 20 days later and were therefore under conditions that are not strictly comparable. The same lack of precision applies to the interpretation of differences noted during Phase III.

Quite apart from the responses to photoperiod, differences in rate of development between cultivars which are very closely correlated to the spring or winter form are very clear during

Phase I, and to a lesser extent in Phase III; the differences between cultivars during Phase II, although of the same order during both photoperiods, do not form any obvious pattern that can be confidently related to seasonal growth adaptation.

Height differences appear between the cultivars only from the 14 h photoperiod, whereas leaf number gave clearly different values under both photoperiods (Table 2). It is clear, therefore, that increased height was achieved under 14 h day length as a result of increased internode length. At 8 h photoperiod all genotypes remained short and active node development conferred a rosette habit to winter types. The large difference in internode extension in the winter forms was only expressed under conditions of long day.

Differences in total leaf area were very marked especially between the extreme spring form, Kievskij Mutant, and the others. The very much greater total leaf area of the later, winter forms, is thus compounded of changes in leaf number and increased area of individual leaves.

The effect of vernalisation

In one experiment (Table 3) three cultivars were subjected to vernalisation treatments of 5°C for 21 days at three different stages of seed and seedling development. Chilling was timed to ensure that plants were simultaneously at the two leaf stage in all treatments before transfer to the glasshouses. The plants were grown in a heated glasshouse in natural daylight supplemented with fluorescent light to provide a constant 12 h photoperiod.

A second experiment (Tables 4 and 5) using Kievskij Mutant involved cold treatment at 2°C for 21 days and the plants were grown in an unheated greenhouse with supplementary light from High Pressure Mercury Vapour lamps providing a 14 h day length.

As with photoperiod the effect of vernalisation on rate of development is clear only during Phase I. No differences can be detected between treatments in the duration of Phase II and the differences recorded for two of the cultivars in Phase III are almost certainly due to temperature effects during flowering

TABLE 3

EFFECT OF VERNALISATION (21 DAYS AT 5°C) ON DURATION (DAYS) OF PHASES OF DEVELOPMENT IN THREE CULTIVARS OF *L. albus*

Vernalisation stage	Kievskij Mutant				Tapioszele				VLS			
	Phase											
	I	II	III	Total	I	II	III	Total	I	II	III	Total
Imbibed seed	9	54	18	81	14	60	13	87	14	60	20	94
Expanded cotyledon	9	56	18	83	14	70	11	95	14	63	18	95
2 emerged leaves	17	59	11	87	22	64	10	96	22	62	14	98
Unvernalised	23	59	14	96	25	63	13	101	23	65	14	102

TABLE 4

EFFECT OF STAGE OF VERNALISATION ON DURATION (DAYS) OF DEVELOPMENTAL PHASES IN KIEVSKIJ MUTANT

Treatment stage	Phase			Total
	I	II	III	
Unvernalised	26	48	11	85
4 Primordia (seed fully imbibed)	15	53	8	76
8 Primordia (Cotyledons emerged)	18	53	8	79
10 Primordia (Primary leaves visible)	26	48	9	83
SED - 9DF	1.2	1.5	1.3	1.3

and pod set. Of particular interest in Table 3 is the effect of chilling related to morphogenesis of the apex; maximal developmental rates were expressed in seedlings treated at or before the expanded cotyledon stage when less than ten primordia had been initiated. The response when two true leaves had emerged was much reduced in the early cultivar Kievskij Mutant whereas in the later forms there was no effect. The response to chilling at a lower temperature (Table 4) confirms those of the first experiment except that in this test even the early

TABLE 5

EFFECT OF STAGE OF VERNALISATION ON STEM AND FLOWER CHARACTERS IN KIEVSKIJ MUTANT

Treatment stage	Mainstem height (cm)	Node number	Flower number	Pod number	% pod set
Unvernalised	30	11.8	6.8	3.7	54
4 Primordia (seed fully imbibed)	25	10.3	7.0	4.3	60
8 Primordia (Cotyledons emerged)	30	11.8	9.5	5.3	56
10 Primordia (Primary leaves visible)	31	11.3	8.8	4.8	55
SED - 9DF	1.7	0.4	0.6	0.4	3.8

cultivar, Kievskij Mutant, showed no response at the true leaf stage. Plant height was affected only by the earliest chilling treatment (Table 5) but there was an unexpected increase in flower and pod number on plants vernalised at the later stage of development.

The effect of vernalisation on field grown plants (Tables 6, 7 and 8)

Vernalisation at 2°C was imposed on imbibed seed for 21 days, and the seedlings grown to emergence under glass before transplanting to the field at 36 plants per m^2 in randomised complete blocks with three replicates, each plot containing 36 plants. Records on Phases I and II were confined to mainstem development; for primary and secondary branches the records cover days to the appearance of the first flower on the respective branch orders.

The effect of vernalisation on Phase I of the mainstems, leading to 17 days earlier flowering, was the largest recorded in any of the experiments. The effect of vernalisation on development of primary and secondary branches, paralleled almost precisely the mainstem effects, indicating that the later flowering of the higher order branches was a carry-over from the mainstem response which persisted throughout the observation period.

TABLE 6

EFFECT OF VERNALISATION ON DURATION (DAYS) OF DEVELOPMENTAL PHASES OF FIELD-GROWN PLANTS OF KIEVSKIJ MUTANT

Stem part	Treatment	Phase I	Phase II	Phase III	Total to maturity
Mainstem	Unvernalised	27	41	11	180
	Vernalised	10	45	11	168
		(2.2)*	(1.4)	(0.7)	(1.1)
Primary branches	Unvernalised	80		14	179
	Vernalised	67		13	170
		(2.8)		(1.6)	(1.5)
Secondary branches	Unvernalised	98		9	189
	Vernalised	85		9	174
		(1.9)		(1.0)	(1.0)

* ( ) = SED, 2DF

TABLE 7

EFFECT OF VERNALISATION ON MAINSTEM HEIGHT, NODE AND BRANCH NUMBER IN KIEVSKIJ MUTANT GROWN IN THE FIELD

	Unvernalised	Vernalised	
Mainstem height (cm)	20	16	(1.6)*
Node number	13.8	9.7	(0.5)
Primary branch number	3.7	3.1	(0.1)
Secondary branch number	8.6	5.3	(0.6)
Basal branch number	2.8	0	(0.3)

( )* = SED, 2DF

Mainstem height, node and branch number (Table 7) were all reduced by chilling, causing a more determinate habit and an agronomically more acceptable level of vegetative growth. It is noteworthy, however, that many components of yield (Table 8) including flower, pod and seed number and seed yield per plant were all adversely affected by vernalisation suggesting that for comparable yields under ideal ripening conditions, crops grown from vernalised seed should be grown at a higher population density.

TABLE 8

EFFECT OF VERNALISATION ON COMPONENTS OF YIELD PER PLANT IN KIEVSKIJ MUTANT GROWN IN THE FIELD

	Unvernalised	Vernalised	
Flower number	42	36	(1.8)
Pod number	21	15	(1.3)
Total seed number	65	49	(1.3)
Total seed weight (g)	20	15	(0.9)
Weight/100 seed (g)	30.8	31.2	(0.6)
Dry weight (g)	40	31	(1.8)
Harvest index	50	50	(2.6)

The effect of photoperiod and vernalisation (Tables 9 and 10)

The combined effects of vernalisation of imbibed seed at 2°C for 21 days and varying photoperiods were tested on Kievskij Mutant. The plants were grown under glass, all receiving a minimum of 8 h natural daylight; longer photoperiods were applied in light-proof compartments by means of tungsten filament lamps. Each photoperiod was replicated twice within which vernalised and unvernalised lots were randomised in single blocks.

The effects of treatments on rate of development were again largely confined to Phase I, and predominantly to plants which had been chilled; there was some indication of a slight effect of photoperiod at 14 h on unvernalised plants. The combination of the photoperiod and vernalisation reduced the duration of Phase I by 30%, 6 days.

Mainstem height increased with day length as a result of lengthening of the internodes showing that plant form responded to photoperiod irrespective of chilling treatment. Vernalised plants had fewer internodes and therefore reduced height at all photoperiods. The attainment of maximal height at 12 h with a reduction at 14 h which seems to be indicated by the data in Table 10 cannot unfortunately be checked with the results in Table 2.

TABLE 9

EFFECT OF VERNALISATION AND PHOTOPERIOD ON DURATION (DAYS) OF DEVELOPMENTAL PHASES IN KIEVSKIJ MUTANT

Treatment	Phase			Total
	I	II	III	
Unvernalised				
8 h	25	42	14	81
10 h	24	43	14	81
12 h	25	38	14	77
14 h	22	40	13	75
Vernalised				
8 h	20	41	10	71
10 h	17	43	12	72
12 h	16	40	11	67
14 h	14	43	13	70
SED - 24DF	1.4	1.8	1.1	1.5

TABLE 10

EFFECT OF VERNALISATION AND PHOTOPERIOD ON MAINSTEM HEIGHT (cm), INTERNODE LENGTH (cm) AND NODE NUMBER IN KIEVSKIJ MUTANT

Treatment	Plant height	Internode length	Node number
Unvernalised			
8 h	24.6	1.7	14.9
10 h	28.5	2.0	14.5
12 h	45.3	3.1	14.2
14 h	37.6	2.8	13.5
Vernalised			
8 h	18.3	1.5	12.3
10 h	27.8	2.1	12.4
12 h	36.2	3.0	12.0
14 h	34.0	3.0	11.5
SED - 24DF	3.7	0.7	0.3

CONCLUSIONS AND DISCUSSION

Vernalisation and photoperiod treatments separately increased rate of development and altered plant height in all cultivars, but the photoperiod effect was more marked in plants that had been vernalised than in those subjected to higher growing temperatures throughout their life cycle. Vernalisation of plants grown in the field under identical photoperiod increased the rate of development of the mainstem apex very markedly as compared with unvernalised control and the effect persisted throughout the life of the plant. Increased rate of development was invariably associated with reduced stem length and greater dominance of the mainstem apex giving reduced branching at all nodes and a more acceptable level of determinacy and of vegetative growth. These effects were unfortunately correlated with reduced pod and seed number per plant and with an unchanged harvest index.

The most important results from these experiments are the relationships between the stage of the apex and the degree of response to vernalisation. At the temperatures applied, the vernalisation effect is greatest and most consistent at stages of seedling development between full seed imbibition and cotyledon emergence. At cotyledon emergence the apical meristem has eight primordia microscopically visible, so that the complete morphogenetic pattern of only 12 - 14 vegetative primordia in the early cultivar Kievskij Mutant is probably physiologically finally determined at this time, thus rendering vernalisation treatments at later stages ineffective.

It is a common experience with spring-sown crops of *Lupinus albus* that the rate and form of plant development cannot be accurately related to sowing date. In some seasons the earlier sowings condition rapid development and shorter plant height while in others this pattern is reversed. The results presented here suggest that the reason for the variation in response to sowing date relates to the temperature during the earliest stage of germination; early spring-sown seed when followed by relatively high temperature during germination will experience minimal vernalisation effects as compared with later sown seed exposed to cold spells.

Since the response to vernalisation of even the earliest cultivars now available determines earliness and plant habit in *L. albus*, the elimination of sensitivity to temperature differences at the seedling stage ranks as a major priority for improving the adaptation of the species to a wider range of geographic and climatic conditions.

REFERENCES

Almeida, J.R.M. de, 1966. Mutantes Doces No Melhoramento de Forragens. Agronomia Lusitanica. Vol. XXV 311-323.

Rahman, M.S. and Gladstones, J.S., 1972. Control of lupin flower initiation by vernalisation, photoperiod and temperature under controlled environment. Australian Journal of Experimental Agriculture and Animal Husbandry. 12, 638-645.

DISCUSSION

M. Lenoble *(France)* Vernalisation of the Kievskij mutant occurs in response to relatively high temperatures which explains why the same number of nodes occurred on vernalised and non-vernalised plants. A cultivar which requires a relatively low temperature for effective vernalisation responds with a much reduced number of nodes, number of flowers, and so on.

W. Williams *(UK)* That is absolutely true. The surprising thing was that when Kievskij mutant was unvernalised, at a short photoperiod, it produced the same number of nodes as cv. Hope which is a very high vernalisation requiring winter genotype. Another surprising result was that Kievskij mutant, which is the truest spring type that we have, has still so much of a residual requirement for vernalisation. This is why it is as unpredictable in its temperature response in the spring as intermediate and winter types. Even the so-called spring types do have a vernalisation requirement which militates against obtaining a repeatable result with lupin. The only way to increase predictability in this respect is through genetic control. However, we now know that we can look for response at the early seedling stage.

I also think it is necessary genetically to eliminate the photoperiodic response. In this material a photoperiodic response is obvious but it is not the positive response seen in winter cereals where the apex becomes competent to flower, or remains incompetent to flower. In Kievskij it is rather a quantitative effect of increasing extension and increasing the rate of development. A winter crop of *L. albus* in southern Spain, Portugal or Italy, will respond to short day photoperiod with excessive vegetative growth. There will be failure of apical dominance.

R. Thompson *(UK)* Exactly how were the photoperiods given?

W. Williams Using light tight compartments natural daylength was supplimented as necessary by mercury vapour lamps.

R. Thompson This means that photoperiodic effects and photosynthesis and growth are confounded.

R. Hardwick *(UK)* I agree. I wonder whether the Reading workers have considered the possibility of fluid drilling vernalised seed. Is it necessary to remodel *Lupinus* by changing the genotype? Would it not be possible to change the phenotype by vernalising the seed and then using the fluid drill to sow the germinating seed into the soil?

W. Williams I am sure that would be possible.

J. Picard *(France)* Once a genetic solution has been achieved it is usually the easiest one to use in practical terms.

W. Williams A major consideration with technology such as fluid drilling is whether the returns on the crop bear the cost.

M. Frauen *(FRG)* Are autumn sown crops more predictable than spring sown, and how much winter hardiness is available in *Lupinus albus*?

W. Williams My impression based on limited experience of the winter sown crop is that it also is unpredictable; under winter conditions in Spain and Portugal, for example, very little vernalisation occurs under the high winter temperatures.

Regarding the presence of winter hardiness, results have been conflicting. In England a winter sown crop would be unsuccessful whilst in southern Italy there would be every chance of success. Between these countries, in Portugal, Spain and France, exchange of material has shown variable results depending on the precise location.

INDUCED DUPLICATIONS IN CHROMOSOME SEGMENTS OF PEA AND THEIR SIGNIFICANCE FOR SEED PROTEIN STUDIES

S. Grillo[1] and F. Saccardo[2]

[1]Istituto di Agronomia - Cattedra di Miglioramento Genetico, Facoltà di Agraria, Portici - Naples,

[2]Laboratorio Valorizzazione Colture Industriali, CNEN, CSN della Casaccia, S. Maria di Galeria, Rome, Italy.

INTRODUCTION

Several mutants with dicentric chromosomes have been induced by mutagenic treatments in peas, tomato, durum wheat (Saccardo and Monti, 1974) and field beans (Errico et al., 1982).

Studies have been made on the phenotypic effect of these chromosome aberrations in peas, on their mitotic and meiotic behaviour and on their transmission through different generations (Monti and Saccardo, 1969, 1979; Saccardo, 1971).

Lines with duplications in some chromosome segments were obtained in the progenies of pea mutants with dicentric chromosomes and this paper reports the mechanism of their induction. The paper also reports on the protein content and on the electrophoretic pattern of the seeds of one of these pea lines, to determine whether the induced duplication affected the seed proteins.

MECHANISM OF INDUCTION OF CHROMOSOME SEGMENT DUPLICATIONS

In previous research on the fodder pea cv. Parvus, several Ml variegated plants were isolated after X-irradiation of pollen, which were characterised by the presence of dicentric chromosomes. The variegated plants showed a mixochimeric structure due to the 'breakage-fusion-bridge' cycles of the dicentric chromosomes.

These aberrant chromosomes were not only persistant but also transmissible to subsequent generations. Studies at meiosis showed the heterodicentric nature of these chromosomes, which associate with their homologues giving origin to rings of 4 - 5 chromosomes. In the disjunction of such associations, by breakage of the double bridges in anaphase, gametes with deleted or duplicated chromosome segments can also be hypothesised (Figure 1), with the production in the following generations of not-variegated plants in which a segment of one of the two chromosomes involved in the dicentric could be duplicated.

On the basis of such findings and considerations, mutagenic treatments of pollen by X-radiation (750 Rads) on the cv. Sprinter produced three variegated mutants with dicentric chromosomes from 183 M1 plants, corresponding to 1.63% which is similar frequency to that found in the cv. Parvus (1.80%).

The progenies from the M1 variegated plants were examined with particular care and among the M2 plants only those with normal phenotypes were taken to the next generation. From the M3 and subsequent generations only the best lines were selected taking into account only yield and seed quality traits. In the M6 generation two of these lines (7631 and 76124) in replicated trials had high yield and better seed quality than the mother line cv. Sprinter. Cytological analysis of these mutants (Table 1) showed the presence of two longer chromosomes (Figure 2) in their genomes, normal meiosis and pollen fertility (98 - 99%).

TABLE 1

CYTOLOGICAL ANALYSIS, POLLEN FERTILITY AND SEED SETTING IN TWO MUTANT PEA LINES AND IN THEIR CROSSES WITH cv. SPRINTER

Material	Chromosome number at mitotic metaphase 2n=	Bivalents at meiotic MI	Pollen fertility (%)	Ovules per pod no.	Seeds per pod no.
<u>Lines</u>					
Sprinter	14	7	99.00	5.0	3.1
7631	12 + 2 with dupl.*	7	98.00	6.2	4.6
76124	12 + 2 with dupl.*	7	99.00	6.2	5.0
<u>F1</u>					
Sprinter x 7631	13 + 1 with dupl.*	7	95.00	6.0	4.1
7631 x Sprinter	13 + 1 with dupl.*	7	80.03	6.6	4.0
Sprinter x 76124	13 + 1 with dupl.*	7	96.02	6.2	3.8

* Chromosomes characterised by duplicated segments.

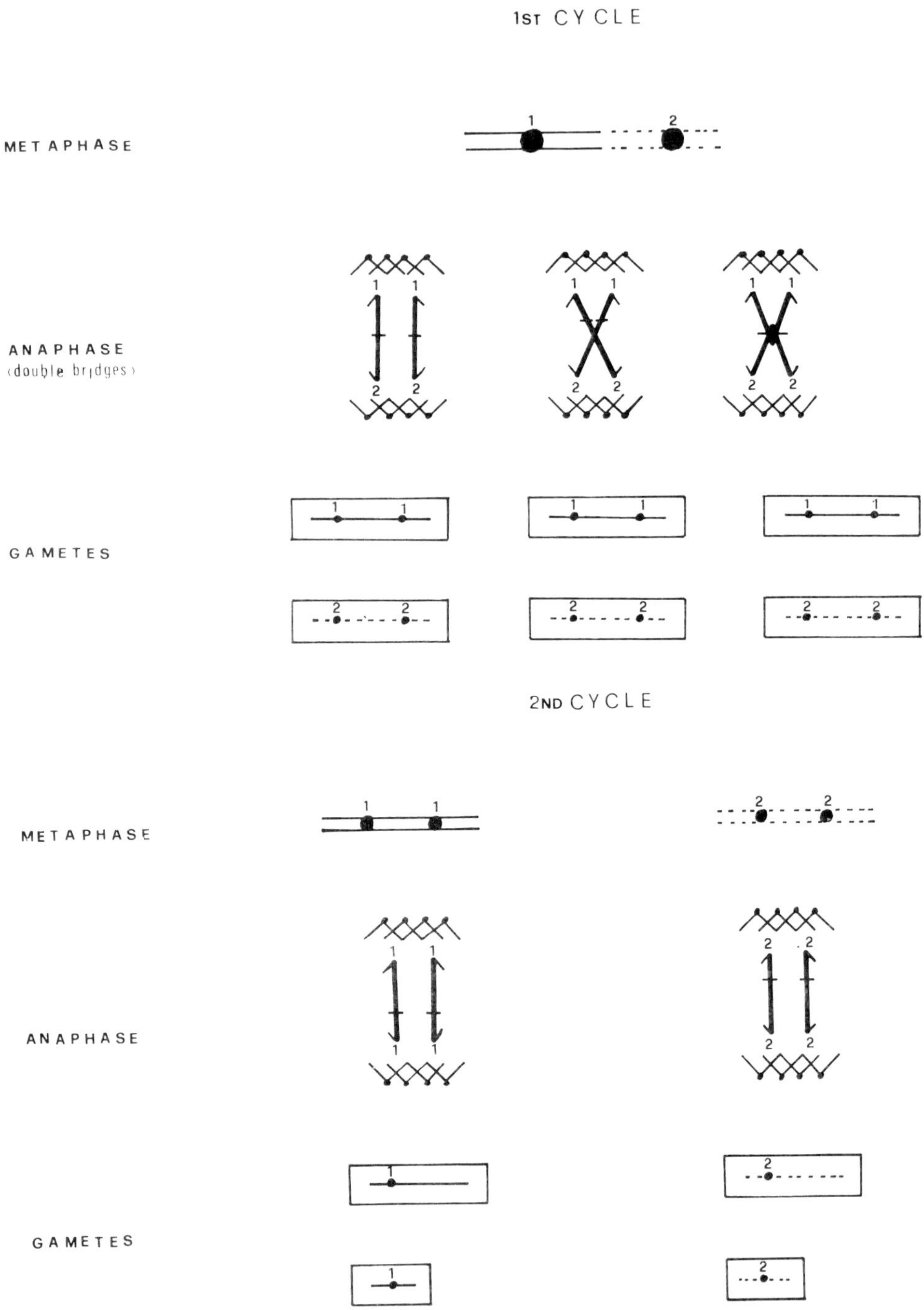

Fig. 1. Behaviour of dicentric chromosomes through the meiotic cycles.

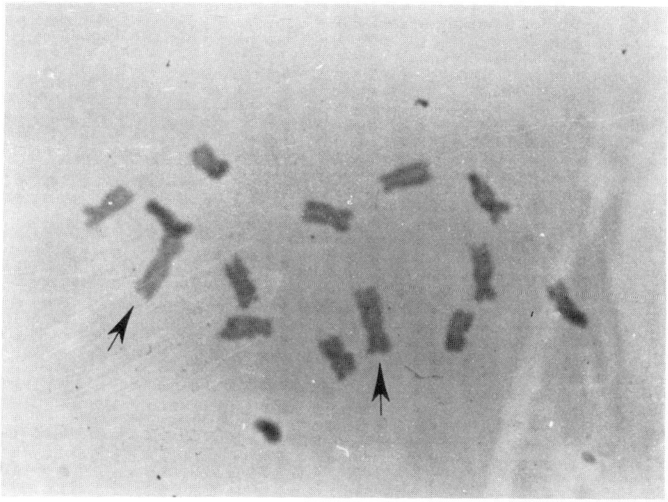

Fig. 2. Line 7631 with 2n = 14; the two abberrant chromosomes are indicated.

In the F1 obtained between the mutant line 7631 and the cv. Sprinter, only one longer chromosome, normal meiosis and pollen fertility (80 - 95%) were found. It was then confirmed that the two pea mutant lines selected in M6 are characterised by the presence of a duplicated chromosome segment. Preliminary results on line 7631 seem to indicate that the duplication is on chromosome number 6.

SEED PROTEIN ANALYSIS

Line 7631 was analysed to see if the cytological changes influenced the amount and quality of seed protein. Total N contents of the ripe seeds were determined by the micro-Kjeldahl method and line 7631 had a value of 5.1% N compared with 4.7% for the mother line Sprinter. The determination of total N at different stages of seed maturity confirmed these data (Figure 3) and a constant difference between the two genotypes was observed.

No differences in non-protein-nitrogen were found between the two lines and this suggests that the differences in total N is usually due to differences in protein nitrogen which in the mutant line was 3% higher than that in the mother line (32.3% against 29.3%).

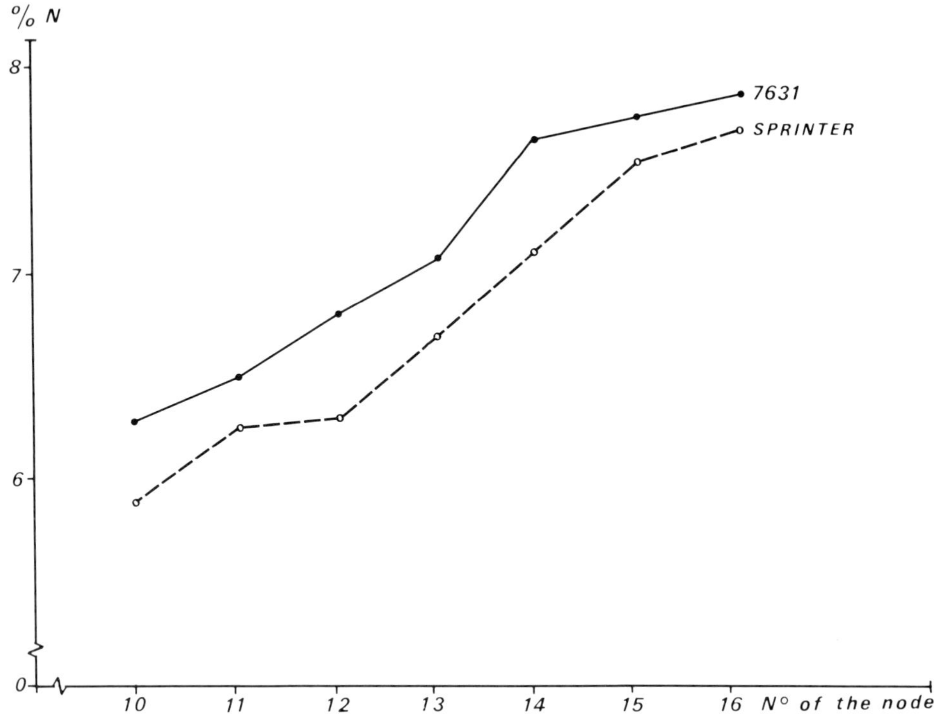

Fig. 3. Relationship between N in the seed and the node of origin of the seed.

On the basis of these results, research was started to explore possible differences in the protein electrophoretic patterns between the two lines. The crude extracts and total globulins were prepared according to the method described by Rao and Pernollet (1981). The samples were analysed on 4% stacking and 12% running polyacrylamide gel in the presence of SDS using the Laemmli discontinuous buffers system (Laemmli, 1970).

No difference between 7631 and the mother line Sprinter was observed in the electrophoretic patterns of the crude extracts and total globulins (Figure 4). No data have yet been obtained on the fractionation and characterisation of vicilin and legumin and on quantitative analysis of the albumin and globulin fractions.

CONCLUSIONS

The mutation technique proved to be more appropriate and

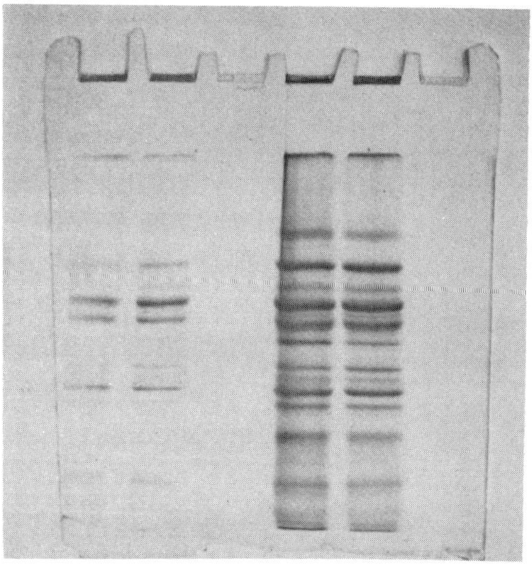

Fig. 4. SDS-elecrophoretic patterns. From left to right: crude extracts of Sprinter (a) and 7631 (b); globulins of Sprinter (d) and 7631 (c).

faster than other techniques (Hagberg et al., 1972) adopted to induce duplications. This increases the prospects of widening the genetic variability for pea breeding as exemplified by line 7631 which showed a 3% increase in the protein content in comparison with its mother line.

Loci for vicilin and legumin fractions have been localised on chromosomes 7 and 2 (Casey, 1982) while the chromosome modified in line 7631 is probably chromosome 6.

Mutants with duplications in other chromosomes will be isolated by the same technique and studied, in order to find lines with different electrophoretic patterns, which should increase our understanding of the synthesis and structure of pea proteins.

REFERENCES

Casey, R., 1982. The genes for *Pisum* legumin. In: Perspectives for peas and lupins as protein crops (Eds. R. Thompson and R. Casey). This publication.

Errico, A., Saccardo, F., Monti, L.M. and Conicella, C., 1982. Comportamento di un cromosoma dicentrico radioindotta in *Vicia faba* L., XXV Convegno SIGA. Gardone Riviera - Brescia (Italy). May 28-30, 1981.

Hagberg, A., Persson, G. and Hagberg, G., 1972. Utilization of induced chromosome abberrations. Translocations, duplications and trisomics in barley. IAEA Meeting, 'Induced Mutations and Plant Improvement'. Buenos Aires, 1970.

Laemmli, U.K., 1970. Cleavage of structural proteins during the assembly of the head of Bacteriophage T4. Nature, 277-680.

Monti, L.M. and Saccardo, F., 1969. Mutations induced in pea by X-irradiation of pollen and the significance of induced unstable chromosomes in mutagenic experiments. Caryologia, 22, 81-96.

Monti, L.M. and Saccardo, F., 1979. The gamete irradiation technique in pea genetics and breeding. Monografia IV di Genetica Agraria: 141-152.

Rao, R. and Pernollet, J.C., 1981. *Pisum sativum* seed storage globulins as genetic markers for wrinkled and smooth seed character: comparison of genotypes through different protein extraction and electrophoretic procedures. Agronomie, Vol. 1.

Saccardo, F., 1971. Behaviour of dicentric chromosomes in peas. Caryologia, 24, 71-84.

Saccardo, F. and Monti, L.M., 1974. Isolation of aneuploids originating from induced unstable chromosome abberrations. 'Polyploidy and Induced mutations in plant breeding'. IAEA, Bari: 277-287.

DISCUSSION

W. Williams *(UK)* Can you give an estimate of the standard deviation or the standard error of the value for percentage protein that you quoted?

F. Saccardo *(Italy)* Not yet because these are preliminary results.

W. Williams This is very important in relation to protein percentages. They interact so violently with the environment with regard to maturity, amount of nitrogen and so on. I would like to ask the editors of the proceedings of this meeting if they would obtain from the contributors estimates of the variations of these mean figures for publication. Otherwise it will be very difficult to assess the real value of the figures. (Whilst sympathetic with the view expressed, this would cause unacceptable delay in publication and the reader must therefore attach the appropriate interpretation to means quoted without errors. Editors).

R. Casey *(UK)* Measures of protein content are obtained in several different ways including ninhydrin analysis by alpha-amino nitrogen, and total nitrogen in association with a conversion factor usually 6.25. The value of 6.25 has been shown by Sosulski in Saskatoon to be inappropriate to pea protein and a value of 5.5 would be better. Some standardisation is obviously needed. It may well be that some of the very high protein contents quoted are not as high as they appear to be.

J. Picard *(France)* Perhaps reference should be made to total nitrogen without use of a correction factor. Alternatively, the correction factor used should be clearly stated.

IN VITRO BREEDING OF LUPINS

Christine Sator

Institute of Crop Science and Plant Breeding,
Federal Research Center of Agriculture,
Braunschweig-Völkenrode (FAL),
Bundesallee 50, 3300 Braunschweig,
Federal Republic of Germany.

INTRODUCTION

There is an increasing need for plant protein for human and animal consumption and legumes are an obvious choice. Among the temperate legumes lupins have the highest protein content (Table 1).

Lupins were used as crop plants before von Sengbusch selected the sweet varieties in the 1920s. Lupin seeds have been used both as animal feed-stuff and also in human diet. As can be seen from Figure 1, lupin seed can be prepared for a wide variety of culinary uses. During, and after, the first World War lupins were used industrially in large amounts. This culminated in a lupin banquet given by the German Association of Applied Botany in 1918 in Hamburg when lupin soup and lupin steak, roasted in lupin oil and spiced with lupin extract, were served on a tablecloth made of lupin fibre. Lupin margarine, with 20% lupin constituents, was served as well as cheese made from lupin protein, lupin liquor and lupin coffee. Lupin soap, lupin paper and lupin adhesives were also available at that time (Becker-Dillingen, 1929). This wide range of uses for lupin and its high protein content and good quality suggests little need to adapt soyabean for the Central European climate, but rather that it would be more worthwhile to breed greater stability and higher yields into lupin.

Conventional breeding of lupins has been successful through intraspecific crosses and the selection of mutants. X-ray treatment of seeds has resulted in improved total dry matter and flowering but it has not led to better seed quality (Hackbarth, 1957) (Table 2). Interspecific crosses (Roemer, 1916; Gollmick, 1937; Jaranowski, 1962; Williams, 1980) have all been unsuccessful.

TABLE 1

NUTRIENTS IN LEGUME SEEDS

Plant species	Proteins in % t m	Oil in % t m	Lysin in %	Meth. protein	Cyst.
Vicia faba	25.5 - 35.4	1.2 - 2	5.9	0.8	1.4
Pisum sativum	18.3 - 28.7	0.6 - 5.5	7.0	0.7	-
Lupinus luteus	36.0 - 47.6	4.0 - 7.1	2.8 - 6.4	0.3 - 1.0	0.7 - 4.8
L. albus	34.3 - 44.9	9.9 - 14.5	5.0 - 5.8	0.3 - 0.5	1.0 - 1.6
L. angustifolius	28.0 - 37.9	5.3 - 6.6	3.8 - 6.0	0.4 - 1.0	0.5 - 0.9
L. mutabilis	31.7 - 45.9	13.1 - 23.1	5.3	0.4 - 1.4	1.4
Glycine max	40	21	6.6	1.2	1.2
		FAO Standard	5.4	3.5	

From Brouwer, W., 1976, Hill, G.D., 1977 and Anon, 1965.

TABLE 2

THE TYPE AND ORIGIN OF MUTATIONS

Plant species	Mutation	Plant part Leaf	Stem	Flower	Pulse	Seed
Lupinus luteus	spontaneous	1	-	2	2	6
	X-ray	11	2	1	2	-
Lupinus angustifolius	spontaneous	2	-	5	-	9
	X-ray	5	-	-	-	-

The recently developed technique of protoplast fusion has led to 18 different interspecific and intergeneric hybrid plant varieties. The hybridisations have been done mainly within the family Solanaceae with Datura, Atropa, Nicotiania, Petunia, Solanum and Lycopersicum species. Fusion also has been done successfully with two Daucus species and between Daucus and Aegopodium within the Umbelliferae. Satisfactory results have been achieved with Arabidopsis and Brassica within the Cruciferae (Harms, 1980). Even though positive results in fusing protoplasts are limited to these three plant families, this new breeding method provides the possibility of overcoming the

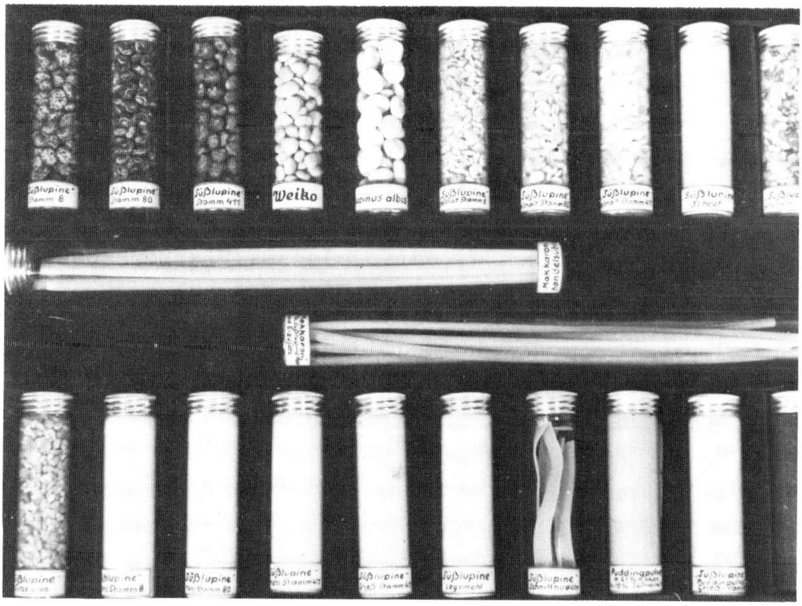

Fig. 1. Multiple uses of lupin seeds.

incompatibility which may exist between plant species. It may, therefore, be useful for producing interspecific lupin crosses to obtain desirable characteristics from two species. For example, the previously mentioned high protein content of the yellow lupin *Lupinus luteus* and the high seed yield of *Lupinus polyphyllus* or *Lupinus hartwegii*, or the technique may be used for exploring crosses of *Lupinus albus* with *Lupinus mutabilis* to combine high protein and high oil content with good yields.

Protoplasts may be fused in two different ways (Table 3). Protoplasts may be isolated from leaf tissue or callus from the heterozygous cross-parents. This will lead to a tetraploid fusion product. The second possibility may be to produce haploid plant material of the two cross-parents and to fuse haploid protoplasts which will lead to a diploid fusion product. This second possibility might simplify subsequent breeding.

TABLE 3

POSSIBILITIES OF FUSING PROTOPLASTS OF *Lupinus luteus* AND *Lupinus polyphyllus*

	Plant	Protoplast	Fusion product	Hybridplant
I	*lut* 2n →	*lut* 2n	*lut*/*poly* 4n →	4n
	poly 2n →	*poly* 2n		
II	*lut* n →	*lut* n	*lut*/*poly* 2n →	2n
	poly n →	*poly* n		

Techniques for successful culture of *Lupinus* callus are restricted to those described for *Lupinus polyphyllus* by Wink et al. (1980). Moreover, neither isolation nor regeneration of protoplasts from *Lupinus luteus*, *Lupinus polyphyllus* or *Lupinus hartwegii* has been described.

ANTHER CULTURE

Since Guha and Maheshwari (1964) succeeded in producing haploid Datura plants from cultured anthers, the method has been used successfully with many plant species to produce haploids from microspores. Using the pollen of lupins at the stage known to be most suitable for callus/embryo induction for many other plant species and with sugar, 2, 4 D and kinetin in the culture medium, satisfactory levels of callus formation were obtained for *Lupinus polyphyllus*.

The phases of pollen development were found to correspond to changes in both anther size and bud colour. The microspores change from the uninucleate to binucleate stage as the anthers change from green to yellow, and whilst the petals are still green. It is from pollen at this stage of development that haploid callus is most readily obtained. Sixty percent of anthers at this stage, of about 2.5 mm in length, produced callus.

After callus induction in the dark, green callus is formed in the presence of light (Figures 2 and 3), first growing soft and amorphously, later becoming hard. At this stage globular structures may arise which may differentiate into roots or shoots (Figures 4, 5, 6). Callus with shoots is induced to produce roots after transfer to a rooting medium (Figure 7).

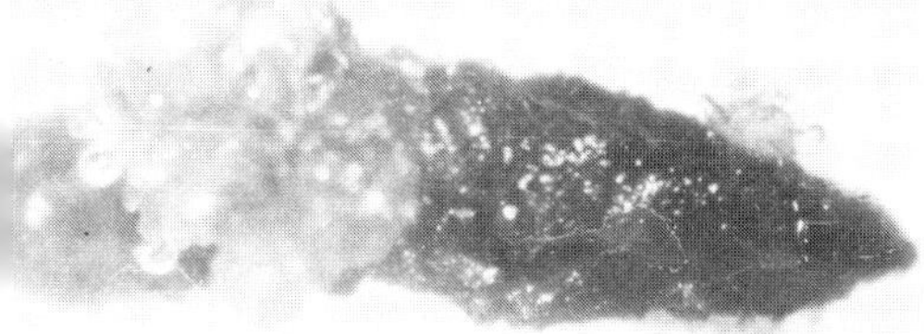

Fig. 2. Anther of *Lupinus polyphyllus* with callus.

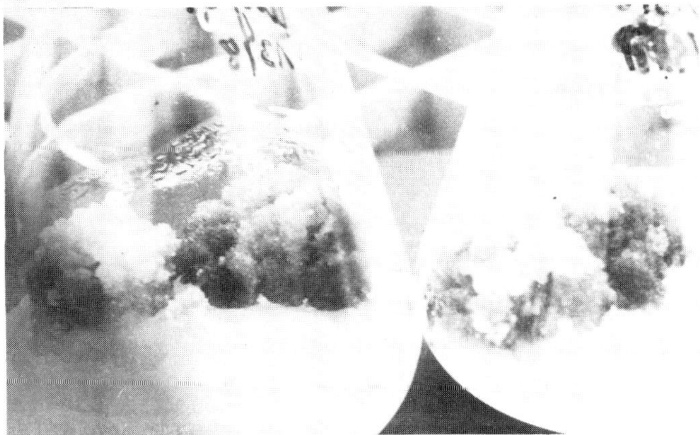

Fig. 3. Soft amorphous callus of *Lupinus polyphyllus*.

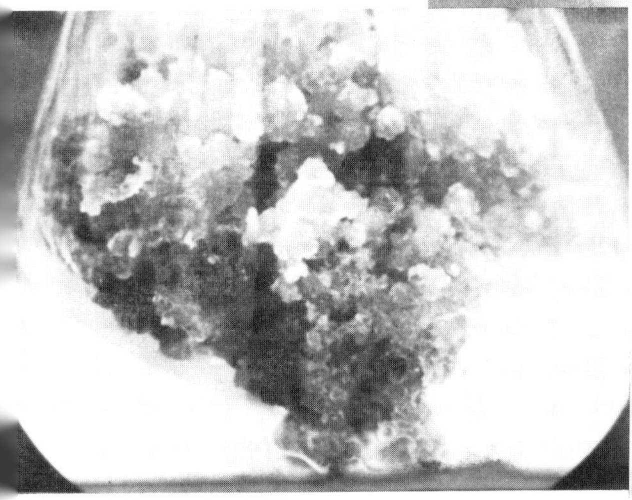

Fig. 4. Hardened callus of *Lupinus polyphyllus* with globular structures.

Fig. 5. Shoot from *Lupinus polyphyllus* of anther derived callus.

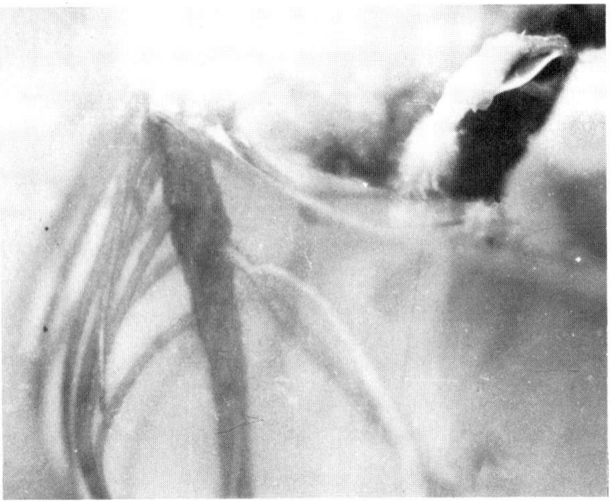

Fig. 6. Roots from *Lupinus polyphyllus* of anther derived callus.

Fig. 7. Shoots and roots from *Lupinus polyphyllus* differentiated after callus formation of anthers.

It has not yet been determined whether plants produced from anthers of *Lupinus polyphyllus* are haploid or diploid. If they turn out to be haploid these plantlets may serve as material from which to isolate haploid protoplasts. Regeneration from *Lupinus hartwegii* or *Lupinus luteus* has not been achieved. Callus of *Lupinus luteus* is difficult to grow but current results are promising.

PROTOPLAST CULTURE

There are difficulties in isolating protoplasts from lupin species. The only experiment describing their isolation (Binding et al., 1981) failed to obtain live dividing protoplasts. But since isolation and regeneration of protoplasts from other leguminous species has been done successfully, for instance Dos Santos et al. (1980) regenerated *Medicago sativa* and Gresshoff (1980) regenerated *Trifolium repens* from protoplasts, success might reasonably be anticipated for lupins. Moreover, prospects exist for successful protoplast fusion between different lupin species because success has been obtained between legumes and species from other families. For example soyabean and tobacco protoplasts have been successfully fused (Kao, 1977) and Binding (1978) achieved fusion between *Vicia faba* and *Petunia hybrida*, but in neither experiment did hybrid plants result.

Our experiments on the isolation and culture of protoplasts have included different enzyme mixtures and concentrations as well as changes in the pH value or the addition of $CaCl_2$, which is said to protect the protoplasts after isolation, as well as changing the composition of the washing solution for the isolated protoplasts. Isolation of protoplasts from *Lupinus hartwegii* proved to be the easiest followed by protoplast isolation from *Lupinus polyphyllus*. Protoplast yields from *Lupinus luteus* were very poor. None of the isolated protoplasts showed cell division. Nevertheless it seems worthwhile continuing with the work because of the potential advantages of protoplast fusion for breeding new lupin varieties.

MERISTEM CULTURE

In addition to the above new method for breeding, tissue culture may be useful for the rapid multiplication of valuable plant material. So far, about 700 plant species have been regenerated by vegetative propagation *in vitro* (Pierik, 1979). Although lupins are not mentioned in Pierik's review it is clear from our own experiments that this is possible. Seed may also be used as a source for callus, which means that examples with attractive qualities could be multiplied rapidly, after analysis of a small sample of the seed itself (Figure 8).

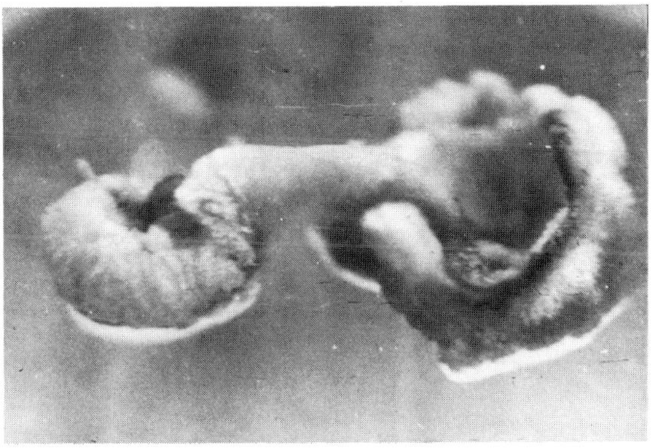

Fig. 8. Callus formation at half of a seed from *Lupinus polyphyllus*.

REFERENCES

Anon, 1965. Handbuch der Lebensmittelchemie BDI. Springer Verlag Berlin, Hamburg, New York.

Becker-Dillingen, J., 1929. Handbuch des Hülsenfruchtbaues und Futterbaues, Berlin, Paul Parey.

Binding, H., Nehls, R., Kock, R., Finger, J. and Mordhorst, G., 1981. Comparative studies on protoplast regeneration in herbaceous species of the dicotyledoneae class. Z. Pflanzenphysiol. 101, 119-130.

Binding, H. and Nehls, R., 1978. Perspectives in Plant Cell and Tissue Culture. Ed. I.K. Vasil. International Review of Cytology Suppl. 11 B. Academic Press. Mol. Gen. Genet. 164, 137-143.

Brouwer, W., 1976. Handbuch des speziellen Pflanzenbaues Bd II. Paul Parey Berlin-Hamburg.

Dos Santos, A.V.P., Outka, D.E., Cocking, E.C. and Davey, M.R., 1980. Organogenesis and somatic embryogenesis in tissue derived from leaf protoplasts and leaf explants of *Medicago sativa*. Z. Pflanzenphysiol. Bd 99, 261-270.

Gresshoff, P.M., 1980. *In vitro* culture of white clover: callus, suspension, protoplast culture and plant regeneration. Bot. Gaz. 141/2, 157-164.

Gollmick, F., 1937. Über die Artkreuzungen bei Lupinen. Züchter 9, 65-68 In· Handbuch der Pflanzenzuchtung II Aufl. Bd IV. Paul Parey, 1959 Berlin-Hamburg.

Guha, S. and Maheshwari, S.C., 1964. *In vitro* production of embryos from anthers of Datura. Nature 204, 497 ff.

Hackbarth, J., 1957. Die Gene der Lupinenarten I, II, III. Z. Pflanzenzücht. 37, 1 ff, 81 ff, 185 ff.

Harms, Ch., 1980. Pflanzliche Zell- und Gewebekulturen. Umschau 80, 707-712.

Hill, G.D., 1977. The composition and nutritive value of lupin. Commonwealth bureau of nutrition 47/8, 511-529.

Jaranowski, J., 1962. Fertilization and embryo development in the genus *Lupinus*. Genetica Polonica 3, 209 ff, 333 ff In: Lehrbuch der Züchtung landwirtschaftlicher Kulturpflanzen Bd 2 Paul Parey 1970.

Kao, K.N., 1977. Perspectives in Plant Cell and Tissue Culture. Ed. I.K. Vasil. International Review of Cytology Suppl. 11 B. Mol. Gen. Genet. 150, 225-230. Academic Press 1980.

Pierik, R.L.M., 1979. *In vitro* culture of higher plants. Wageningen, The Netherlands.

Roemer, Th., 1916. Züchtung alkaloidarmer Lupinen Landw. Jahrbuch 50, 433-443. Handbuch der Pflanzenzüchtung Bd. IV II. Aufl. Paul Parey 1959.

Williams, W., Akhtar, M.A. and Fauyi, M., 1980. Cross compatability between European and American *Lupinus* species. Botanical Journal of the Linnean Society 81/3, 225-232.

Wink, M. et al., 1980. Alkaloid pattern of cell suspension cultures and differentiated plants of *Lupinus polyphyllus*. Planta medica 3/138, 238-245.

DISCUSSION

W. Williams *(UK)* Is anther culture successful and if so are the resulting plants haploid?

C. Sator *(FRG)* Anther culture is possible but we have not yet determined whether the plantlets are haploid.

W. Williams It would be very interesting if haploid plants are produced. All European lupins have 32 or more chromosomes and even though they behave genetically like diploids, they are almost certainly polyploids.

RATES OF POD AND SEED DEVELOPMENT IN *LUPINUS ALBUS*, *L. MUTABILIS*, *VICIA FABA*, *PISUM SATIVUM* AND *LATHYRUS LATIFOLIUS*

S. Davies and W. Williams
University of Reading, Department of Agricultural Botany, Whiteknights, Reading RG6 2AS, UK.

INTRODUCTION

The slow rate of maturation of some grain legume species constitutes a serious limitation to their success at the more northerly latitudes where summers are cool and the humidity level during harvest months is high. The very large and fleshy pod walls characteristic of cultivars of *Lupinus albus* mature and desiccate very slowly. Preliminary observations suggest that there is a wide range of variation in seed and pod maturation between the two major temperate grain legumes *Pisum sativum*, *Vicia faba* and lupin species.

In this investigation the development of tissue of the pods was recorded at successive periods after pollination in five species representing four genera of grain legumes. The relationships between pod growth and water content and morphogenesis of the embryo were examined in detail.

MATERIALS AND METHOD

The species and cultivars used in the study and their origin were as follows:

Lupinus albus cv. Kievskij Mutant - early type ex USSR.
Lupinus mutabilis - population Jauja Market - early unselected type, ex Peru.
Pisum sativum cv. Little Marvel - rapid growth, dwarf, ex UK.
Vicia faba cv. Maris Bead - spring type, moderately early, ex UK.
Lathyrus latifolius sp. *albus* - ornamental and wild - European distribution.

All the material was grown under glass during the spring and summer months from early March to the end of September. The reference date for records was anthesis, and values for different intervals are the mean of 5 observations on different plants selected at random.

Fresh weight and moisture content of pods

Changes in fresh weight and moisture content of pods are given in Figures 1 and 2.

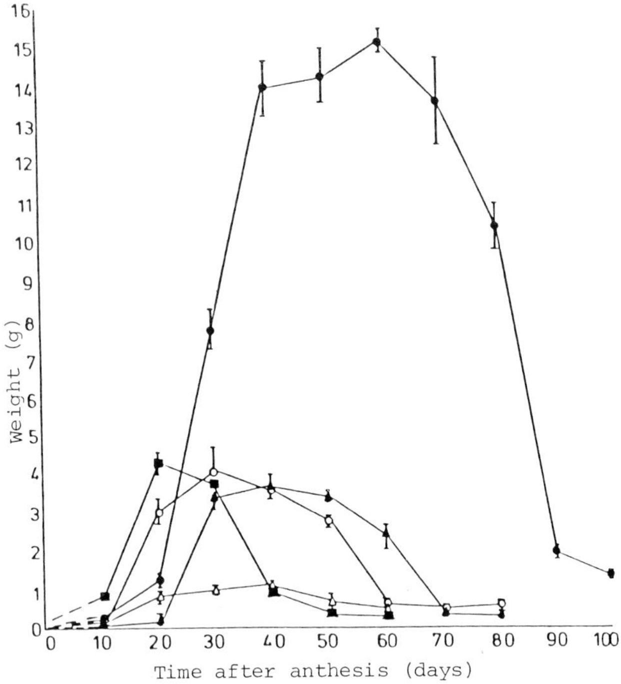

Key: *L. albus* ●
L. mutabilis ○
P. sativum ■
V. faba ▲
Lathyrus sp. △

Fig. 1. Comparison of pod fresh weight of chosen legumes. Each point is the mean of 5 replicates.

The lowest maximum pod weight was recorded in the wild species *Lathyrus latifolius*. The maximum pod weights were not significantly different in *Vicia faba*, *Lupinus mutabilis* and *Pisum sativum*, but time taken to achieve these weights differed, *Pisum sativum* and *Vicia faba* being respectively the earliest and latest

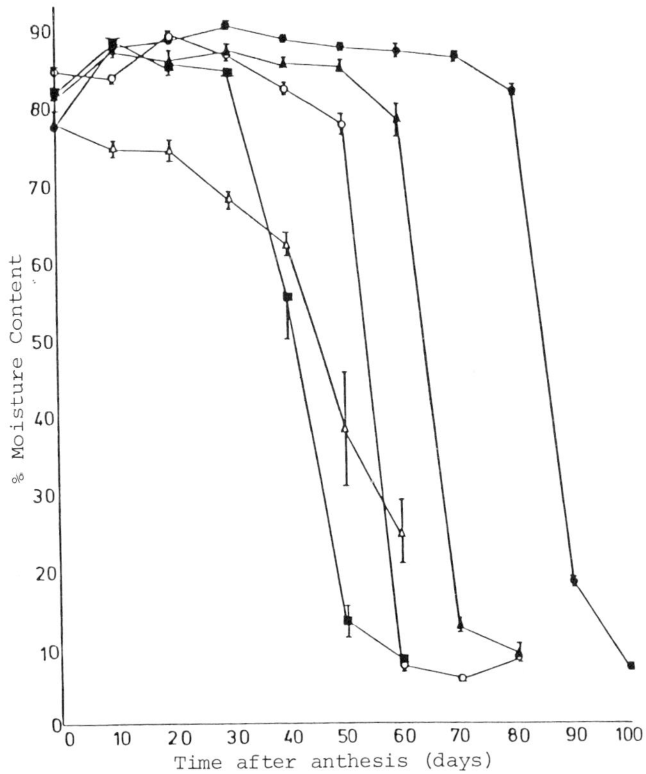

Key: *L. albus* ●
L. mutabilis ○
P. sativum ■
V. faba ▲
Lathyrus sp. △

$$\% \text{ MC} = \frac{\text{Fresh wt.} - \text{Dry wt.}}{\text{Fresh wt.}} \times 100$$

Fig. 2. Comparison of pod moisture content of chosen legumes. Each point is the mean of 5 replicates.

with *Lupinus mutabilis* intermediate.

The pattern of weight accumulation in pods of *Lupinus albus* was outstandingly different from the other four species, both in respect of the maximum weight attained and in the very long interval during which increases in weight occurred.

The point at which the pods ceased to lose weight was the time when the mature harvest stage had been reached. *Pisum sativum* and *Lathyrus latifolius* were the earliest to achieve constant weight at about 50 days from anthesis while *Lupinus mutabilis* and *Vicia faba* were at the same stage 10 and 20 days later respectively, and *Lupinus albus* only reached a constant pod weight at day 90 after anthesis, fully 40 days later than *Pisum sativum*.

The water content of the pods remained fairly constant, after an initial rise during the first ten days after anthesis, until the final stage of maturation when water loss was rapid. Significant losses in fresh weight were observed in all species except *Lathyrus latifolius* prior to loss of moisture. Thus the pods were in a state of negative balance in respect of photosynthate some considerable time before tissue desiccation had commenced.

Rates of embryo and seed development

The records of seed dry weights in Table 1 show several very significant differences in the time taken to achieve maximum seed weights in the different species. At day 40 after anthesis, *Pisum sativum* had completed seed filling following a very rapid rate of dry matter accumulation between days 30 and 40. *Lupinus mutabilis*, the second species in order of rate of dry matter accumulation in the seed, had achieved approximately 70% of its maximum weight at this stage, whereas the equivalent for *Vicia faba* and *Lupinus albus* were less than 46% and 35% respectively. The delay in active seed development and seed filling in *Vicia faba* and *Lupinus albus* extends over a surprisingly long period of at least 30 days after anthesis, and in the latter seed weight increases are relatively low until after day 40. The onset of rapid accumulation of dry matter in the seed is closely related to the achievement of maximum fresh weight (Figure 1) of the pods.

TABLE 1

AVERAGE EMBRYO DRY WEIGHT (mg)

Species	Days after anthesis				
	10	30	40	50	60
L. albus	0.90	5.0	93	267	409
L. mutabilis	0.70	25	187	267	243
Pisum sativum	0.30	44	303	279	294
Vicia faba	0.20	3.0	218	393	473
Lathyrus spp.	0.03	1.0	32	47	54

The germination tests given in Tables 2 and 3 confirm in other ways the relative rate of maturation of seeds in the species. Seeds of *Pisum sativum* are fully mature and competent to germinate at day 25, followed by *Lupinus mutabilis* and *Vicia faba* at day 45, whereas full seed maturity in *Lupinus albus* was achieved only after the excessively long period of 70 days.

TABLE 2

GERMINATION OF EMBRYOS EXCISED ON TO FILTER PAPER (Max No. = 10)

Days after anthesis	Species *L. albus*	*L. mutabilis*	*P. sativum*	*V. faba*
15	0	0	1	0
25	0	3	10	0
35	0	0	10	2
45	0	10	10	8
55	5	10	10	8
70	9	9	9	7
80	10			9

TABLE 3

GERMINATION ON FILTER PAPER OF EMBRYOS MATURED IN EXCISED PODS (Max No. = 10)

Days after anthesis	Species *L. albus*	*L. mutabilis*	*P. sativum*	*V. faba*
15	0	0	1	0
25	0	1	10	1
35	2	10	10	10
45	10	10	10	10
55	10	10	10	10

Maturation of the seeds in pods (Table 3) had no effect on germination in *Pisum sativum* whereas in the late maturing species, maximum germination figures were attained 10 - 25 days earlier. *Lupinus albus*, which showed the most significant improvement in germination following pod maturation, significantly had the heaviest pods indicating that a substantial proportion of the dry matter in the seed of this species is normally derived from the pods.

DISCUSSION AND CONCLUSION

These experiments have revealed important variations in pod development and in the rate of seed formation in some important species of grain legumes. These differences directly affect the range of climatic adaptation of the species as crop plants and thus deserve special consideration.

Although preliminary observations suggest that late maturation of seed and fruit in some of these species is conditioned by excessive tissue development of the pods, it is clear from the results that pod growth and desiccation is closely related to morphogenesis of the embryo. Changes towards early maturity need therefore to be considered from the standpoint of interdependent processes in both the embryo and the pod wall. Thus, it is probable that it is the dependence of the embryo on the moisture balance and photosynthate supply from the pod tissues that condition the greater pod development in *Lupinus albus* and, also, to a slightly lesser extent, in *Vicia faba*. Final desiccation of the fruit tissue seems to be triggered by rapid seed filling and the demands of advanced states of seed maturity.

The difference between *Pisum sativum* and the other species in respect of seed maturation within prematurely excised pods is further evidence of differences in the contribution of dry matter by the pods during seed development. In *Lupinus albus* because of the apparent large contribution of the pods during development, seed maturation in excised pods conditioned full capacity for germination 40 - 45 days before normal pod desiccation. At this stage the plant had achieved the stage when field crops are sufficiently dry for direct machine harvesting of the seed. Thus direct threshing of standing crops must frequently be a less satisfactory alternative than ground drying of the cut crop before harvesting, especially in areas where there is a premium on early harvest to avoid the higher humidity of the autumn months.

DISCUSSION

C.L. Hedley *(UK)* Lupin is the only one of those plants you described in which the pod makes a net contribution to the carbon economy of the plant during seed development. What effect do you think altering the development of the pod may have on carbon economy?

Also, what temperature was used for the studies on pod development?

W. Williams *(UK)* I can't give the mean temperature but the plants were grown in the glasshouse during the summer months, and therefore at relatively high temperatures.

There is a net contribution of carbon from the pods but it is quite small. On the basis of published work by Pate and our own experiences with defoliation experiments, the contribution of the pods is not likely to exceed 10% of total assimilation. The position of the pods relative to the canopy is obviously important, for following full canopy development, the pods will be at or near compensation point.

C.L. Hedley Even 10% could be important.

M. Lenoble *(France)* It may well be that a prolonged ripening period in *Lupinus albus* is necessary to ensure a high oil content. If this is so then selection for earliness could reduce oil content.

W. Williams This could well be important but we have not yet explored the relationship between maturity and oil content. Seeds on secondary branches which develop late have a higher oil content. The secondary and tertiary pods are borne high in the canopy and therefore intercept more light during the later stages of development and our evidence suggests that this results in their higher oil content.

Figures I have given relate to pod and embryo development on the main stem. We now have some indication that there is less delay in embryo formation on the secondary and tertiary branches. There seems to be a stage of plant development when in response to a stimulus, possibly triggered by day length, there is an increase in the rate of embryo growth and rapid mobilisation of assimilate, including alkaloids, from the leaves to the pod.

J. Picard *(France)* A similar situation occurs in *Vicia faba* where there is a period of arrested development of the pods at the bottom of the plant.

R.C. Hardwick *(UK)* Observations of quiescent organs are often shown to correlate with 'competition' for carbohydrates and photosynthates from some other part of the plant. Correlations of growth patterns within the plant may be very fruitful in this respect.

PHYSIOLOGICAL PERSPECTIVES ON YIELD COMPONENTS IN PEAS

R.C. Hardwick
National Vegetable Research Station,
Wellesbourne, Warwick, UK.

In the early years of this century, when the concept of yield components was introduced, botany was taking its first steps towards scientific maturity. Much published work was still descriptive and phenomenological, but an increasing number of workers were exploring more rigorous quantitative approaches to their subject. Yield analysis was conceived by Balls and Holton in 1915, and taken up after the War by Engledow and Wadham (1923) and others. At about the same time, growth analysis was introduced (Briggs et al., 1920). Both yield analysis and growth analysis start with a major parameter (yield, growth rate) and express it as the product of a series of components. This 'top down' approach has a strong intuitive appeal and both techniques have continued in vogue among plant physiologists for more than sixty years. Yet I would argue that growth analysis and yield analysis have contributed little of practical value to programmes of plant breeding and crop improvement, and it seems to me that the need for simple descriptive techniques is almost at an end. As we come to understand more of the biochemistry and biology of plant cells and cell organelles, plant breeders will seek to replace traditional breeding methods by genetic manipulation; they will want to handle nucleotide sequences, not components of yield. Similarly, the advances now in plant and crop physiology are in crop models, not growth analysis. If the 'bottom-up' approaches are successful botany will have moved out of the descriptive phase and become a mature, predictive science. The difficulty I foresee is that whether the approach is bottom-up or top-down, all paths lead to the same central conundrum - the nature of the integration of plant activities. This problem was encountered by yield analysis in the 1960s and it led to a certain amount of disillusion. It will pose equal difficulties for crop modellers and for genetic engineers, but it has not, so far as I am aware, been faced up to by those disciplines. Some progress has been made on the yield analysis front and in this paper I wish to explore how, at the level of yield components, we can put some aspects of the phenomenon of the integration of plant activities onto a quantitative footing.

Consider the basic yield analysis equation. In the case of *Pisum sativum* (Wellensiek, 1941) the yield components are

Weight per pea	(w)
Number of peas per pod	(p)
Number of pods per raceme	(r)
Number of racemes per plant	(n)
Number of plants per unit area	(d)

The algebraic product of these components is numerically equal to yield

$$Y = w.p.r.n.d.$$

Since the racemes of *Pisum* are produced in a monopodial sequence, meaningful extra information can be gained by making a separate yield analysis at each node and writing the yield components as column vectors. The yield equation is then expressed using matrix algebra (Hardwick and Milbourn, 1967). In either case, the yield analysis equation is a valid description, *post hoc*, of what happened in a given experiment, in any given treatment or cultivar. But if it is to be useful, yield analysis needs to be valid in a predictive sense. Can we predict the effect on Y of changing one of the yield components r? The equation might seem to predict that doubling the value of r (the number of pods per raceme) would lead to a doubling of yield - but our experience is that this does not happen (Hardwick et al., 1979). Genetic analysis shows that the expression of some yield components is controlled quite simply (see, for example, Krarup and Davis (1970) on the genetic control of ovule number in *Pisum*, and Hole (1977) on the control of flower number); yet selection for increased expression of particular yield components has not in general led to increases in yield (Geadelman and Peterson, 1976; see also Kuhn and Stucker (1976) and Langer (1967)). 'The limitation of yield analysis is that the components are the ultimate results of underlying growth and developmental processes falling out of the field of view' (Bruinsma, 1966). More specifically 'Components of seed yield are to some extent interdependent in their development' (Adams, 1967). Since the interdependence is generally negative, the effects of any increase in one yield component are reduced by countervailing decreases in other components. Many of the plant breeders who sought to increase yields by manipulating yield components soon discovered that it is difficult to bring about any changes in a heavily damped system.

This poses the physiologist with two problems; (i) how is the degree of interdependence to be measured and (ii) how is it to be explained? To quantify the degree of interdependence of the yield components we follow other workers (Dewey and Lu, 1959; Driscoll and Abel, 1976; Fisher, 1980) and use a logarithmic transformation of the yield equation. If C_i and C_j are the logarithms of the ith and jth yield components and if σ_k^2 is the variance of the logarithm of the kth yield component, then the situation in crops is (generally) that the covariances do not equal zero; i.e. that

$$Q = \sum_i \sum_j \mathrm{cov}\,(C_i\, C_j) \neq 0$$

The minimum possible value of Q is $-\sum_k \sigma_k^2$ and the maximum is $\sum_i \sum_j \sigma_i\, \sigma_k$. It would be more convenient to have a scale running from (say) 0 to 1. Consider the parameter W

$$W = \frac{\sum_k \sigma_k^2 - Q}{\sum_k \sigma_k^2 - \sum_i \sum_j \sigma_i\, \sigma_j} = \frac{\sigma^2}{(\sum_k \sigma_k)^2}$$

where σ^2 is the variance of log yield. We now have the required scale on which to measure the degree of interdependence of yield components. It runs from complete compensation of components (W = 0) through complete independence (W = 0.5) to complete additivity (W = 1.0). The condition that W = 0.5 when the components are independent (Q = 0) requires a minor transformation of the parameter; details are given by Hardwick and Andrews (1980). These authors also report some empirical tests of the parameter and discuss its use in plant breeding projects.

If the amount of interdependence between components can now be measured, the reasons for its occurrence are still a matter for debate. In some cases the associations between yield components may be due to genetic linkage. Rasmusson and Cannell (1970) advanced the linkage theory to account for observations in some experiments on barley; but this explanation was challenged by Adams and Grafius (1971). They proposed that the interdependence between components reflected a common dependence on some limited environmental resource i.e. that it was phenotypic in origin. The theory was subsequently elaborated (Grafius and Thomas, 1971), and some mathematical reasoning was corrected (Driscoll and Abel, 1976). Recently Lloyd (1980) has discussed the notion of the serial adjustment of maternal investment during fruit growth - a very similar idea to Grafius' 'oscillatory response of sequential components to limited resources'.

There is some evidence that yields of legumes are source limited (Sinclair and de Wit, 1975). In peas, Meadley and Milbourn (1971) concluded from a series of shading experiments that, 'The limitation to yield in peas lies less in the ability of the plant to produce pods than in the capacity of the crop-stand to produce dry matter'. On the other hand Hedley and Harvey (1979) found that pea yields were able to compensate for levels of shade up to 30%, and there is much recent evidence that rates of photosynthesis are regulated by sink activity (Herold, 1980; Keener et al., 1979; Neales and Incoll, 1968). This makes the simple Grafius theory (a cake can be cut up in many different ways, but the sizes of the slices will be interdependent if the size of the cake is fixed) more difficult to sustain.

An alternative theory for the interdependence of yield components is available. A picture is beginning to emerge from recent research of the plant as a homeostatic organism, strongly regulated by feedback pathways (see, for example, Gersani et al. (1980)). Homeostasis provides an alternative explanation for the interdependence of yield components, which on this view results from the regulation of development and is part of the overall integration of the plant's activities. If a homeostatic system is perturbed, the effect on yield will be small - just as in a source-limited system. But if the perturbation affects the homeostatic system's control point, the perturbation in yield may be unexpectedly large. A possible example of a shift in the system's control point has recently come to hand in the report by Thompson and Taylor (1980) of an experiment on *Vicia*. By growing plants in pots and altering the interplant spacing during the season these authors found some rather large effects on yield. They concluded that, 'The canopy may be able to support to maturity a considerably greater load than would normally develop, providing constraints to development are removed during a short but critical stage of growth'. We have preliminary indications of a similar situation in a transplanting experiment with *Phaseolus* (Andrews and Hardwick, 1981).

At the end of the road down which botany is travelling is the prospect of a plant science as predictive and powerful as metallurgy or hydrodynamics. Against this perspective, yield analysis and other descriptive techniques have an ever dwindling importance. In recent work on yield analysis, as well as in recent work on crop models, there is an implicit assumption that plants are machines which are driven by the environment. It would be unfortunate if this view were to persist for very much longer. A much more tenable concept is that the plant is a richly interconnected, self-regulating system in which the various components of growth and yield are strongly coupled and all are habitually controlled at some level of operation that is below their maximum. To pursue this concept further we need methods of estimating the degree of coupling between components. Estimates of yield components, analysed by the mathematical approach outlined in this paper, may have some part to play in the development of this idea.

REFERENCES

Abel, G.H. and Driscoll, M.F., 1976. Sequential trait development and breeding for high yields in safflower. Crop Science 16: 213-216.

Adams, M.W., 1967. Basis of yield component compensation in crop plants with special reference to the field bean, *Phaseolus vulgaris*. Crop Science 7: 505-510.

Adams, M.W. and Grafius, J.E., 1971. Yield component compensation - alternative interpretations. Crop Science 11: 33-35.

Andrews, D.J. and Hardwick, R.C., 1981. The effects of temperature and other variables on the contribution of lateral branches to yield in *Phaseolus vulgaris*. Annals of Botany 48: 15-23.

Balls, L. and Holton, F.S., 1915. Analyses of agricultural yield. 1. The spacing experiment with Egyptian cotton. Philosophical Transactions of the Royal Society B 206: 327-332.

Briggs, G.E., Kidd, F. and West, C., 1920. A quantitative analysis of plant growth. Annals of applied Biology 7: 103-123.

Bruinsma, J., 1966. Analysis of growth, development and yield in a spacing experiment with winter rye. Netherlands Journal of agricultural Science 14: 198-214.

Dewey, D.R. and Lu, K.H., 1959. A correlation and path coefficient analysis of components of crested wheat grass seed production. Agronomy Journal 51: 515-518.

Driscoll, M.F. and Abel, G.H., 1976. A correct logarithmic transformation for standardizing multiplicative trait variables. Crop Science 16: 301-303.

Engledow, F.L. and Wadham, S.M., 1923. Investigations on yield in the cereals. Journal of agricultural Science 13: 390-439.

Fisher, N.M., 1980. The effect of time of planting on four bean *(Phaseolus vulgaris)* genotypes in Kenya. Journal of agricultural Science 95: 401-408.

Gersani, M., Lips, S.H. and Sachs, T., 1980. The influence of shoots, roots and hormones on sucrose distribution. Journal of experimental Botany 31: 177-184.

Geadelmann, J. and Peterson, R.H., 1978. Effects of two yield component selection procedures on maize. Crop Science 18: 387-390.

Grafius, J.E. and Thomas, R.L., 1971. The case for indirect genetic control of sequential traits and the strategy of deployment of environmental resources by the plant. Heredity 26: 433-442.

Hardwick, R.C. and Milbourn, G.M., 1967. Yield analysis in the vining pea. Agricultural Progress 42: 24-31.

Hardwick, R.C., Andrews, D.J., Hole, C.C. and Salter, P.J., 1979. Variability in number of pods and yield in commercial crops of vining peas *(Pisum sativum* L.*)*. Journal of agricultural Science 92: 675-681.

Hardwick, R.C. and Andrews, D.J., 1980. Genotypic and environmental variation in crop yield. A method of estimating the interdependence of the components of yield. Euphytica 29: 177-188.

Hedley, C.L. and Harvey, M.J., 1979. The effects of shading on the yield components of six 'leafless' pea genotypes. Annals of Botany 44: 469-478.

Herold, A., 1980. Regulation of photosynthesis by sink activity - the missing link. New Phytologist 86: 131-144.

Hole, C.C., 1977. Environmental control of flower number in multi-flowered cultivars of *Pisum sativum* L. Annals of Botany 41: 1217-1223.

Keener, M.E., de Michele, D.W. and Sharpes, P.J.H., 1979. Sink metabolism: A conceptual framework for analysis. Annals of Botany 44: 659-669.

Krarup, H.A. and Davis, D.W., 1970. Genetic control of ovule number in peas *(Pisum sativum* L.*)*. Crop Science 10: 517-519.

Kuhn, W.E. and Stucker, R.E., 1976. Effect of increasing morphological component expression on yield in corn. Crop Science 16: 270-274.

Langer, R.H.M., 1967. Physiological approaches to yield determination in wheat and barley. Field Crop Abstracts 20: 101-106.

Lloyd, D.G., 1980. Sexual strategies in plants. I. An hypothesis of serial adjustment of maternal investment during one reproductive session. New Phytologist 86: 69-79.

Meadley, J.T. and Milbourn, G.M., 1971. The growth of vining peas. III. The effect of shading on abscission of flowers and pods. Journal of agricultural Science 77: 103-108.

Neales, T.F. and Incoll, L.D., 1968. The control of leaf photosynthesis rate by the level of assimilate concentration in the leaf: a review of the hypothesis. Botanical Reviews 34: 107-125.

Rasmusson, D.C. and Cannell, R.Q., 1970. Selection for grain yield and components of yield in barley. Crop Science 10: 51-54.

Sinclair, T.R. and de Wit, C.T., 1975. Photosynthate and nitrogen requirements for seed production by various crops. Science 189: 565-567.

Thompson, R. and Taylor, H., 1981. Factors limiting growth and yield of *Vicia faba*. In *Vicia faba*: physiology and breeding. Edited by R. Thompson. Published by Martinus Nijhoff The Hague.

Walton, P.D., 1972. Factor analysis of yield in spring wheat. Crop Science 12: 731-733.

Wellensiek, S.J., 1941. Oogstanalyse. 1 Orientatie bij erwten. Mededlingenen Landbouwhoogeschool, Wageningen 45 (1) 1-29.

DISCUSSION

Plancquaert *(France)* In our country the majority of peas and beans are cultivated in the cereal area. Variability in leguminous crop yield is a problem for our farmers. We have studied the effect of soil and climate on these plants but we have not been able to explain the effect of climate because we do not know which are the most important factors. Last year we started an experimental programme to study the effects of meteorological data on yield factors - number of flowers, number of pods, and so on. In our opinion knowledge of the physiology of these legumes is very important; perhaps the next congress could be on this subject.

R.C. Hardwick *(UK)* I feel that much of the variability that we see in pea yields is associated with sub-clinical pathology. The failure of yield production may be because of rotting of leaves and pods. Would you agree that is an important component?

Plancquaert In our work we try to get the best conditions for cultivation. We spray against disease, against aphids, and so on, so that we can isolate the effects of climate.

R.C. Hardwick Was there any rot in your crops due to *Botrytis*?

J. Picard *(France)* I am familiar with this programme. There is some rot, of course, but the point Plancquaert is making is that they have tried to get the best possible conditions for growth.

IMPROVING THE PEA CROP

B. Snoad

John Innes Institute,
Colney Lane, Norwich, UK.

INTRODUCTION

The demand to improve the dried pea crop as a source of vegetable protein has never been stronger than it is now and within the EC this demand is particularly strong because of our almost total dependence upon imported soya.

The problems that stand in the way of crop improvement are many and varied and are bound to differ from region to region as functions of the environment. At the John Innes Institute we can only speak with confidence about growing peas in the UK but in all probability our experiences are similar to those which are encountered in northern Europe generally. In our relatively cool environment with its prolonged growing season we can usually guarantee to grow good crops of dried peas with acceptable yields. We would obviously like to improve these yields but at the same time we have two other equally important tasks; to reduce the significant year-to-year yield variation which we experience and to weather-proof the crop which so often collapses prior to being harvested. At the time of lodging the ground is frequently wet so that combine harvesting is difficult and the quality of the end product is continuously being reduced by pathogen attack and even premature germination. Presumably the yields of crops grown further south in the EC will be limited more by high temperatures and by water deficiency but the harvesting process itself is less likely to raise problems in these environments. The other big difference between regions will probably be associated with the spectra of pests and diseases with the aphid-transmitted viruses being more important in warmer climes.

What exactly is meant by improving the pea crop? Put most simply we need to increase the yield of seed and of protein per unit area, we need to improve the quality of that protein and we need to introduce maximum stability of response to environment.

Because of our regional differences within the EC it is impossible to generalise very widely about the specific plant improvements that are necessary but there are a number of areas of research which are worth considering as being common to the majority of us. This paper aims, therefore, to define the main attributes of a good dried pea crop, to summarise and compare the most recent attempts that have been made to achieve progress and then, finally, to discuss these areas of research which appear to be of common interest. It is difficult to cite references to many of the statements made, partly because plant breeders are not prolific writers and partly because many of these statements have come from discussions with other breeders of from observations made at trial sites. References can be made, however, to the Annual Reports of the John Innes Institute and of the Processors and Growers Research Organisation as well as to two review articles (Hedley and Ambrose, 1982; Snoad, 1980).

THE IDEAL PEA CROP PLANT

Such a plant must possess a number of specific attributes and must respond to a number of accepted agricultural practices.

1. It must be quickly and easily established when sown with conventional farm drills. This implies tolerance of temperature and soil moisture variation and the development of a fast-growing and efficient root system, areas of research which have been somewhat neglected.
2. There must be efficient inoculation of the root system by effective strains of Rhizobium and mycorrhiza to ensure adequate supplies of nitrogen and phosphates.
3. Resistance to or protection from fungal pathogens is necessary during germination and crop establishment, as well as tolerance of the herbicides that are usually required.
4. A growth rate is required that is sufficient for the crop to compete effectively with any weeds that may still occur but not high enough to result in the excessive inter-plant competition which is so characteristic of present day cultivars and which is associated with a high coefficient of variation and yield instability.

5. There must be continuing tolerance of variation in soil water content, particularly as plants begin to develop pods.

6. A very low frequency of basal branch development in competitive sward conditions is necessary since there is ample evidence to show that the secondary branches of pea crop plants contribute very little towards economic yield.

7. A constant yield of some 6 - 7 g of seed per plant is required which implies that each plant need only bear four or five pods. (This statement is based upon the fact that plant density in a pea crop will differ from one variety to another, principally as a function of seed size. Thus if seed of 250 mg weight is sown at 200 - 250 kg/ha and a density of 80 plants/m^2 is achieved the yield would be expected to be in the region of 5 t/ha which is 1.75 t above the UK national average and therefore very acceptable.)

8. There should be resistance to or protection from those pests and diseases which are known to be of economic importance.

9. Pods that are kept well clear of the ground are desirable in order to speed crop drying, avoid pathogen attack and aid machine harvesting.

10. When dry, pods should not split open and thus shed their seed, either spontaneously during the drying process or during harvesting itself.

11. When dried for storage seeds should be uniform in size, shape and colour for ease of handling and for eye appeal in selling.

12. Seed protein content and quality should be as high as possible and remain so when the crop is grown over a number of seasons and sites.

RECENT APPROACHES TO CROP IMPROVEMENT

A number of attempts to improve the pea crop over the past 10 - 15 years have met with varying degrees of success and are of importance to us in reaching conclusions about the best ways to pursue our objectives.

Fasciation

This character can easily be introduced by one or two

recessive genes and the objectives have been to strengthen the stem by means of the multiplication of its vascular strands which gives it a ribbed effect and to increase the uniformity of seed development and seed size by having the majority of the flowers developing at one point on the stem at a multiple node.

The standing ability of a crop of some fasciated plants is improved but there are a number of significant disadvantages that outweigh any gain in this direction.

There has been insufficient research to indicate whether it is due to poor penetrance of the gene or due to genotype/environment interaction but the end result is significant plant-to-plant variation for the degree of fasciation within a crop and quite considerable differences between crops grown in different countries.

The centre of gravity of the plant is unacceptably high when the seed has developed and this contributes towards collapse of the crop.

The sink/source relationship is excessively strained by the development of so many pods at one level so that the amount of seed abortion is very high.

Finally, the crown-like top of the plant acts as a moisture trap and the amount of damage due to *Botrytis* is significantly increased.

Winter peas

Little seems to be known about the genetic control of winter hardiness, which in all probability is due to a combination of cold tolerance and disease resistance, but it does seem possible to introduce it from the so-called Austrian Winter Pea types. The reasons for developing these forms of pea would include the possibility of drilling during the autumn which is more popular with many farmers and the hope that the date for harvesting would be brought forward sufficiently to escape some of the rainfall which is often experienced.

In practice, the number of problems and uncertainties associated with the varieties tested in northern Europe are more significant than the gain of some degree of winter hardiness and so it is perhaps a crop for areas in which the winters can be guaranteed to be mild rather than severe.

Bird and cold damage always occur, although they vary considerably, and lead to a high incidence of basal branching from stems which tend to be weakened by pathogen attack in the region of the hypocotyl.

Weed control is especially difficult and the use of post-emergence herbicides would be hard to justify on a damaged crop and at the low temperatures experienced early in the spring.

The date of harvest does not seem to be significantly earlier than can be obtained with a spring-sown crop and neither is the yield an impressive one.

Rogue peas

This is a general term for off-type plants found in crops and one of these, the so-called rabbit-eared or tare-leafed rogue, has been the subject of intermittent research since the early part of the century. It is particularly fascinating because segregation in the F2 from a cross between such a rogue pea and a normal pea may or may not occur and the extent to which it does occur is totally unpredictable. Indeed in some crosses the F2, F3, F4 etc. may be composed exclusively of rogue plants. There is so far no conclusive evidence to show that this particular character is controlled by any Mendelian genetic system and it is therefore a far from satisfactory character to employ in a plant breeding programme.

Rabbit-eared rogues occur with different frequencies in a large number of pea cultivars. One popular variety grown in the UK, Maro, contains these rogues quite frequently and it was observed by personnel at the Processors and Growers Research Organisation that the off-types tended to remain more upright than the normal forms when the crop began to lodge. It was decided to select these rogue forms of Maro and by selfing them build up enough seed to test on a field scale. These tests were sufficiently convincing for the rogue forms to be registered as a variety in their own right, a variety called Progreta.

Plants of Progreta have significantly smaller leaflets and stipules than many peas so that a distinctly modified canopy is built up in which there is more light penetration and presumably more air movement. They also have somewhat stiffer stems which gives them a more erect habit and certainly keeps the crop standing longer than Maro which is notorious for its inability to remain upright.

Seed of Progreta is a little smaller than that of Maro but the yields of the two varieties are not significantly different. One unexpected disadvantage of Progreta is that it has proved to be far more susceptible to downy mildew than Maro.

Peas with modified leaves or stipules

A number of major modifications of leaves, tendrils and stipules have been examined during the past ten or more years as possible tools for crop improvement. Two recessive genes in particular have survived these studies, one which converts all the leaflets to tendrils and another which brings about a significant reduction in the size of the stipule.

Plants in which only the leaflets were modified became known in the UK as 'semi-leafless' peas whereas those which additionally possessed the very small stipules were called 'leafless' peas. Now that we can produce, by other genetic manipulation, 'leafless' peas with large stipules and 'semi-leafless' peas with small stipules it is becoming evident that these terms have lost their meaning. It is perhaps better to refer to them all as 'leafless' peas and note the degree of stipule reduction that each specific phenotype exhibits.

Plant breeders in most countries are now using these foliar modifications in their search for better crop plants because the reduction in foliage results in better light penetration of the canopy, more air movement and possibly faster drying of the crop. At the same time these plants have an improved standing ability, due to tangling of the extra tendrils, which in some instances lasts right up to the time of machine harvesting. Coincidentally it has been shown that one of the small stipule forms tends to overwinter successfully and also that it exhibits a marked resistance to damage in waterlogged soil.

These types of plant are not without their problems; for example the standing ability of a particular line may be unpredictable and can vary considerably from site to site or season to season. Too little material has been developed yet, however, for maximum potential to be assessed.

Leafless plants with very small stipules have a low growth rate and are only likely to give high yields when grown at higher densities, densities which are verging on the uneconomic because of seed costs. Leafless plants with normal sized stipules have growth rates on the other hand which do not differ very much from those of leafed peas which it is felt makes them too competitive for a good crop plant. Since we know that we have the two extremes there is reasonable optimism that the right balance can be detected somewhere in the range between them.

Attempts have been made to combine the 'leafless' and fasciated habits but the criticisms levelled at fasciated plants unfortunately still apply to the combination.

Mixed crops

There have been suggestions and even some small scale experiments about growing mixtures of small stipule 'leafless' with leafed peas and of these 'leafless' peas with cereals, the aim being to improve the standing ability of the weaker partner. Growing mixtures of plants which differ so markedly in growth rate and therefore in competitiveness seems fraught with danger and it appears most unlikely that the 'leafless' peas would survive long in the population or contribute very much to the overall yield.

FUTURE PROGRESS

It seems that fasciation, with or without the 'leafless' habit, and mixed cropping are two areas of research which should have low priority. Winter peas too seem to be of less interest because of their poorer showing in the colder parts of northern Europe.

Unless newer and better ideas are forthcoming it would seem then that our effort should be put into the study and exploitation of peas with foliage modifications and we are left, therefore, with rogue peas and 'leafless' peas, bearing in mind that the 'genetic' control of the rogue character is still totally unexplained.

Having narrowed down the plant models with which to work there are some five main, but highly interactive, areas of research to be considered.

Plant morphology

In the rather more mechanical sense this relates to the size of the plant and of its component parts, its competitive nature in regard to its neighbours whether weeds or other peas, its potential yield components, its tendril and stem structure as they affect standing ability and the ease with which it can be machine harvested. The type of root system developed could well be important here too.

Both the rogue and the 'leafless' pea are of particular importance under this heading and as soon as enough material becomes available all new forms of plant should be tested in a much wider range of environments.

Physiological aspects of plant behaviour

The advent of the 'leafless' pea has had a catalytic effect upon some physiologists and, to a more limited extent, agronomists because of the wide range of entirely new models for experimentation. As a crop plant in particular the pea has been very much neglected and so an excellent opportunity now exists for breeders, physiologists and agronomists to work together in the study and improvement of the plant model that we are seeking to develop.

Among the areas of research currently in progress, some of which are to be referred to in more detail by my colleague, Dr. Hedley during this meeting, are:

The importance of seed size and of the patterns of seed development in relation to plant establishment, growth rate and the provision of economic yield.

The significance of growth rate differences between phenotypes as they influence competition between plants in a crop and the partitioning of yield into its economic and non-economic components. This area of research is especially important in helping to define the extent to which foliage reduction needs to be practised in the development of a better crop plant.

Degrees of intra-plant competition for resources, between ovules in a pod, between pods at a node, between nodes on a plant etc., as modified by phenotype and environment.

The relative importance of specific photosynthetically

active components of plants in a wide range of phenotypes during the various phases of crop development.

The role of the pod and of the seeds themselves in the development of economic yield, bearing in mind the importance that many mutant pod forms differing in size, colour and wall thickness could assume when grown in much lighter crop environments than would normally exist in crops of leafed peas.

Resistance to pests and diseases

Distinct regional needs will influence the type of research to be undertaken. Ideally, guidance should come from reports on the economic importance of the pests and diseases but unfortunately information of this kind seems rarely to be available. It would be wise to decide if this information is not forthcoming because the research has not been undertaken or because the pests and diseases do not always cause measurable economic losses.

Peas and symbiotic bacteria and fungi

The role of Rhizobium in the fixation of nitrogen has been researched extensively in peas but usually with individual, non-competing plants in mind rather than crop plants. There is no evidence to suggest that pea crops respond to extra nitrogen and therefore increased or prolonged nitrogen fixation may not lead to increases in crop yield. In other words there are other factors limiting crop yield that rank higher than nitrogen and so, as long as efficient strains of Rhizobium are present in the soil, the emphasis would be better placed on trying to reduce the energy requirements of the nitrogen fixation process.

The part played by the mycorrhizal population of the soil in the three-way symbiotic relationship between it, the plant and the nitrogen-fixing bacteria is possibly underestimated in peas. Nutrient uptake is improved by the increased surface area which the fungal hyphae provide, particularly in relation to the less mobile nutrients such as zinc and especially of phosphate which is essential not only for plant growth but also for nodulation and nitrogen fixation.

Protein

The biochemistry of seed protein production as it affects protein quality and quantity in peas has also received considerable attention but so far no ways of obtaining improvements have been reported.

It is generally accepted that protein yield per unit area needs to be increased and it seems that initially this is more likely to be achieved by increasing the yield of pea seed than by increasing seed protein content. Unfortunately there is considerable evidence of very significant genotype-environment interactions affecting seed protein content, not only within varieties but also within plants of a single variety. To demonstrate increases in seed protein content or to show correlations between seed yield and protein content with any degree of statistical reliability is therefore exceedingly difficult. The importance of screening our material for any signs of stability of response of protein content to environmental influence cannot be over-emphasised if progress is going to be made in our breeding programmes.

Ideally the quality of pea seed protein should also be improved, but that could equally be said about many legumes including soya. In practice, it appears that the potential users of pea seed are more interested in a guaranteed supply of seed of as high a protein content as possible than they are in the quality of that protein; perhaps that is just as well for the plant breeder.

CONCLUSIONS

It seems reasonable to conclude that remodelling the pea plant by restructuring its leaves and by appropriately reducing its stipule size is the most effective way of modifying the crop canopy and of converting a good garden vegetable into a more efficient crop plant. This view is obviously shared because there are many other pea breeders, including those in the private sector in most parts of the world, who are working with similar concepts. Providing that we can keep the backing of the physiologists and the agronomists, an ideal opportunity exists for some good fundamental research work and for significan progress in the development of a much better crop. At least we are in a position to have the physiologists working with the breeders in examining crop models rather than following along behind which has so often been the case with other crop plants.

Fortunately a significant amount of genetic variation exists in peas and has been carefully preserved in a number of genebanks. The amount of naturally occurring variation is extensive and is probably sufficient for any consideration of induced mutation work to be deemed unnecessary.

The breeding techniques available are few but generally well tested and reliable and include pedigree breeding, back-crossing and, to a smaller extent, single seed descent. Because the multiplication rate of peas is low, the provision of illuminated glasshouse space or the co-operation of contacts in the southern hemisphere are essential for speeding up the progress of any programme.

It is frequently assumed that trialling techniques for dried peas are well established and reliable but the fact that there are so many ways that different organisations approach the same problem casts some doubt. As long as the methods do not take into account the differences between hand and machine harvesting, the extent to which edge effects can modify yield differentially according to phenotype, the proportion of the plot which is influenced by the edges and which can be altered by adjusting the dimensions of that plot and the need to harvest the crop at the optimum time as the farmer would, then problems of making meaningful comparisons will continue to exist.

Sufficient quantities of seed of new pea plant models will soon be available for trials on a wider scale and the EC could play an important role in their co-ordination. The need to establish and maintain an efficient and reliable trial system cannot be over-emphasised, especially if the opportunity is taken to monitor the protein content of all crops that are grown in the search for some degree of stability.

REFERENCES

Hedley, C.L. and Ambrose, M.J., 1982. Designing 'leafless' plants for improving the yields of the dried pea crop. Advances in Agronomy 34 (in press).

Snoad, B., 1980. The origin, performance and breeding of leafless peas. ADAS Q. Rev. 37: 69-86.

DISCUSSION

R.C. Hardwick *(UK)* I would like to pursue the idea of competitive ability. It is usually thought that in yield/density studies there are two important parameters. One is the ceiling yield per unit area and the other is the yield per plant not subject to competition from neighbours. In most cases these two parameters describe the increase in yield associated with increase in density, up to the point of inflexion where values start to level off. Do the John Innes workers have any evidence that a third parameter which one could call 'competitive ability', is necessary to explain their data, or is the variation between varieties explained by one or both of the parameters I have mentioned?

C.L. Hedley *(UK)* The main problem with the 'leafless' pea is biological yield. With the conventional crop the optimal planting density for biological yield and for seed yield tend to be similar whereas with 'leafless' crops they tend to be different.

R.C. Hardwick I take your point that the variation you are getting with density is variation in 'harvest index'. But perhaps this does not preclude the existence of a third parameter which would explain genetic variation in 'competitive ability'.

FOUR YEARS OF EXPERIMENTAL RESEARCH ON LUPINS

L. Postiglione

Universita' di Napoli,
Istituto di Agronomia Generale Ecoltivazioni Erbacee,
Portici - Naples, Italy,

CULTIVATION OF LUPINS IN ITALY: CURRENT SITUATION AND FUTURE PROSPECTS

Lupins are cultivated in Italy at present as a crop intended mainly for ploughing in as green manure and only to a limited extent for use as animal forage. The reason for this is that lupins are not very palatable because they contain certain alkaloids and because their nutritional value as forage is low. The seed crop is grown mainly to produce seed for the green manure crops and only a very small amount of seed is used for human or animal consumption. Seed is suitable for consumption only after lengthy immersion in salt water to eliminate the bitter taste caused by alkaloids (lupanine, lupinine, sparteine, oxylupanine).

The species normally grown in Italy is *Lupinus albus* L., which formerly included *L. vulgaris* (Common lupin) and *L. termis* (Egyptian lupin). The different types or local variants which are now grown are generally of the *L. vulgaris* type and are known by the name of the area in which they are most commonly found. It is therefore most likely that similar types are known by different names (for example, Vairano lupin and Cassino lupin.

As lupins follow an autumn - winter cycle* and grow well in fairly poor soils without too much attention, they became a major crop in some areas of southern Italy where they were grown as a source of green manure on account of their considerable organic enrichment of the soil. But as the practice of ploughing-in green manure has become less widespread, especially in the poorer areas and where crop yields were not so great, the cultivation of lupins has steadily diminished over recent years.

* As already stressed in a previous paper (Postiglione, 1979), to which readers are referred for a critical examination of the principal agronomic problems and for the bibliography.

Consequently, the area under cultivation for seed slumped from 46 931 ha in 1950 to 4 654 ha in 1980.

Nowadays, however, the cultivation of lupins must be considered from quite a different aspect. Lupins can be grown to produce seed which may be used as a source of protein and to some extent also as a source of seed oil. It is indeed common knowledge, that the protein content of lupins is higher than that of broad beans and that the amino acid composition of *Lupinus albus* is not very different from that of the soyabean (Hudson et al., 1976). Lupin seeds have an oil content of 7 - 16%.

There is also the fact that lupins are grown in the south in the autumn and winter. As a result, they do not require irrigation and so do not compete with other high yielding irrigated crops (whether legumes or not). They also grow well in poor soil, provided it is well drained, and can even grow on marginal land.

Finally there is the fact that lupins require little attention. In view of all this, it is clear that lupins can be a source of low-cost protein. In fact, for the reasons outlined above, lupins can simply alternate with wheat, and can indeed improve the wheat crop by enhancing soil structure and nutrition.

The presence of alkaloids in lupins has hitherto been a major barrier to their use as a source of protein. Fortunately there are now two solutions to encourage their use for this purpose. The technological solution consists of the chemical extraction of the oil and proteins and the elimination by similar means of the alkaloids. The second solution, using genetic methods, involves the production of alkaloid-free, or more accurately, low alkaloid varieties of lupin.

The latter solution actually seems to be less expensive, as it eliminates extraction costs and, where the oil content is very low, the seed could be used directly without being processed. Also this is the most common method used in countries where lupins are grown on a large scale as a source of protein.

In the light of these facts and with a view to promoting the cultivation of lupins as a source of protein, the University of Naples' Institute of Agronomy in Portici carried out the research on lupins described here.

SELECTION AND ADAPTATION

In 1976 - 77 and 1977 - 78 plants obtained from seed samples received from France and Germany were grown on experimental farms at Torre Lama in the Sele plain (province of Salerno) and in the Caiazzo plain (province of Caserta), and the main biological and biometric characteristics were noted.

Work continued at the Torre Lama farm in 1978 - 79 and in subsequent years. In addition to the lupin types already mentioned, which belonged exclusively to the *Lupinus albus* species, work also began on other alkaloid-free cultivars and lines from Germany and Australia. These belonged to the *Lupinus luteus* and *L. angustifolius* species respectively. In 1979 - 80 fourteen types of South American *L. mutabilis*, containing alkaloids and with a high oil content, were added. Finally, 1980 saw the introduction of several types of alkaloid-free *L. albus* with five from Portugal, four from Australia and one from Chile.

Certain local types of *L. albus*, commonly grown in the south of Italy, were used as control plants. A selection process involved the elimination of the types which, during the previous two years, had proved unsuitable for experimental purposes on account of their lack of resistance to winter cold, exceptionally slow growth, inadequate flowering and seed production or high alkaloid content as revealed by analysis. In the case of the fourteen types of *L. mutabilis* from South America, only two survived the winter of 1980 which was characterised by moderate temperatures during the initial growth period and unusually cold spells in January and February.

As a result of this procedure only 20 types (Table 2) remained of the 42 original (Table 1), but at the same time a number of families were found which seemed particularly interesting.

A problem which arose between 1979 and 1980 was that of nodulation. It was noted, particularly for the cv. Populor

TABLE 1

CULTIVARS AND TYPES TESTED BY THE INSTITUTE OF AGRONOMY, PORTICI 1976 - 81

Species	Type or cultivar	Source
	Without alkaloids	
Lupinus albus	Type 7	INRA 1976
"	Type 10	"
"	Type 18	"
"	Type 35	"
"	Type 38	"
"	Type 39	"
"	Type 40	"
"	Type 41	"
"	Type 42	"
"	Type 45	"
"	Type 53	"
"	Type 55	"
"	Type 57	"
"	LA Lupinus albus	INRA selection
"	Multolupa	Forschungsstelle von Sengbusch, Hamburg, 1976
"	Sweet Hungarian lupin	Hungary, 1979
"	Red Hamburg	Australia, 1980
"	Ultra lupinus	"
"	930 - 3	Portugal, 1980
"	895 - 2	"
"	968 - 2	"
"	551 - 3	"
"	804 - 4	"
"	Multolupa	Chile, 1980
Lupinus angustifolius	Unicrop	Department of Agriculture, Western Australia, 1978
"	Uniharvest	Western Australia, 1978
"	Unicrop	CSIRO, Perth, Western Australia, 1978
Lupinus luteus	Populor	Institut für Kleintierzucht, Celle, Germany, 1978
	With alkaloids	
Lupinus mutabilis	Tiraque	Bolivia CLP-10, 1979
"	Copacabana	" CLP-11 "
"	Falsuri	" CLP-12 "
"	Colomi	" CLP-13 "
"	Larati	" CLP-14 "
"	Copacabana	" CLP-15 "
"	"	" CLP-16 "
"	Quillacollo	" CLP-17 "
"	Potosi	" CLP-19 "
"	Cuzco	Peru CLP-20 "
"	Yungas	Bolivia CLP-21 "
"	Huancayo	Peru CLP-22 "
"	"	" CLP-23 "
"	Tiraque	" CLP-24 "

TABLE 1 CONTINUED

CULTIVARS AND TYPES TESTED BY THE INSTITUTE OF AGRONOMY, PORTICI 1976 - 81

Species	Type or cultivar	Source
	Local control plants with alkaloids	
Lupinus albus	A (seed crop)	Zorzi Company, Padua
"	B (green manure crop)	"
"	C	Institute of Agronomy, Pisa
"	Cassino lupin	Local plant
"	Vairano lupin	"
"	Common Vairano lupin	"
"	Bitter Cortale lupin	"
"	Philadelphia lupin	"

TABLE 2

CULTIVARS AND TYPES TESTED BY THE INSTITUTE OF AGRONOMY, PORTICI 1980 - 81

Species	Type or cultivar	Source
	Without alkaloids	
Lupinus albus	Type 10	INRA 1976
"	Type 18	"
"	Type 39	"
"	Type 53	"
"	LA Lupinus albus	INRA selection
"	Multolupa	Forschungsstelle von Sengbusch, Hamburg, 1976
"	Sweet Hungarian lupin	Hungary, 1979
"	Red Hamburg	Australia, 1980
"	Ultra lupinus	"
"	930 - 3	Portugal, 1980
"	895 - 2	"
"	968 - 2	"
"	551 - 3	"
"	804 - 4	"
"	Multolupa	Chile, 1980
Lupinus angustifolius	Unicrop	Department of Agriculture, Western Australia, 1978
"	Uniharvest	Western Australia, 1978
Lupinus luteus	Populor	Institut für Kleintierzucht Celle, Germany, 1978
	With alkaloids	
Lupinus albus	Vairano lupin	Local plant
Lupinus mutabilis	Quillacollo	Bolivia CLP-17, 1979
"	Potosi	" CLP-19 "

(L. luteus), that growth was stunted and the plants produced few flowers and pods, as a result of a failure of the roots to nodulate. The Institute of Microbiology of the Faculty of Agriculture in Portici is now studying the problem in collaboration with the INRA microbiology laboratory in Dijon.

COMPARISON OF VARIETIES

During the four successive years from 1977 - 78 to 1980 - 81 comparative tests were carried out on the types and varieties under cultivation at the experimental farm at Torre Lama in the Sele plain in the province of Salerno.

Methods and materials

The Torre Lama experimental farm enjoys a Mediterranean climate of temperate and rainy winters and hot, dry summers. Figure 1 gives the temperature and rainfall data for the four years during which the experimental work was carried out.

The soil is of alluvial origin with small deposits of pyroclastic material from Vesuvius. It is good neutral clayey soil with an adequate amount of phosphorus and a large amount of potassium. The properties of the soil in which the experimental work was carried out are given in Table 3.

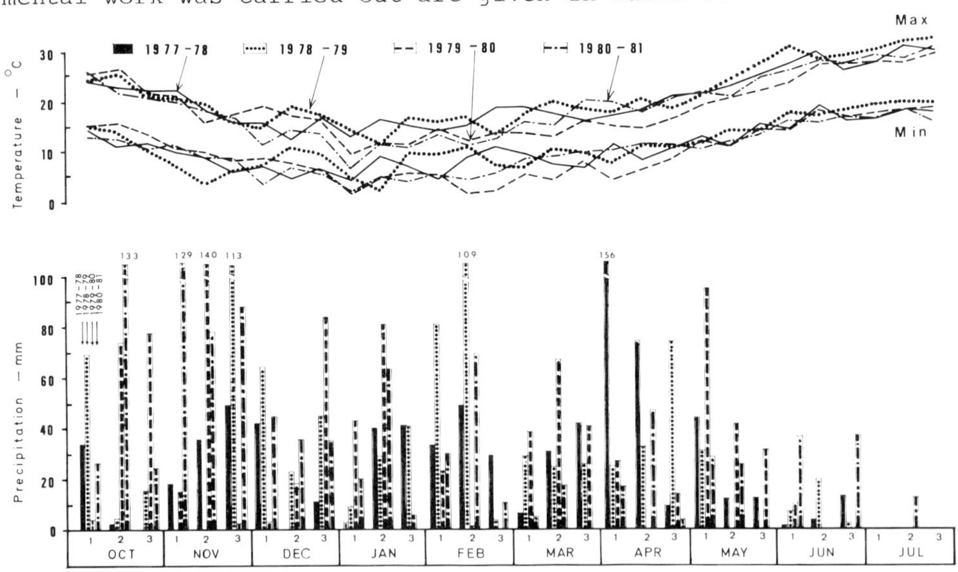

Fig. 1. Meteorological conditions

TABLE 3

SOIL PROPERTIES OF PLOTS USED FOR TESTS

Year	pH reaction
1977/78	7.0
1978/79	6.8
1979/80	7.2
1980/81	6.9

Plot size was 12 m^2 and three replicates were used in a randomised block design. Culture of the crops was based on good local practice and are outlined in Table 4. In the first three years the plant density was the same as for local lupins (approximately 16.7 plants/m^2) and the distance between rows was 30 cm. In the final year, when the results of the density experiments were known, the lupins were planted in rows 25 cm apart with a density of 26.7 plants/m^2.

TABLE 4

CULTIVATION PROGRAMME

	1977-78	1978-79	1979-80	1980-81
Previous crop	wheat	maize	wheat	wheat
Fertiliser:				
nitrate (kg/ha)	100	100	150	150
phosphoric acid (kg/ha)	600	600	600	600
Sowing	6 Dec 77	7 Nov 78	6 Sep 79	17 Dec 80
Weeding	1 Mar 78 10 Apr 78 26 Apr 78	15 Jan 79 6 Mar 79	8 Jan 80 16 Jan 80 19 Feb 80	16 Feb 81 6 Apr 81
Harvesting	20 Jul 78	28 Jul 79	24 Jul 80	15 Jul 81

When the lupins were harvested a sample of ten plants was taken to determine average plant size, the height of the first pod, the number of pods per plant (and whether borne individually or in groups of two, three, four or more) and, in the 1979 - 80 season, the amount of lateral branching. Data were also collected from each plot on the number of plants at harvest, the total fresh weight and estimates of dry weight, the weight of the seed and 1 000 seed weight.

Results

Seed yield was very poor in the first three years but gave very high values in 1980 - 81 (Table 5). When yields were generally poor, local ecotypes tended to be ranked high in terms of productivity and their yield was satisfactory in 1977 - 78 and 1979 - 80. The yield of all cultivars was low in 1978 - 79 mainly because of the high rainfall between December and March and especially because the experimental area was often waterlogged. During the first three years the yield from the French types was generally poor, although some of these types revealed characteristics of considerable interest (e.g. high protein content and low alkaloid content). The cv. Multolupa from Germany produced acceptable results, apart from those in the second year. Poor results were obtained from the cultivars of the *L. angustifolius* and *L. luteus* species.

In 1980 - 81, in the light of the results of the density experiments, a density of 267 000 plants/ha was adopted. The new density, together with more favourable environmental conditions and the elimination of the less promising types, resulted in an overall average yield of 3.97 t/ha. The yield of some cultivars, such as the Multolupa, both the selections from Chile (6.56 tonnes) and Germany (5.66 tonnes), the Red Hamburg and INRA's Type 53, exceeded that of the Vairano lupin, the local variety used as a control plant (5.08 tonnes). These levels of productivity were confirmed by field scale cultivation in 1980 - 81 of cv. Multolupa *(Lupinus albus)* for the Food Technology Institute of Milan. A plot of 700 m^2 on the same experimental farm at Torre Lama produced a total yield of 272 kg, which is equivalent to a yield of 3.9 t/ha. The cv. Unicrop also gave acceptable results but CLP 17 (the only type of the *L. mutabilis* species to be used in the tests) and cv. Populor produced very poor results.

The total yield of dry matter showed the same trend as the seed yield, and throughout the four year period there was a significant correlation between the two parameters (Figure 2).

TABLE 5

YIELD RESULTS

Type/variety	1978 seed yield t/ha	1978 total plant yield t/ha	1978 1 000 seed weight	1979 seed yield t/ha	1979 total plant yield t/ha	1979 1 000 seed weight
Type 7	1.327	5.633	380	0.294	1.049	307.00
10	0.900	3.327	420	0.385	1.597	345.66
18	0.380	1.357	300	0.181	0.465	260.32
35	1.560	6.577	330	0.470	2.010	323.73
38	0.907	3.807	360	0.519	1.752	292.84
39	0.316	1.273	310	0.337	1.257	301.90
40	1.320	7.440	410	0.431	1.299	323.23
41	0.980	3.773	390	0.513	1.427	335.66
42	0.670	2.303	340	0.363	1.073	290.85
45	1.363	3.860	370	0.547	1.668	322.76
53	0.917	3.997	420	0.504	1.816	334.46
55	1.353	5.280	390	0.520	1.521	321.84
57	0.717	2.683	270	0.596	1.615	259.27
Lupinus albus	1.087	5.320	380	0.367	1.454	341.79
Multolupa (Germany)	1.287	4.993	560	0.336	1.887	416.30
Sweet Hungarian lupin	-	-	-	-	-	-
Red Hamburg	-	-	-	-	-	-
Ultra Lupinus	-	-	-	-	-	-
Multolupa (Chile)	-	-	-	-	-	-
Lupinus angustifolius Unicrop	-	-	-	0.184	0.616	125.37
Lupinus angustifolius Uniharvest	-	-	-	0.258	0.809	125.99
Lupinus angustifolius Unicrop (CSIRO)	-	-	-	-	-	-
Populor	-	-	-	-	-	-
A	-	-	-	0.402	1.738	-
B	-	-	-	0.479	1.853	-
C	-	-	-	0.418	1.534	-
Cassino lupin	3.913	13.863	380	0.377	1.661	350.60
Vairano lupin	2.187	7.580	791	0.428	1.930	610.65
Common Vairano lupin	2.260	7.787	390	0.284	1.509	341.53
Bitter Cortale lupin	2.767	8.587	350	0.299	1.229	327.09
Philadelphia lupin	3.223	9.850	320	0.576	1.289	307.68
ELP - 17	-	-	-	-	-	-
P = 0.05	0.568	1.741		0.162	0.670	
P = 0.01	0.760	2.330		0.217	0.895	

In the year which produced the best results (1980 - 81) the yield of biomass ranged from 3.55 t/ha for Populor to 16.07 t/ha for Vairano.

TABLE 5 (cont.)

Type/variety	1980 seed yield t/ha	1980 total plant yield t/ha	1980 1 000 seed weight	1981 seed yield t/ha	1981 total plant yield t/ha	1981 1 000 seed weight
Type 7	0.118	2.933	233.99	-	-	-
10	0.656	13.733	279.06	4.762	12.967	337.92
18	0.282	5.200	208.74	2.407	5.013	272.21
35	0.385	7.233	317.46	-	-	-
38	0.528	8.333	225.68	-	-	-
39	0.855	10.333	266.97	4.807	12.833	328.72
40	0.522	11.067	236.45	-	-	-
41	0.362	5.167	281.32	-	-	-
42	0.477	7.867	241.69	-	-	-
45	0.799	9.333	275.51	-	-	-
53	0.470	10.400	244.25	5.533	12.433	339.13
55	1.297	13.667	253.67	-	-	-
57	0.463	8.400	248.53	-	-	-
Lupinus albus	0.933	10.267	304.29	4.637	9.500	341.09
Multolupa (Germany)	1.727	15.733	324.24	5.657	12.200	364.65
Sweet Hungarian lupin	0.095	3.933	201.19	-	-	-
Red Hamburg	-	-	-	5.553	11.667	351.94
Ultra lupinus	-	-	-	2.273	3.793	288.05
Multolupa (Chile)	-	-	-	6.557	13.000	393.22
Lupinus angustifolius Unicrop	1.220	15.800	120.23	3.263	8.033	130.51
Lupinus angustifolius Uniharvest	0.262	6.600	119.39	1.673	4.247	132.58
Lupinus angustifolius Unicrop (CSIRO)	1.055	17.864	133.13	-	-	-
Populor	0.837	7.400	107.96	1.383	3.547	119.23
A	2.467	23.000	518.11	-	-	-
B	2.390	20.933	327.76	-	-	-
C	2.383	24.833	543.48	-	-	-
Cassino lupin	1.470	24.467	321.19	-	-	-
Vairano lupin	1.568	26.300	445.00	5.077	16.067	595.95
Common Vairano lupin	0.762	18.333	293.05	-	-	-
Bitter Cortale lupin	1.298	22.133	304.17	-	-	-
Philadelphia lupin	1.930	25.400	295.35	-	-	-
ELP - 17	-	-	-	1.953	16.000	150.32
P = 0.05	0.727	5.027	49.12	0.566	3.343	34.25
P = 0.01	0.971	6.715	65.61	0.765	4.519	46.30

There were also significant correlations between 1 000 seed weight and grain yield in 1979 - 80 and 1980 - 81. In the latter year 1 000 seed weight fluctuated between 595.95 g for Vairano lupin and 119.23 g for Populor. The cv. Multolupa proved to be the most productive with 1 000 seed weight of 364.65 g for the variety from Germany and 393.22 g for the Chilean variety.

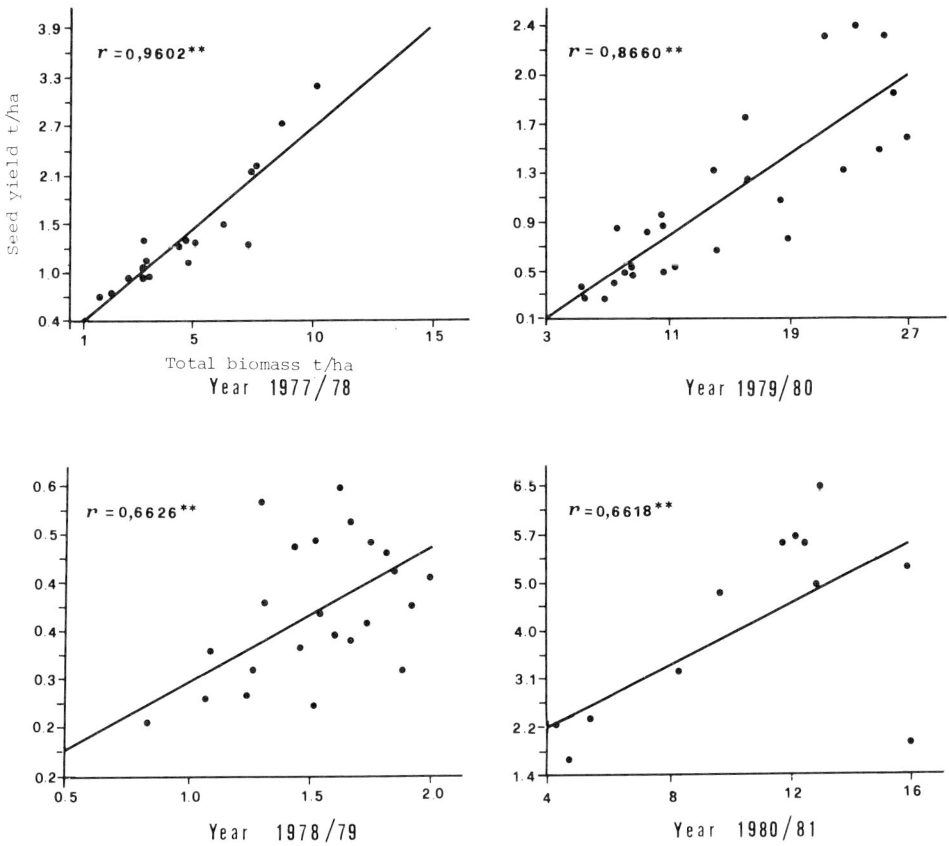

Fig. 2. Correlation between seed production and total production.

Local cultivars were taller (Tables 6 - 9) than the others, usually exceeding 1 m in height. Indeed in 1980, all accessions exceeded 1 m in height, with the local cultivars reaching 2 m. This was probably the result of the particular weather conditions. Mild temperatures in the early stages of growth, until early February, meant that most of the cultivars were developing flowers at this time. The plants then suffered during the rest of February as a result of a severe cold spell, which was followed in March by rising temperatures and frequent rain which encouraged new extension growth. Plant height was correlated with total plant production for three years (Figure 3) but in 1978 - 1979 growth was poor and there was no significant correlation.

TABLE 6

DETAILS OF MORPHOLOGICAL CHARACTERISTICS FOR 1978 EXPERIMENTS

Type or variety	Height of plant in cm	Height (cm) of first pod	Number of pods per plant				
			Single	Twin	3-pod	4-pod	More than four pods
Type 7	88.33	11.37	6.10	4.57	3.63	2.23	1.57
10	89.33	19.03	3.73	2.53	3.10	2.40	2.50
18	39.67	12.90	2.67	2.07	1.67	1.23	1.50
35	75.17	16.33	3.93	3.77	3.30	1.47	1.63
38	82.17	13.23	5.67	2.87	2.30	1.97	3.30
39	33.63	12.17	1.50	1.93	1.50	1.50	1.67
40	78.67	15.73	6.97	5.07	5.07	1.67	1.63
41	71.50	11.03	6.60	4.57	2.70	1.70	1.67
42	60.83	16.70	4.70	2.73	1.93	1.47	1.77
45	73.07	15.47	4.53	3.37	2.37	1.63	2.90
53	86.13	16.93	5.27	3.03	2.60	1.67	2.07
55	66.67	16.83	7.40	3.83	3.47	1.87	4.40
57	55.83	14.90	2.93	2.73	1.63	2.03	2.87
Lupinus albus	98.37	17.93	4.20	2.80	2.97	1.17	1.83
Multolupa (tedesca)	89.83	15.03	2.63	2.57	2.33	0.87	2.47
Cassino lupin	168.50	67.50	2.67	1.67	1.47	1.23	4.60
Vairano lupin	139.13	38.67	3.97	2.03	1.47	1.30	3.30
Common Vairano lupin	127.67	36.17	2.90	2.17	1.27	0.80	3.90
Bitter Cortale lupin	124.07	36.37	3.17	1.47	1.30	1.83	6.90
Philadelphia lupin	131.83	43.17	2.00	1.17	0.70	0.57	4.93
P = 0.05	7.80	4.40	1.43	1.08	0.96	0.71	1.00
P = 0.01	10.25	5.79	1.88	1.42	1.26	0.94	1.32

The position of the first pod (a factor of great importance in relation to mechanical harvesting) varied between years, and was greatest in 1979 - 80 as also was total plant height. The relative performance of the types and varieties grown in the various years was nevertheless similar. In 1980 - 81 the cultivar with the highest position of its first pod was CLP 17 (39.73 cm), followed by the Vairano lupin (37.57 cm), and the cultivar with the lowest position of its first pod was Populor (9.83 cm).

TABLE 7

DETAILS OF MORPHOLOGICAL CHARACTERISTICS FOR 1979 EXPERIMENTS

Type or variety	Height of plant in cm	Height (cm) of first pod	Number of pods per plant				
			Single	Twin	3-pod	4-pod	More than four pods
Type 7	79.03	13.93	10.40	7.60	3.30	1.87	1.47
10	85.47	23.50	7.37	5.07	3.30	2.03	3.13
18	53.53	14.07	3.03	2.40	2.37	1.87	3.20
35	88.47	22.13	14.40	6.83	2.67	1.83	2.33
38	91.10	19.97	6.33	5.63	3.63	2.23	6.03
39	87.67	20.20	6.77	4.57	3.10	2.60	4.77
40	90.93	20.10	9.93	8.57	4.57	2.37	3.33
41	89.20	14.20	12.67	9.80	4.93	3.07	2.77
42	80.07	18.60	6.67	4.73	4.17	3.20	4.63
45	86.53	18.73	9.33	7.73	5.57	2.83	4.83
53	74.00	25.23	13.70	9.30	3.67	1.97	3.73
55	82.97	22.57	7.73	5.67	4.17	2.47	5.80
57	86.73	21.40	7.63	5.63	5.20	3.70	8.43
Lupinus albus	161.00	23.43	8.70	6.79	3.50	1.57	2.33
Multolupa (tedesca)	155.30	31.40	6.30	4.03	3.00	2.03	4.63
Lupinus angustifolius Unicrop	85.77	15.30	13.83	6.00	3.30	1.77	1.20
Lupinus angustifolius Uniharvest	54.80	17.03	15.23	6.77	3.87	1.77	2.33
A	92.07	37.50	3.43	2.17	1.07	0.93	2.40
B	82.07	39.17	3.10	2.93	2.33	1.77	4.33
C	91.77	26.07	11.97	1.83	1.20	0.50	2.70
Cassino lupin	134.43	55.27	1.60	0.80	0.77	0.87	4.27
Vairano lupin	144.27	51.30	3.33	2.67	1.43	1.00	2.80
Common Vairano lupin	177.73	61.23	1.77	1.63	0.87	0.87	3.20
Bitter Cortale lupin	167.33	59.83	0.60	1.07	0.77	0.97	3.37
Philadelphia lupin	141.40	52.87	0.57	0.57	0.80	0.73	3.90
P = 0.05	8.47	5.51	3.19	2.01	1.25	0.89	1.47
P = 0.01	11.14	7.25	4.19	2.64	1.64	1.16	1.93

For all four years the height of the first pod was directly correlated with the plant height (Figure 4).

Of particular interest in respect of crop production is the average number of pods per plant and also the number of pods per raceme. In this regard the local types were outstanding as also were the best of the more recent

TABLE 8

DETAILS OF MORPHOLOGICAL CHARACTERISTICS FOR 1980 EXPERIMENTS

Type or variety	Height of plant in cm	Height (cm) of first pod	Number of pods per plant				
			Single	Twin	3-pod	4-pod	More than four pods
Type 7	128.67	81.00	2.93	4.40	2.00	1.73	1.07
10	167.33	96.00	5.27	4.20	2.80	1.27	2.80
18	105.00	68.00	4.47	3.93	2.13	1.53	1.00
35	165.33	89.67	7.87	4.67	2.40	1.07	1.07
38	151.27	77.53	2.87	1.73	1.53	0.87	3.87
39	147.47	79.33	6.73	6.47	4.00	2.20	1.60
40	170.47	86.40	4.40	5.60	3.60	1.27	0.47
41	146.33	75.33	3.40	4.20	3.20	1.13	2.07
42	130.00	69.33	6.13	6.27	5.60	1.80	2.20
45	159.33	86.67	4.80	3.87	3.07	1.93	2.20
53	164.80	77.67	7.20	6.80	3.40	1.73	3.07
55	154.93	83.00	4.40	4.47	3.00	2.53	4.67
57	168.33	85.33	3.13	1.93	2.60	1.53	3.13
Lupinus albus	166.33	85.33	4.00	3.40	1.87	1.47	3.47
Multolupa (tedesca)	170.00	79.00	2.00	3.87	1.87	2.07	4.27
Sweet Hungarian lupin	135.33	90.33	3.93	3.60	2.40	0.73	0.27
Lupinus angustifolius Unicrop	138.67	96.33	6.33	4.93	5.80	4.27	5.73
Lupinus angustifolius Uniharvest	123.80	92.00	5.00	3.60	2.67	1.27	1.07
Lupinus angustifolius Unicrop CSIRO	146.47	68.33	4.13	3.73	4.33	2.47	3.13
Populor	98.13	56.33	5.67	2.00	1.93	0.93	5.87
A	199.00	79.87	2.00	1.73	1.80	0.87	3.33
B	195.67	92.53	5.53	4.00	3.07	0.73	2.07
C	197.67	96.00	3.40	2.20	2.33	0.47	1.67
Cassino lupin	167.33	89.67	3.60	2.20	1.80	1.53	4.53
Vairano lupin	202.67	95.33	3.47	1.93	1.67	0.80	1.87
Common Vàirano lupin	168.67	95.13	1.93	2.93	1.87	1.53	4.13
Bitter Cortale lupin	169.13	82.67	3.00	2.00	1.80	1.73	5.13
Philadelphia lupin	171.33	80.27	3.27	2.13	1.40	0.87	4.27
P = 0.05	13.91	9.47	2.24	1.93	1.34	1.15	1.52
P = 0.01	18.28	12.80	2.95	2.53	1.76	1.50	2.00

acquisitions. In 1980 - 81, when a higher plant density was used, there was a sharp drop in the number of racemes with more than four pods (Figure 5). These findings will be taken into consideration in the work on genetic improvement.

TABLE 9

DETAILS OF MORPHOLOGICAL CHARACTERISTICS FOR 1981 EXPERIMENTS

Type or variety	Height of plant in cm	Height (cm) of first pod	Number of pods per plant: Single	Twin	3-pod	4-pod	More than four pods
Type 10	91.83	22.23	0.80	1.00	0.63	0.73	1.67
18	47.03	13.23	1.13	1.53	1.67	0.93	1.80
39	92.40	23.10	1.03	1.47	1.23	1.03	2.30
53	85.67	24.63	0.57	0.87	0.80	0.57	1.20
Lupinus albus	71.23	14.83	0.93	1.53	1.43	1.20	1.50
Multolupa (tedesca)	82.50	18.60	0.67	1.23	1.20	1.00	1.63
Red Hamburg	74.83	20.23	0.87	1.23	1.37	0.77	1.93
Ultra lupinus	41.40	14.70	0.80	0.77	0.50	0.67	2.47
Multolupa (Chile)	86.63	22,87	0.60	1.23	0.90	1.27	1.33
Lupinus angustifolius Unicrop	55.30	15.10	6.57	7.33	3.07	1.03	0.77
Lupinus angustifolius Uniharvest	60.37	13.00	8.57	10.17	6.73	2.33	1.10
Populor	38.97	9,83	0.93	0.60	0.77	0.40	1.90
Vairano lupin	107.50	37.57	0.70	0.77	0.77	0.47	0.93
CLP-17	91.23	39.73	0.20	0.47	0.37	0.33	2.83
P = 0.05	4.67	2.21	1.10	0.91	0.76	0.49	0.65
P = 0.01	6.13	2.90	1.44	1.20	1.00	0.64	0.85

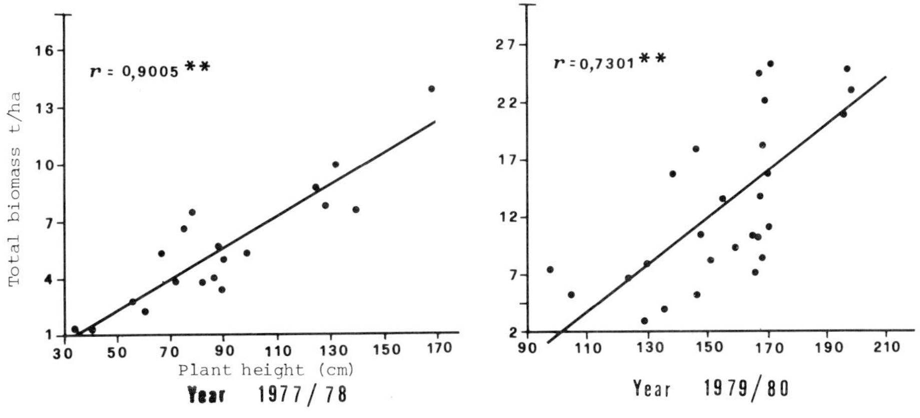

Fig. 3. Correlation between total production and plant height.

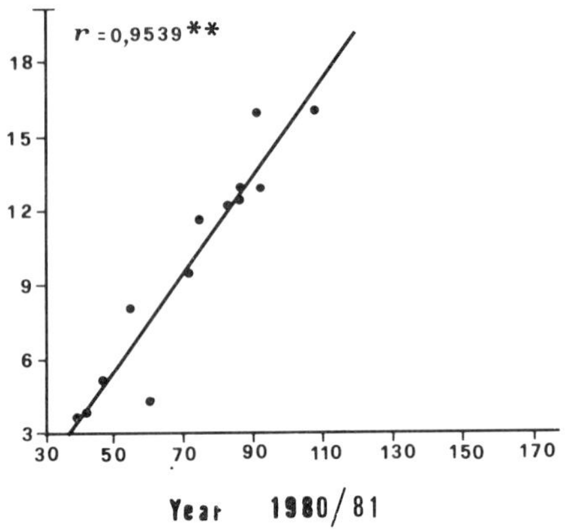

Fig. 3. (cont.)

Correlation between total production and plant height.

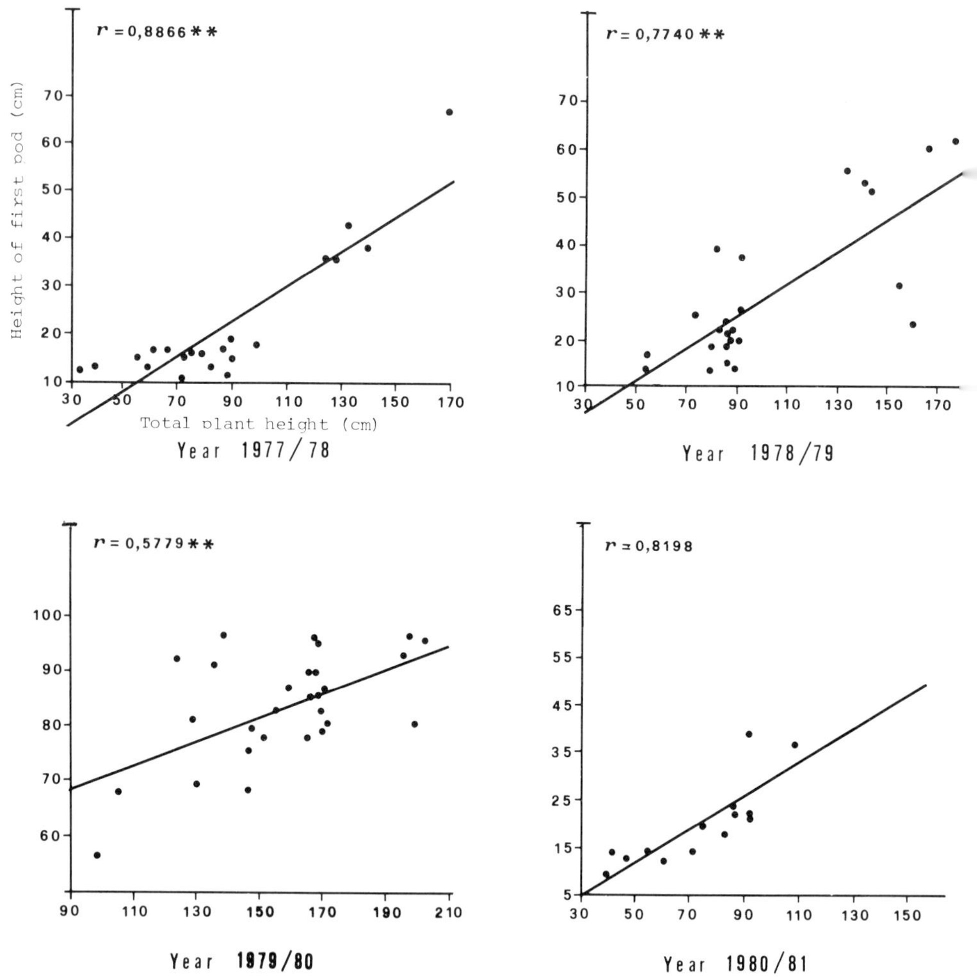

Fig. 4. Correlation between height of first pod and plant height.

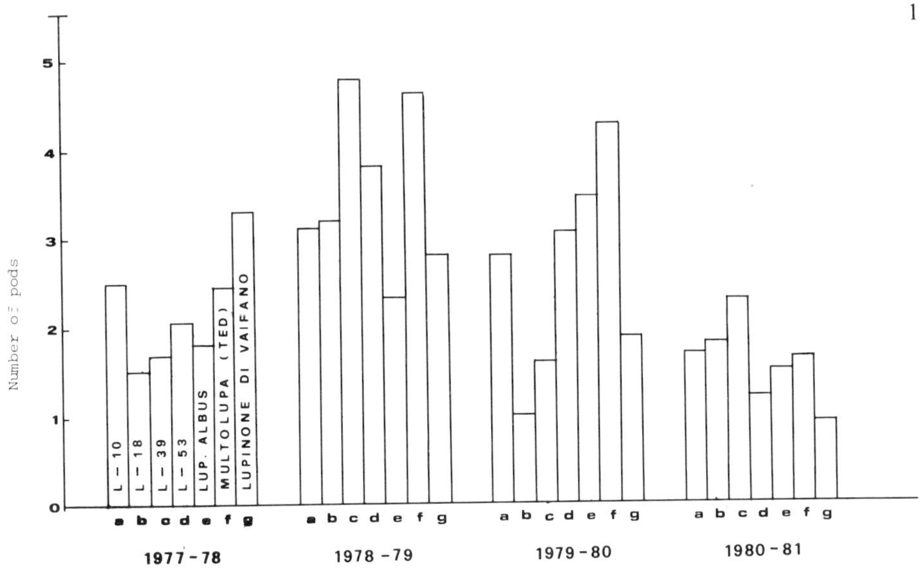

Fig. 5. Number of pods on racemes with more than four pods.

Lateral branching was limited, and in some types absent, perhaps as a result of the planting density used.

PLANTING DENSITY EXPERIMENTS

Because production in the previous years had been somewhat poor, especially for recently introduced types and varieties, a density experiment was carried out in 1979 - 80 with the aim of improving the productivity of the sweet lupin by improved cultural technique (Figure 6).

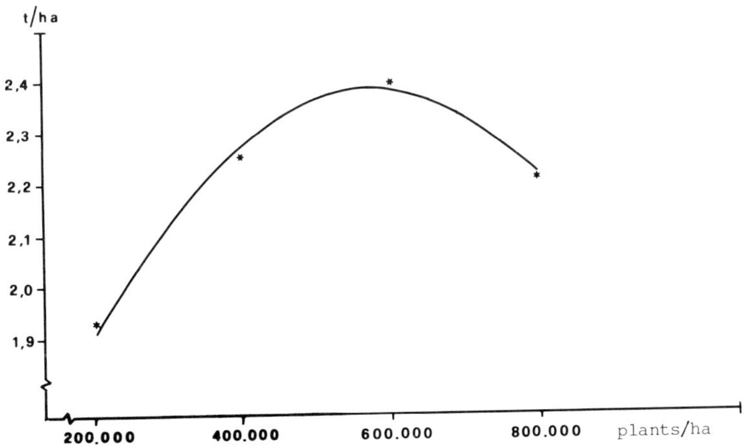

Fig. 6. Relation between plant density and yield.

Methods and materials

The experiment was carried out at the Torre Lama experimental farm on the same land as that used in 1979 - 80, for comparisons between varieties. Three cultivars were selected for the tests: the German cv. Multolupa and INRA's Type 10 for the *Lupinus albus* species and cv. Unicrop for the *Lupinus angustifolius* species. The Vairano lupin *(Lupinus albus)* was used as a control. Each cultivar was grown at 200 000, 400 000, 600 000 and 800 000 plants/ha in a randomised block design with three replicates. Seeds were sown in rows 20 cm apart and the various densities were achieved by subsequent hand thinning. Plot sizes were 15 m^2 and each was surrounded by a border of the same density. Cultural details are given in Table 10.

TABLE 10

CULTIVATION PROGRAMME 1979-80

Previous crop	wheat
Ploughing	3 Sep 1979
Harrowing	4 Sep 1979
Fertilising	5 Sep 1979 - P_2O_5 600 kg/ha N 150 kg/ha K_2O 100 kg/ha
Rotary tilling	5 Sep 1979
Sowing	2 Oct 1979
Thinning out	14 Nov 1979
Weeding	19 Nov 1979 - 16 Jan 1980 - 11 Feb 1980 - 27 Mar 1980
Harvesting	7 Aug 1980

Results

The findings showed that density affected yield (Table 11) The amount of seed obtained increased up to a density of 600 000 plants/ha, whereas the biomass increased linearly up to the maximum density examined. There were no significant variations, however, in the 1 000 seed weight or the height of the plants as a result of the densities used. Densities at harvest were

TABLE 11

PLANTING DENSITY AND YIELD DATA

Variable	Yield t/ha seed	Yield t/ha whole plants	1 000 seed weight g	height cm	Plants existing at harvesting: Number	Plants existing at harvesting: % of theoretical number	Plants existing at harvesting: Angular values
Planting density							
200 000 plants/ha	1.928	7.778	348.81	137.46	107.000	53.5	47.30
400 000 plants/ha	2.255	9.911	333.80	147.97	185.080	46.3	42.69
600 000 plants/ha	2.395	11.626	336.03	140.47	235.420	39.2	38.23
800 000 plants/ha	2.206	11.089	333.39	139.44	284.500	35.6	36.50
linear component	x	x x	n.s.	n.s.	x x		x x
quadratic component	x	n.s.	n.s.	n.s.	n.s.		r.s.
P=0.05	0.285	2.455	n.s.	n.s.	39.07		4.26
P=0.01	n.s.	n.s.	-	-	52.62		5.74
Variety							
Multolopa (Germany)	2.450	9.044	360.79	137.55	184.17	40.0	39.06
Unicrop	1.004	4.611	145.47	99.05	160.00	30.6	32.86
LA 10	2.247	11.467	287.01	155.60	194.08	42.4	40.53
Vairano lupin	3.083	15.281	558.37	173.30	271.67	61.6	52.26
P=0.05	0.285	2.455	33.61	12.44	39.07		4.26
P=0.01	0.384	2.306	38.19	16.36	52.62		5.74

n.s. = not significant

lower than the original densities, the depression in value increasing as the theoretical density increased, and this indicated a lower survival rate at the high densities.

The cultivar with the highest yield was the local Vairano lupin (3.08 t/ha), followed by the cvs. Multolupa and Type 10, while the cv. Unicrop had the lowest yield of 1 t/ha.

The average survival rate of the plants was higher than average for the Vairano lupin and lower for the cv. Unicrop. The general pattern of response of the four cultivars tested was similar, and this is confirmed by the lack of significant interaction for any of the characteristics examined.

CONCLUSIONS

A number of conclusions may be drawn after the four years of experimental work aimed at introducing alkaloid-free lupins to southern Italy as a low-cost protein rich plant which can be grown on marginal land.

It seems that the alkaloid-free *Lupinus albus*, *L. angustifolius* and *L. luteus*, can be grown successfully as an autumn - winter crop in the soil and weather conditions of southern Italy. Some types have been found to be well-adapted, and for others, a selection and improvement programme is in progress. Similarly, the same programme of improvement will be used for the lines with alkaloids but with a high oil content of the species *Lupinus mutabilis* .

As a result of the comparative study of the varieties, it was possible to assess yield and morphology of accessions over a four year period during which the meteorological conditions were quite variable.

A number of alkaloid-free cultivars (for example, the German and Chilean varieties of Multolupa, Red Hamburg and Type 53 *(Lupinus albus)* gave higher yields than the local control plant in the best year and with the most favourable growing techniques. Cultivars of *Lupinus angustifolius*, *L. luteus* and *L. mutabilis* provided poorer yields. From these experiments data was also collected for yield components and characteristics, which will be especially useful in future selection work.

In the plant density experiments carried out in 1979 - 80 it was found that seed yield increased as density was increased up to the theoretical density of 600 000 plants/ha. This was confirmed in experiments with different varieties which were carried out the following year.

These encouraging results indicate that technically, at least, sweet lupins could be grown in the south of Italy. This prospect will be further improved in the near future, as improved growing techniques are developed as well as continued selection for improved types. Currently, bacteria inoculation tests are being carried out, with *Lupinus luteus* in particular, and new research is planned on cultivation techniques (e.g. sowing date, nutrition and weed-control).

If lupin cultivation is to expand markedly, however, a market will have to be created. In this respect, firms specialising in the production of fodder will have to demonstrate their requirement for seed.

REFERENCES

Hudson, B.J.F., Fleetwood, J.G. and Zand-Moghaddam, A., 1976. Lupin: An arable food crop for temperate climates. Plant Foods for Man, 2, 81-90.

Postiglione, L., 1979. Progressi nella coltivazione del lupino - Atti convegno: Prospettive delle proteaginose in Italia, Perugia, 267-296.

THE YIELD POTENTIAL OF DRIED PEAS *(PISUM SATIVUM)* AS A SEED LEGUME IN THE FEDERAL REPUBLIC OF GERMANY

M. Frauen and M.H. Ismail

Institut für Planzenbau und Pflanzenzüchtung,
Universität Göttingen, V. Siebold Strasse 8,
D-3400 Göttingen, Federal Republic of Germany.

INTRODUCTION

The amount of dried peas produced in the Federal Republic of Germany is very small. There are only about 2 000 ha (or 0.03% of the arable land) of peas grown for combining when the seed is dry and mature. The average yield is less than 30 dt/ha, but this figure probably does not represent the yield potential of the crop. Increased interest in home grown protein production and the new EC price policy for dried peas prompted the following trial to assess the yield potential of several varieties of dried peas in German conditions.

MATERIAL AND METHODS

The trial was carried out during 1980 and 1981 at the following three locations:

At the breeding station of the NPZ company, Hohenlieth near Kiel;
at the Institute of Plant Breeding, Göttingen; and
at the Landessatzuchtanstalt, Hohenheim.

The plot size used at all three locations was about 10 m^2 and the number of replications was four in Hohenlieth and Göttingen and three in Hohenheim. The 16 pea varieties tested are given in Table 1. The cultivars Facima, Juwel and Watex are bred for culinary use by the canning industry and all the other varieties are of the dried pea type. Different morphological plant types included were leafless, semi-leafless, tare-leaved and conventional. The seed weight of the varieties ranged between 120 and 370 mg.

TABLE 1

PEA VARIETIES TESTED IN THE YIELD TRIAL AT HOHENLIETH, GÖTTINGEN AND HOHENHEIM IN 1980 AND 1981

Variety	Breeder	Seed weight (mg)	Remarks
Auralia	KWS (D)	260	
Facima	van Waveren (D)	150	
Juwel	"	160	wrinkled seeds
Watex	"	200	
Columba	NPZ (D)	250	
Filby	NSDO (GB)	200	leafless
Progreta	PGRO (GB)	280	small leaflets
Amino	Blondeau (F)	270	
Frisson	INRA (F)	120	winter-pea
Finale	Cebeco (NL)	290	
Imposant	"	370	coloured flowers
Stehgolt	Mansholt (NL)	240	
St. 859	"	310	coloured flowers
Hamil	Rolimpex (PL)	230	semi-leafless
Kwartet	"	260	
Zefir	"	230	

RESULTS AND DISCUSSION

Yields at the three locations for both years are given in Table 2 and indicate a trend of increasing yield from the northern part of Germany (Hohenlieth) to the southern part (Hohenheim).

Not only was the average yield higher at Hohenheim, but there seems also to have been a smaller environmental effect over the years at this location.

The relative yields of all varieties tested in the five environments are given in Table 3. Values from the sixth environment, Hohenheim 1980, are not presented because of very high plot error variance in that experiment in which no significant effects of varieties were detected. The least significant differences (LSD) were high in the environments at Hohenlieth

TABLE 2

AVERAGE YIELD (dt/ha) OF 16 PEA VARIETIES GROWN AT THREE LOCATIONS DURING 1980 AND 1981

Location	Year 1980	1981	$\bar{x}$
Hohenlieth	24.4	29.4	26.9
Göttingen	21.4	34.3	27.9
Hohenheim	31.7	34.1	32.9
$\bar{x}$	25.8	32.6	29.2

TABLE 3

RELATIVE YIELD OF 16 DIFFERENT PEA VARIETIES. ENVIRONMENTAL MEAN = 100, ACTUAL MEAN YIELD OF EACH ENVIRONMENT IN dt/ha

	Environment*					
	Hl 80	Hl 81	Go 80	Go 81	Hh 81	$\bar{x}$
Yield (dt/ha) variety	24.4	29.4	21.4	34.3	34.1	28.7
Finale	141	131	149	159	143	145
St. 859	141	128	107	115	129	124
Amino	123	114	91	137	128	121
Auralia	108	144	104	106	103	113
Stehgolt	136	103	105	117	97	111
Imposant	119	131	94	101	101	109
Hamil	105	105	102	94	123	107
Columba	110	96	108	109	103	105
Kwartet	96	94	107	101	100	100
Progreta	77	98	88	110	78	91
Frisson	82	88	95	101	82	90
Watex	116	66	108	85	55	83
Filby	82	54	103	71	105	82
Facima	59	71	88	80	86	77
Juwel	58	76	78	67	86	74
Zefir	46	101	76	50	79	71
LSD (5%)	24	16	27	8	9	22
LSD (1%)	32	21	36	11	12	30

* Hl = Hohenlieth   80 = year 1980
Go = Göttingen   81 = year 1981
Hh = Hohenheim

1980 (Hl 80), Hohenlieth 1981 (Hl 81) and Göttingen 1980 (Go 80). Only two experiments in Göttingen 1981 (Go 81) and Hohenheim 1981 (Hh 81) have reasonably low LSD values.

The varieties in Table 3 are ranked according to their mean yield performance. The highest mean yield was produced by the variety Finale with 41.6 dt/ha which corresponds to 145% of the overall average yield at 28.7 dt/ha.

There was a big gap of 20% or more between the yield of the best variety and a group of eight varieties which ranged from 100% to 124% of the average. These varieties are all of the conventional plant type, with the exception of Hamil which is a semi-leafless type. A third and lower yielding group of cultivars can be distinguished with yield levels below the average and range from 71% to 91%. This last group includes the culinary varieties, the leafless type Filby, the tare-leaved Progreta, and the very small seeded type Frisson.

Interactions of variety x environment are also evident in Table 3. In all the five environments Finale was either the highest or second highest yielding. This could be interpreted as a high level of yield stability. In contrast, there are some varieties with a marked environmental response for example Amino, which at Göttingen in 1980 yielded only 91% of the environmental mean, whilst at Göttingen in 1981 it yielded 137% of the mean. Likewise the variety Filby, which at Hohenlieth in 1981 yielded only 54%, whilst at Hohenheim in 1981 it yielded 105% of the mean. However, all these data which are based on single environment experiments include high error variances and when interpreting the result this should be taken into consideration. It is suggested that the technique and/or layout of yield trials with dried peas requires improvement.

CONCLUSIONS

The following conclusions can be drawn from the experiments in these two years:

1) There are some locations which seem more suitable than others for growing dried peas. This should be seen, however, in the context of competition from other seed legumes such as faba beans. Other experiments have shown that the yield of

faba beans is 20% to 30% higher than that of dried peas at Hohenlieth, and 10% to 20% at Göttingen, whereas in Hohenheim the seed yield of dried peas is often equal to or even higher than that of faba beans.

2) There are large differences in the potential yield of different varieties. The yield of the best variety tested at slightly more than 40 dt/ha is high enough, at least in suitable locations, to give the farmers a good net income from dried peas, and a new and interesting crop for Germany.

ACKNOWLEDGEMENTS

The authors are grateful to Dr. G. Rakow, Norddeutsche Pflanzenzucht (NPZ), Hohenlieth, and Dr. E. von Kittlitz, Landessatzuchtanstalt, Hohenheim, for their interest and care in carrying out the yield trials. Thanks are also due to Prof. Dr. Dr. G. Röbbelen, Göttingen, for his initiative and help in organising this experiment.

DISCUSSION

Planquaert *(France)* Did you compare peas and field beans under the same conditions and if so what were the results? Secondly, did you compare plot yields of peas with those under farm conditions of growing and harvesting.

M. Frauen *(FRG)* Precise comparisons between the yields of peas and field beans are not possible because the variety of field beans grown differed between sites. However, in the north field beans yield up to 50% more than dried peas, and at Göttingen the yield of field beans was about 20% higher than dried peas. The most interesting location was Hohenheim where peas gave higher yields than beans.

With regard to your second question, the difference varies from variety to variety - it may be about 10%.

E.J. Walsh *(Ireland)* Did plot size have any effect on the range of genetic expression, in particular on heritability. There is some evidence that in cereals and using unselected material, heritability in very small plots is higher than in large field plots.

M. Frauen There were no differences in ranking between the different plot sizes in our experiments this year.

CANOPY PERFORMANCE IN CONTRASTING GENOTYPES OF PEA (*PISUM SATIVUM*)

D.K.L. MacKerron and R. Thompson
Scottish Crop Research Institute,
Invergowrie, Dundee DD2 5DA, Scotland.

In Scotland consistency of yield of protein-rich grain crops is jeopardised by poor ripening conditions. The problem is more acute in peas *(Pisum sativum)* than in field beans *(Vicia faba)* because peas lodge more readily and completely than beans, and hence provide little prospect of adequate drying except in prolonged dry conditions. However, the development of 'leafless' peas seems to offer a number of possible advantages in the pea crop. The haulm is generally reduced, the crop stands better by self support, and the canopy, being more open, allows freer air movement with a possible resulting reduction in moulds and improved seed drying. For such a cultivar to gain acceptance commercially, its yield should be comparable with that of conventional cultivars.

Ideally, comparisons of yield between 'leafless' and conventional types would be made between cultivars which were isogenic apart from those genes which control the morphology of the leaf. However, in the absence of such material, comparisons of yield, canopy performance and efficiency, must be made between cultivars which are as closely similar as possible, and the conventional cv. Vedette was chosen for comparison with the 'leafless' cv. Filby (Snoad, personal communication). The two cultivars, which are both round seeded were grown at 120 and 100 plants/m^2 respectively in six replicates. These densities have in the past been shown to give yields on or close to the plateau of the yield v. density relationship.

Light interception was measured using tube solarimeters installed at several heights within the canopies, and growth was followed by taking fortnightly harvests over the growing season. At each harvest the canopy was divided into 10 cm horizontal layers and separated into its components, but this paper will deal only with overall growth. The canopies of both cultivars were supported by horizontal square-mesh netting to reduce lodging especially in Vedette. Total dry matter yields of 10.6 and 8.0 t/ha were obtained for Vedette and Filby respectively ($P < 0.05$) with corresponding yields of dried pea (adjusted to 15% moisture) of 3.6 and 4.3 t/ha ($P < 0.025$).

Canopy heights were measured at approximately weekly intervals. Until the second week of July Vedette formed a taller canopy than Filby but thereafter canopies were of similar height at about 85 cm. Lodging occurred in Vedette despite the support provided, resulting in early August in a reduction of its canopy height to about 55 cm, whilst Filby remained at about 85 cm.

In a pea canopy light is intercepted and used by leaves, stems, petioles, tendrils and pods. So, for relating assimilatory surface area to measurements of light interception, the projected areas of all organs were summed and denoted 'crop area' rather than the more usual 'leaf area'.

The crop area index (crop area/unit area of soil surface) of cv. Vedette reached a maximum of 13 (Table 1) by 9 July then fell sharply but Filby did not reach its maximum crop area of 4 until a fortnight later. Crop areas (Table 2) at their maximum level of development show the contrast between the uniform distribution of the area intercepting light in the Filby canopy and the marked increase in area index especially towards the base in the Vedette canopy. The large proportion of material in the 10 - 20 cm layer of the Vedette canopy was in part a consequence of the canopy collapsing. A reduction in crop area at all levels of Vedette towards the end of July was due to continued collapse and leaf death at the lower levels of the canopy.

The differences in light level within the two canopies are indicated by the profiles of relative light intensity (I/I_0 where I_0 = incident radiation above the canopy and I = incident radiation at any given height within the canopy) (Table 3) in the canopy on 9 July. This shows higher intensities at lower levels in the Filby canopy than in that of Vedette which are partly due to the reduced amount of foliage and partly to the different nature of the canopy. The relationship between relative light intensity and cumulative crop area (Table 2) shows that for a

TABLE 1

CROP AREA INDEX ON SEVERAL OCCASIONS FOR cvs. Filby and Vedette

Days from sowing	Crop area index Filby	Crop area index Vedette
28	0.16	0.72
41	0.67	2.97
55	1.66	8.98
69	3.10	13.30
83	3.85	8.33
98	2.76	5.43
112	1.46	

TABLE 2

CROP AREA INDEX AND RELATIVE LIGHT INTERCEPTION ($I/_{IO}$) OF FILBY AND VEDETTE, AT SEVERAL POSITIONS WITHIN THE CANOPIES, AT THE TIME OF MAXIMUM VALUE OF CROP AREA INDEX OF EACH cv.

	Filby			Vedette		
Height cm	Crop area index	Accumulated crop area index	$I/_{IO}$	Crop area index	Accumulated crop area index	$I/_{IO}$
70-80	0.06	0.06		-	-	
60-70	0.26	0.32		0.48	0.48	
50-60	0.42	0.74		0.69	1.17	
40-50	0.38	1.12	0.38	0.86	2.03	0.14
30-40	0.44	1.56	0.27	1.07	2.10	0.07
20-30	0.53	2.09	0.25	1.48	4.58	0.11
10-20	0.42	2.51	0.20	2.20	6.78	0.07
0-10	0.54	3.05	0.16	4.78	11.56	0.08

given crop area index up to values of about 3 the canopy of Filby intercepted less light. It would be expected of course that above a certain value of crop area index total interception would occur and then both canopies would be equally effective in this respect. Presumably the shape of the tendrils results in more light interception at acute angles, giving greater light scattering into the lower canopy with the cv. Filby. These characteristics ought to signify a more efficient

TABLE 3

LIGHT INTERCEPTION BY CANOPIES OF cvs. Filby and Vedette ON SEVERAL OCCASIONS BETWEEN SOWING AND MATURITY (ACCUMULATED VALUES)

Days from sowing	Acc. light intercepted MJ/m^2	
	Filby	Vedette
28	29	76
41	95	249
55	222	480
69	360	676
83	544	887
98	678	1 026
112	831	

canopy since light is more uniformly distributed over the assimilating surface, resulting in less sub- and supra-optimal levels of illumination. An obvious disadvantage was that the total amount of light intercepted by Filby was less than that intercepted by Vedette (Table 3).

Classical growth analysis of the dry weight and tissue area data was made using Hunt and Parsons (1974) computer program for the method of Hughes and Freeman (1969). Relative growth rate (Table 4) of Filby, which had the smaller seed, was initially the same as that of Vedette but delayed senescence reduced its rate of decline compared with that for Vedette.

Net assimilation rates were higher in Filby than in Vedette throughout the season (Table 4) but a better measure of energy fixing efficiency is the estimate of increases in dry weight per unit amount of light energy intercepted. Canopy efficiency, expressed in this way in g/MJ, is similar for both cultivars but high levels of efficiency are maintained for longer by the Filby canopy (Table 5). The progressive increase in efficiency in the early part of the season may be an artifact accountable for by root growth which was not measured, and which therefore cannot be included in estimates of total dry weight production. High rates of root growth occur until about flowering (Salter and Drew, 1965) which would require a large proportion of total assimilates early in the season.

TABLE 4

RELATIVE GROWTH RATE (RGR g/g/day) AND NET ASSIMILATION RATE (NAR $g/m^2/day$) FOR cvs. Filby and Vedette

Days from sowing	RGR		NAR	
	Filby	Vedette	Filby	Vedette
28	0.114	0.121	11.65	5.26
41	0.092	0.094	9.81	4.11
55	0.069	0.065	8.36	3.29
69	0.046	0.036	6.95	2.45
83	0.023	0.007	4.77	0.76
98	-0.001	-0.024	-0.43	-4.70
112	-0.024		-12.68	

TABLE 5

TOTAL DRY WEIGHTS OF PLANT TISSUE (g) PER MJ OF LIGHT INTERCEPTED ON SEVERAL OCCASIONS FOR cvs. Filby and Vedette

Days from sowing	Growth (g/MJ)	
	Filby	Vedette
28		
41	0.94	0.69
55	1.25	1.25
69	1.64	1.81
83	1.76	0.70
98	0.26	0.78
112	0.05	-

Because of the lack of lamina in Filby, a higher proportion of light was intercepted by the stem and petioles (28%) than by those of Vedette (11%). Thus although the Filby canopy was apparently more efficient than Vedette as assessed by net assimilation rate, its conversion of intercepted light energy into dry matter was not significantly greater than that of Vedette.

The results demonstrate that 'leafless' pea canopies may be as effective as those of conventional cultivars in producing harvestable seed and are as efficient in terms of dry matter production per unit of light intercepted. In this experiment, however, total dry matter production was lower for Filby and, therefore, the harvest index of the 'leafless' canopy was better than that of the conventional one, giving harvest indices of 0.29 and 0.46 for Vedette and Filby respectively.

Total light interception by the conventional cv. Vedette was 1 026 MJ/m^2 which compares with only 831 MJ/m^2 by cv. Filby. There is, therefore, scope for improvement in light interception and if harvest indices in new 'leafless' cultivars can be maintained then there is potential for further increases in yield.

Finally, the time of year when peas ripen in Scotland i.e. August/September generally coincides with deteriorating weather conditions. Substantial amounts of rain may be expected during these two months (25 year average at SCRI = 69.6 and 58.7 mm respectively) while evaporation rates are declining rapidly (6 year average = 76.7 and 44.6 mm respectively). Seeds which are approaching a sufficiently low moisture content may be rehydrated by rain (Table 6) and harvesting delayed considerably. Under such conditions decomposition of the whole plant is rapid. This emphasises the importance of an open canopy which resists lodging and which dries rapidly in northern latitudes.

TABLE 6

CHANGES WITH TIME IN PERCENTAGE MOISTURE CONTENT OF PEA SEEDS OF FOUR cvs.

	Moisture content of seeds as % fresh weight			
Date 1980	Amino	Finale	Columbo	Maro
27 August	30	32	35	50
17 September	36	29	36	33
29 September	33	28	38	35

REFERENCES

Hughes, A.P. and Freeman, R.R., 1969. Growth analysis using frequent small harvests. J. applied Ecol. 4, 553-560.

Hunt, R. and Parsons, I.T., 1974. A computer program for deriving growth functions in plant growth analysis. J. appl. Ecol. 11, 297-307.

Salter, P.J. and Drew, D.H., 1965. Root growth as a factor in the response of *Pisum sativum* L. to irrigation. Nature, London. 206, 1063-1064.

DISCUSSION

R.C. Hardwick *(UK)* I would like to ask about the apparent trade-off between increase in net assimilation rate and, presumably, decrease in leaf area ratio, in the Filby type. It seems that there was a spectacular increase in net assimilation rate but that had no effect on relative growth rate. This must be because of a spectacular decrease in leaf area ratio. Can you speculate on why the two seem to balance so exactly?

R. Thompson *(UK)* I don't know why this should be so, but it is presumably related to light penetration further into the canopy.

I would like to make one comment regarding plant density. I think one possibility for increasing the crop area index in Filby may be to increase the plant density, or as suggested by Dr. Hedley, introduce a small amount of leaf.

BREEDING FOR YIELD AND FOR PROTEIN CONTENT IN PEA

R. Cousin

Station de Génétique et d'Amélioration des Plantes,
INRA, CNRA, Route de St. Cyr, 78000 Versailles, France.

INTRODUCTION

The European Economic Community imports 85% of its protein needs. In France, soyabean oilcake imports are growing constantly, they were 2 000 000 tons in 1976 and 2 500 000 tons in 1979. This year, the total consumption of oilcake will exceed 4 000 000 tons. Given this situation, the authorities in Brussels are taking steps to develop national protein sources such as field beans and peas.

Dried peas contain 23 to 33% protein and 40 to 50% starch. They are a valuable source of high energy for most animal species. They do not contain tannins or alkaloids and can be used direct, without processing. The amino acid content is close to that of soyabean and the lysine content slightly higher. They contain small amounts of trypsin inhibitors and some haemagglutinins the effects of which are not well known. All the experiments on nutrition which have been undertaken in the Zootechnical Department of INRA and ITCF, show that there is no significant difference when one half of the soyabean oilcake is substituted by pea.

Economic problems are the only factors which limit the use of peas in feed rations. There are, however, two possible solutions.

As a short term solution, a subsidy is available from Brussels within the framework of the EC. This subsidy amounts to approximately 40 Fr per quintal of peas incorporated into cattle feed. Consequently the pea crop has increased from 15 000 ha to 80 000 ha in three years.

The long term solution, however, is based on genetics. The purpose is to obtain new varieties with a higher protein yield per hectare, making the crop more competitive in comparison with cereals. Thus, at the Research Station at Versailles, we have undertaken the breeding of peas, with particular emphasis on protein content and yield.

INTERACTION BETWEEN YIELD AND PROTEIN CONTENT IN PEAS

In many species of cereals, yield and protein content are negatively correlated. Thus, initially it is necessary to study that relationship in peas. From our original investigation of 44 pea varieties it seemed that a negative correlation may also exist in peas, the correlation coefficient of P = -0.49 being significant. However, if smooth seeded and wrinkled seeded varieties are treated separately, the correlation coefficient ceases to be significant: for smooth seeded varieties P = -0.27 (not significant) and for wrinkled seeded varieties P = -0.12 (not significant) (Figure 1). Last year a study on 42 smooth seeded varieties gave the same result with a correlation coefficient of P = -0.15 (not significant) (Figure 2). Thus it is possible to breed for a higher protein content without a corresponding decrease in yield.

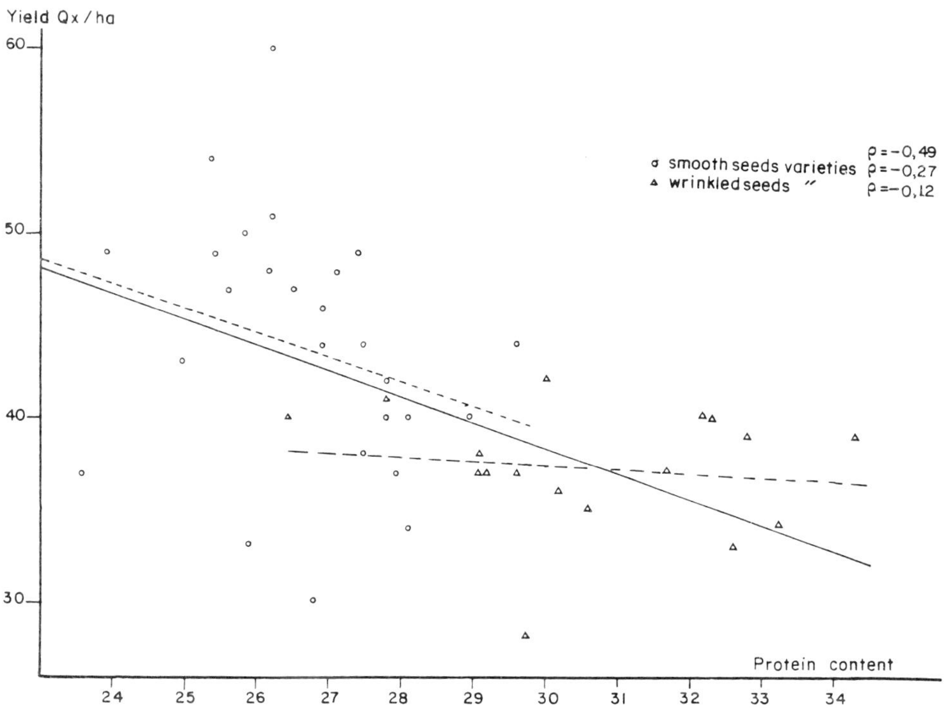

Fig. 1. Relationship between yield and protein content (%) of round and wrinkled seeded peas.

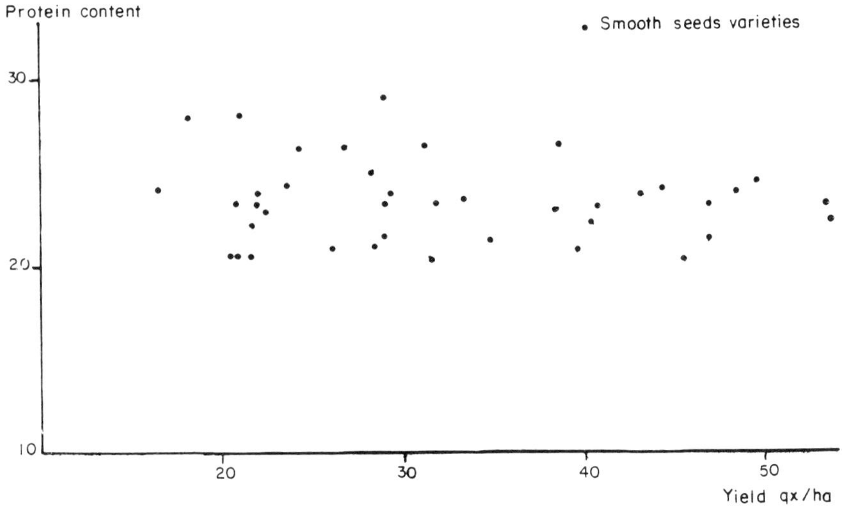

Fig. 2. Relationship between yield and protein content (%) of smooth seeded varieties.

BREEDING FOR PROTEIN CONTENT

Before undertaking breeding for protein content, it is necessary to know the accuracy for the chemical analyses, and variability in protein content.

Accuracy of chemical analyses

The techniques for protein analyses are continually improving. Chemical analyses can be carried out on very small samples. It has been shown that an analysis made on part of a cotyledon gives a good estimate of the protein content of the whole grain. Physical analyses made by an infra-red analyser, especially by 'Neotec' require larger samples (approximately 20 g), but they are much more rapid.

A series of measurements on 26 samples taken from a batch of ground peas of the variety Hybris, gave values with a standard deviation of 0.2 of the percent protein content (Figure 3). Furthermore, when samples of 150 seeds were ground individually and analysed, a standard deviation of 0.3 was obtained. However, when the sample was halved (75 seeds), the standard deviation doubled. This method was suitable for separating breeding lines which differed by 1 to 2% in protein content.

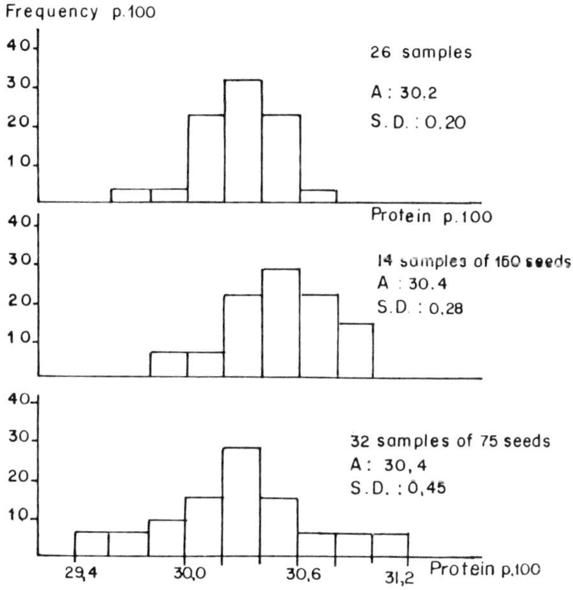

Fig. 3. Range of protein contents in relation to sampling method made on one lot of peas cv. Hybris. A = average; S.D. = standard deviation.

Analyses of the variation in protein content within a variety

We have observed that the protein content of a variety varies widely, for example from 26 to 33% for the cv Minarette, depending on the year and area of production. Moreover, the protein content between individual seeds varied from 20 to 45% for the same genotype (σ = 4.9) (Figure 4). The position of the seeds on the plant was an important determinant of this variation (Figure 5). Protein contents were similar for peas within a single pod, but the protein content decreases from the first fertile node to the top of the plant. A similar pattern exists on branches.

Analyses on about 15 plants from the same plot indicated a standard deviation for protein content of about 2. The protein content also varies with the region in which the crop is grown. In 1976 the average content of ten varieties growing in eight different regions varied from 24.4 to 28.1%, and in 1977 from 23.8 to 29.2% (Table 1). The variation between regions was greater than the genetic variability and ranged from 25.4 to 27.9%. However, there is no correlation between yield in a particular region and protein content (Figure 6).

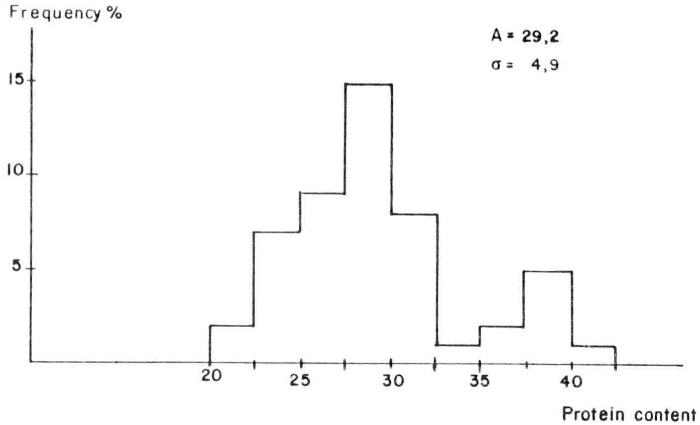

Fig. 4. Variation in the seed protein content (%) within the same lot (50 seeds) of cv. Minarette. A = average; σ = standard deviation.

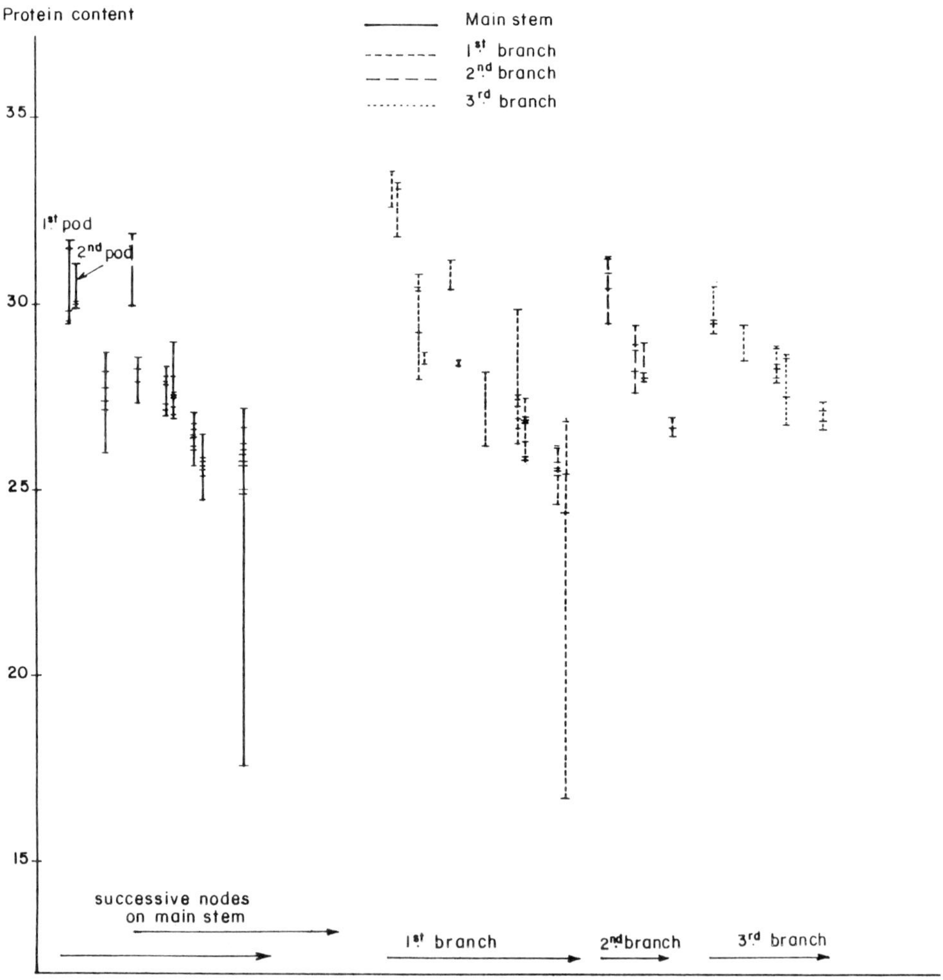

Fig. 5. Variation in protein content (%) of seeds in relation to their position on the plant cv. Primcovert.

TABLE 1

AVERAGE PROTEIN CONTENTS (%) FOR SEVERAL VARIETIES AND DIFFERENT REGIONS IN 1976 AND 1977 (%)

1976 Area	%	Variety	%	1977 Area	%	Variety	%
Somme (Mons)	28.1	Caméor	27.9	Yvelines (Versailles)	29.2	Pilet	27.9
Seine-Maritime	27.3	Pilet	27.8	Sarthe	27.8	Caméor	27.6
Yvelines (Versailles)	26.7	Vercas	27.4	Vienne	27.5	Printana	26.4
Côte-d'Or (Dijon)	26.4	Preclamex	27.0	Somme (Mons)	26.5	Vercas	26.4
Yvelines (La Minière)	26.1	Spiket	25.9	Yvelines (La Minière)	26.2	Finale	26.0
Tarn	26.1	Finale	25.8	Côte-d'Or (Dijon)	25.2	Amino	25.8
Maine-et-Loire	25.1	Printana	25.7	Aube	24.8	Ménivil	25.4
Vienne	24.4	Ménivil	25.2	Tarn	23.8	BL 7-76	25.4
		Colmo	25.2				
		Amino	24.8				

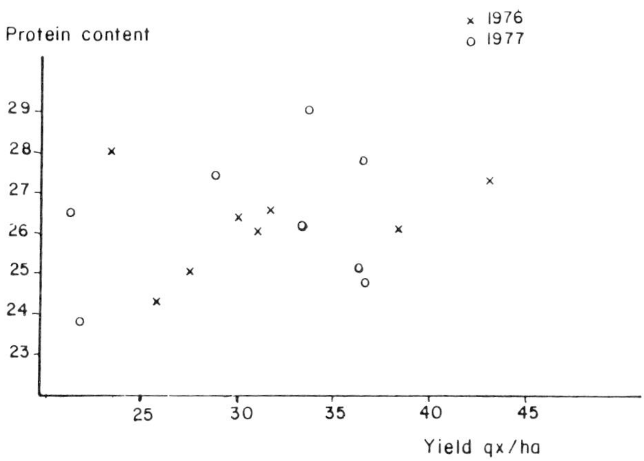

Fig. 6. Relationship between protein content (%) and productivity in different areas.

At present, in co-operation with Picard and Duthion, we are trying to interpret the relationship between protein content and the variation in soil.

Genetic variability

Because of the large variation in protein contents observed, several tests of the varieties under the same conditions are required to provide good estimates of genetic variability. The contents varied from 26 to 33% for wrinkled seeded varieties and from 23 to 31% for smooth seeded varieties (Figure 7). The wrinkled seeded varieties were about 2% richer in protein content than the smooth seeded varieties. This was due to variations in the synthesis of starch. The smooth seeded had a higher starch content and were richer in amylopectin than the wrinkled seeded types.

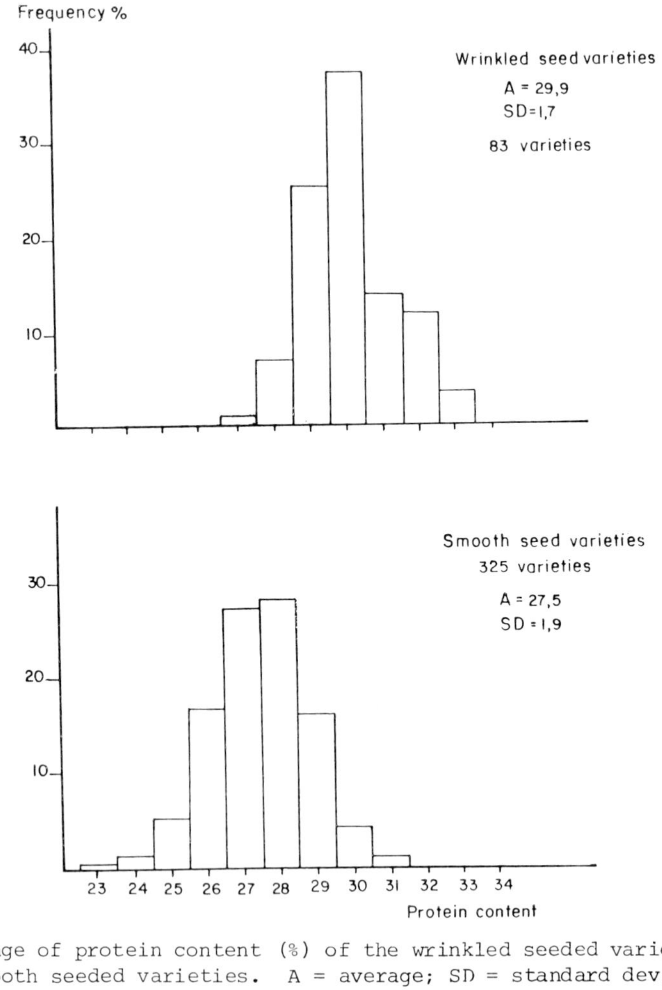

Fig. 7. Range of protein content (%) of the wrinkled seeded varieties and smooth seeded varieties. A = average; SD = standard deviation.

Heredity of protein content

In order to study the heredity of this character we undertook diallel crosses in each group of smooth seeded and wrinkled seeded varieties (Figure 8). The analysis showed that the high protein contents depended on recessive genes. However, among the wrinkled seeded varieties Juwel had the greatest protein content and gave low values for variance and covariance. Its genotype is presumed to possess some dominant genes. The regression line located above the first bissetrix shows the incomplete dominance of this character.

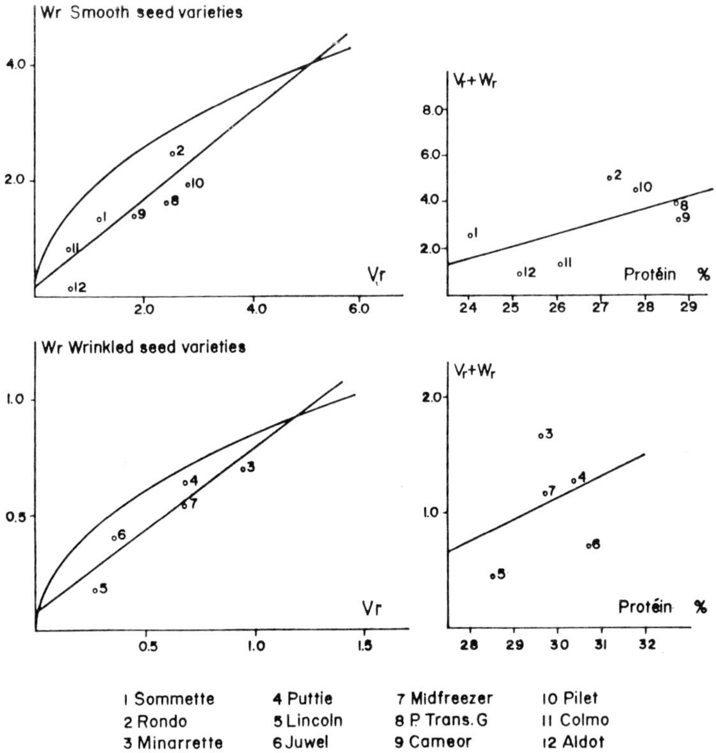

Fig. 8. Wr Vr analyses for the protein content of smooth and wrinkled pea.

Breeding for higher protein content

In the breeding programme selection could only begin at F4 due to the effect of seed position on the plant and the plant effect. This requirement was demonstrated experimentally by a cross Minarette x Frimas (Figure 9). In F2 seeds we observed a wide distribution of protein content from 20 to 45% and 20 to 30% for the F2 plants, and only 27 to 33% for F3 lines. Another study has shown that no correlation exists between protein content and seed size (Figure 10). So it is possible to breed for high protein and small seed size.

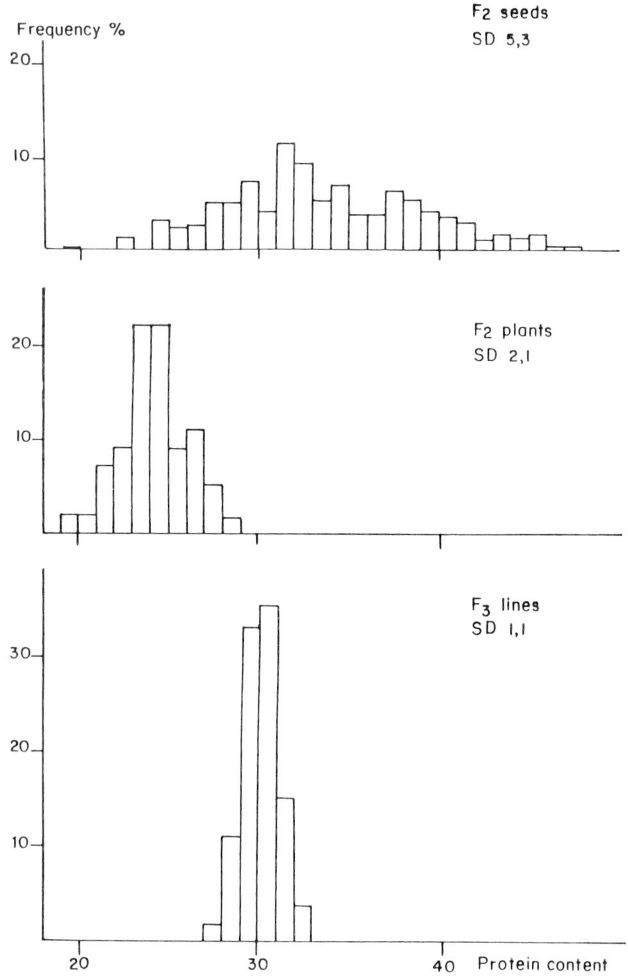

Fig. 9. Protein content (%) of the F2 and F3 progenies from the cross Minarette x Frimas.

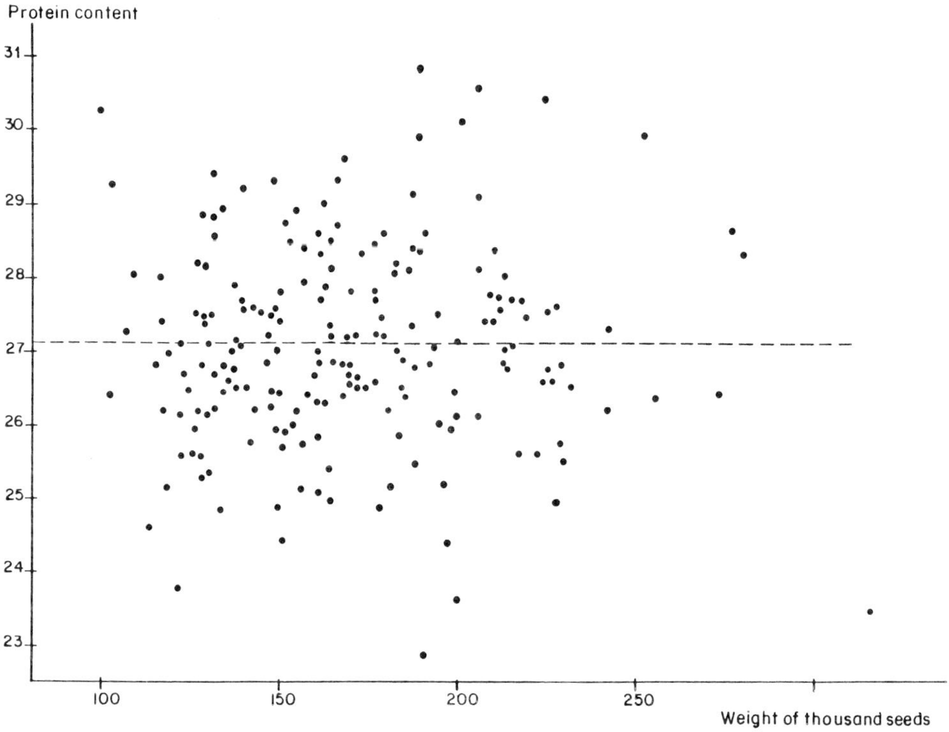

Fig. 10. Relationship between protein content (%) and seed size of the different varieties.

Breeding for quality of protein

In peas, the proteins constitute the complement of starch. We found evidence of several groups of starch with different grain shapes (Figure 11). Smooth seeded varieties have round starch grains, whereas wrinkled seeded varieties have composite grains. This character is controlled by a single recessive gene. It was also found that some wrinkled seeded varieties had round starch granules. This new group was controlled by another recessive gene. The double recessive could be obtained and constituted the 4th group which had composite granules. Every group had a clearly defined starch content as well as amylopectin composition.

Considerably less variation was observed for the amino acid composition which was similar to that of soyabean proteins. Some slight differences may exist between the groups with varying starch levels. For example, smooth seeded varieties seem richer in arginine and slightly poorer in methionine and lysine than wrinkled seeded types (Figure 12).

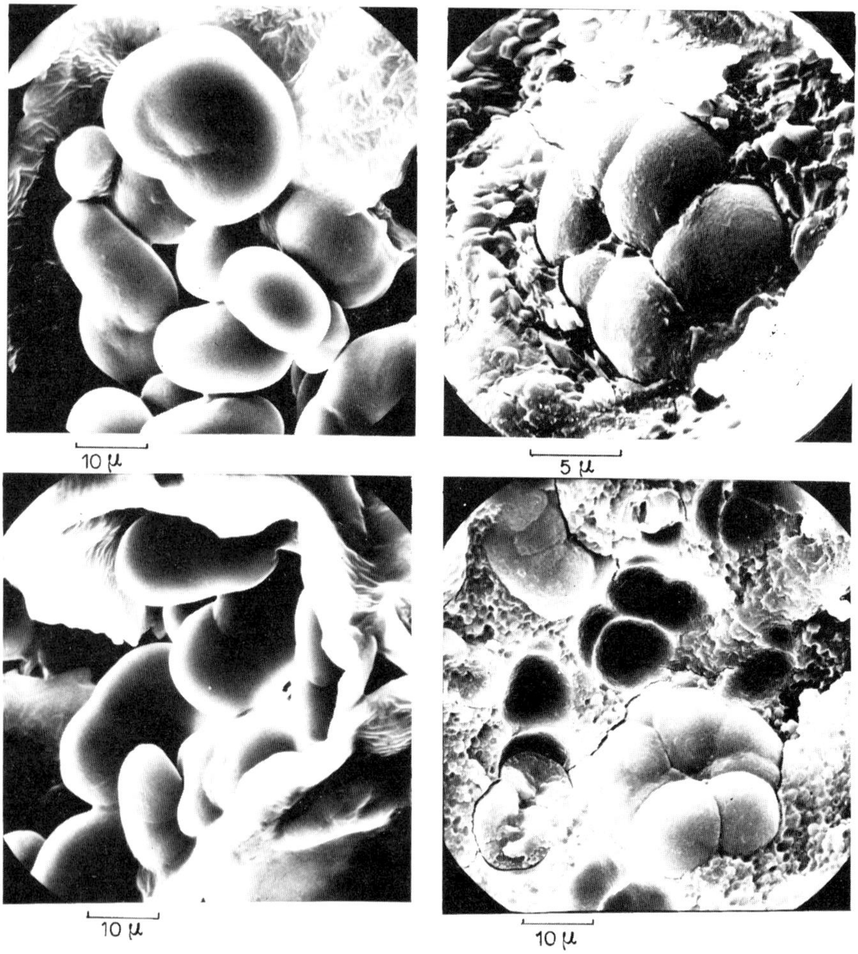

Fig. 11. Different shapes of starch granules:

a) Round starch granules of the smooth seeded varieties (e.g. Aldot)
b) Composite starch granules of the wrinkled seeded varieties (e.g. Kelvedon Wonder).
c) Round starch granules of some wrinkled seeded varieties (e.g. Alaska Sweet).
d) Composite starch granules of the double recessive line (from the cross Kelvedon Wonder x Alaska Sweet).

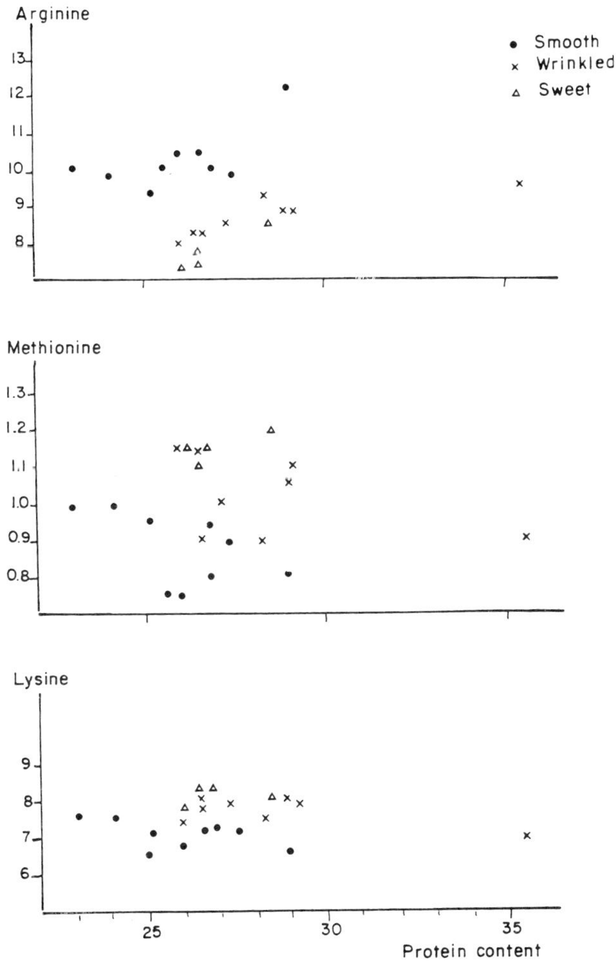

Fig. 12. Arginine, methionine and lysine contents in varieties with different starch groupings.

BREEDING FOR YIELD

About fifteen years ago we introduced winter hardiness from some winter fodder peas into the garden pea. This programme made it possible to obtain the winter pea varieties, Frimas, Floriver and Frisson. These varieties tolerate -15°C when well hardened. In the Rhone Valley and south western France winter peas give about 40% more yield than spring peas. In Versailles and Dijon during the last four years, yield varied from 22 to 62 qx/ha depending on variety and year (Table 2). The new line 22, which has still to be registered in the common catalogue, out-yields Frisson by some 20%.

TABLE 2

PRODUCTIVITY OF WINTER PEA VARIETIES IN VERSAILLES AND DIJON IN THE LAST FOUR YEARS qx/ha

	1978	1979	1980	1981
Versailles				
Frimas	24	12	42	22
Frisson	26	26	50	25
Line 22	32	23	46	45
Dijon				
Frimas	34	47	50	
Frisson	39	50	57	53
Line 22	52	54	62	62

Yield stability

Variation in yield is correlated to damage due to frost and to the subsequent development of pathogens. Thus it is necessary to increase the level of resistance to frost and to the main pathogens and to determine the best cultivation methods.

Winter hardiness

The analysis of diallel crosses showed that cold resistance was a quantitative character with intermediate dominance. In the progeny of some crosses it was possible to obtain transgressive lines for this character. Thus, in the progeny of the cross Frimas x Floriver, we selected lines that were more cold resistant than their parents. This was clearly demonstrated during the winter 1978 - 1979 when temperatures fell below -20°C. To breed for different levels of winter hardiness, a method of selection in controlled conditions was developed. Before exposing the plants to the required minimum temperature, it is necessary progressively to lower the temperature over a period of two to three weeks so that the plants become hardened (Figure 13). When applied to material selected in controlled rooms, this method does not take into account all the factors involved in field observations because winter hardiness is more complex than just frost resistance.

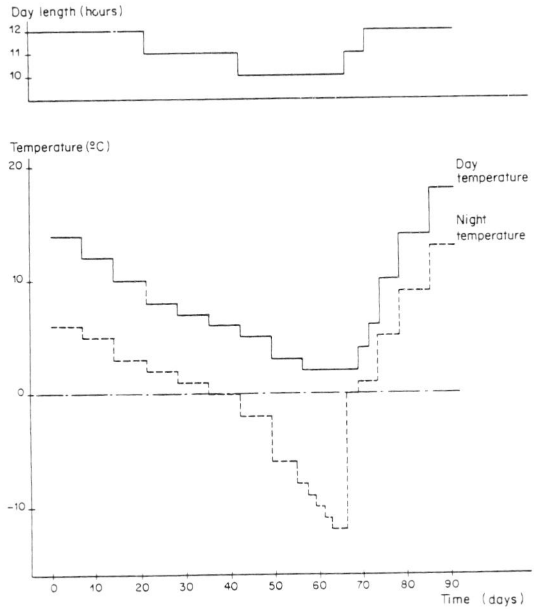

Fig. 13. Progress of temperature during evaluating winter hardiness in controlled conditions.

Disease resistance

The particular weather conditions during recent years have encouraged the development of some pathogens, especially *Ascochyta*. The resistance found in the variety Rondo is effective against nearly all physiological strains of *Ascochyta pisi* and represents a source with good heritability. Resistance to *Ascochyta pinodes* has proved more difficult to obtain.

Yield variation in relation to culture

Sowing date can be an important factor in the success of the crops. Plants from seeds sown in the first fortnight of November acquire better winter hardiness and better resistance to spring frost which was reflected in yield (Figure 14).

Study of yield factors

About 30 characters in winter and spring peas have been examined in relation to yield. The first study revealed slight correlations between yield and the different characters examined.

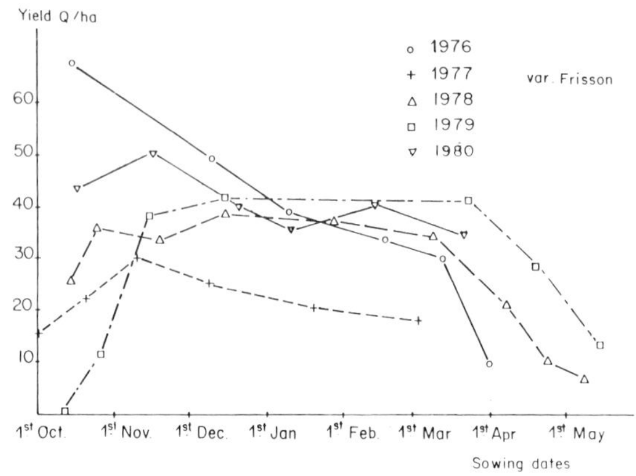

Fig. 14. Relationship between yield and sowing date.

This suggests that the best yields result from the interaction of several phenomena. Progressive multiple regression analysis was used to determine the order of decreasing importance of the characters and their regression coefficients. Initial results show that multiple regression involving about six characters gave a good estimate of yield. In this case between 40 and 70% of the variability was explained by differences between years. Among the other characters studied, height, leaf area, thousand seed weight, and number of branches, appear to be the most important, but their regression coefficients were low.

To be productive, spring peas should be quite short with some branching, small leaf area and large seeds. In contrast, winter peas may be taller but should also have a small leaf area and branching. Seed size seems to be less important; they may be smaller and more numerous.

Analyses showed that the vegetative organs were over-developed in peas. They used too much assimilate in relation to that used for seed growth. Thus, mutants without leaves seem to be promising in pea breeding and especially so in the winter pea

We have transferred the gene afila 'af' by back-cross into the variety Frisson and obtained an isogenic Frisson line. This material enabled us to examine the effect of this gene on photo-synthesis and yield variation. The afila line is 10 to 20% more productive than the normal. The leaf area of the plant was reduced by 40% but was more evenly distributed along the stem especially at the level of fertile nodes (Figure 15). Leaf growth of the semi-leafless line was weaker but persisted for longer (Figure 16).

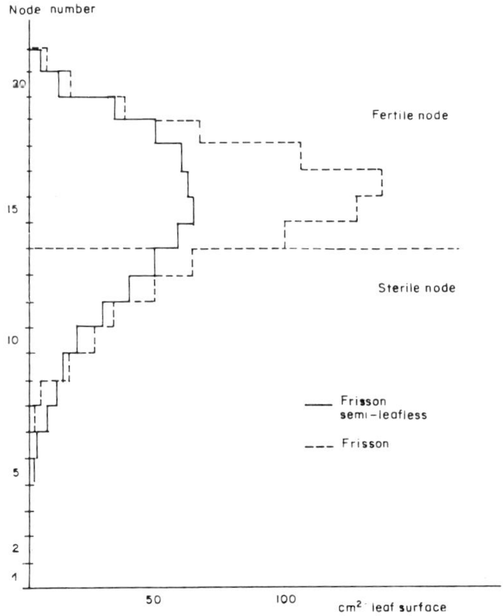

Fig. 15. Leaf area at each node for cv. Frisson and Frisson semi-leafless.

Photosynthetic activity was examined by colleagues Jolivet and Moutot and they found the activity of the semi-leafless stipules during the flowering period was twice that of the variety Frisson. That of tendrils may be compared with that of leaflets (Table 3). Nevertheless, photosynthetic activity in the leaf of a semi-leafless variety is lower than in normal varieties during much of the growth.

The higher productivity of the semi-leafless type may be due to differences in the plant as a whole. It is necessary to establish whether the stipule has prolonged photosynthetic activity resulting from better light penetration through the foliage. The assimilate which is not used for leaflet growth migrates preferentially to the seeds.

Finally, this gene gives the plant better standing ability, the large number of tendrils allowing a stronger link between plants. Thus the absence of lodging facilitates harvesting and leads to increased yield.

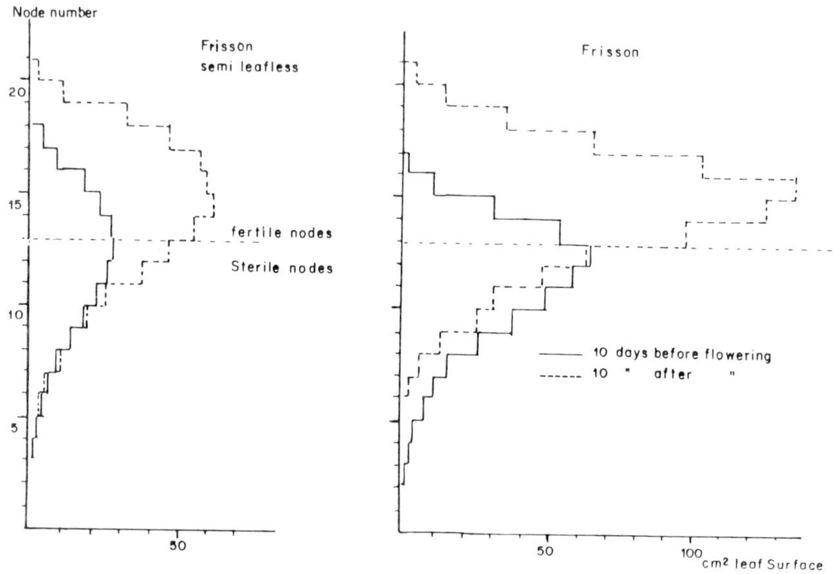

Fig. 16. Leaf area 10 days before and 10 days after flowering for cv. Frisson and Frisson semi-leafless.

TABLE 3

RADIOACTIVITY FIXED PER UNIT AREA OF ASSIMILATING SURFACE

	Leaflet	Tendril	Stipule	
	Frisson	Frisson semi leafless	Frisson	Frisson semi leafless
Plantlet stage	55	73	27	38
Flowering stage	65	54	39	76
Pod stage	82	75	65	68

The relationships obtained with about ten characters explained only about 70% of the variability. This may have been due to the fact that we investigated characters related to the aerial plant parts. Pea is a legume and utilises nitrogen fixed by Rhizobium in the nodules. Together with Picard and Messager we have started a study of nitrogen fixation by Rhizobium in the framework of a research contract with

Elf-Aquitain. We aim to establish whether improved nitrogen fixation would contribute to increased yields or increased protein content.

About 50 varieties have been grown in glasshouses with or without nitrogen fertiliser. Great variability in nitrogen assimilation and atmospheric nitrogen fixation was recorded. Fixation did not seem to be correlated with nitrogen assimilation. However, fixation was negatively correlated with earliness. Nevertheless, with Rhizobium alone, the weight of seeds varied from 40 to 100% of the weight obtained when nitrogen fertiliser was added, depending on variety (Figure 17).

Thus, the analysis of these different characters contributes to the definition of selection criteria and makes breeding increasingly less empirical.

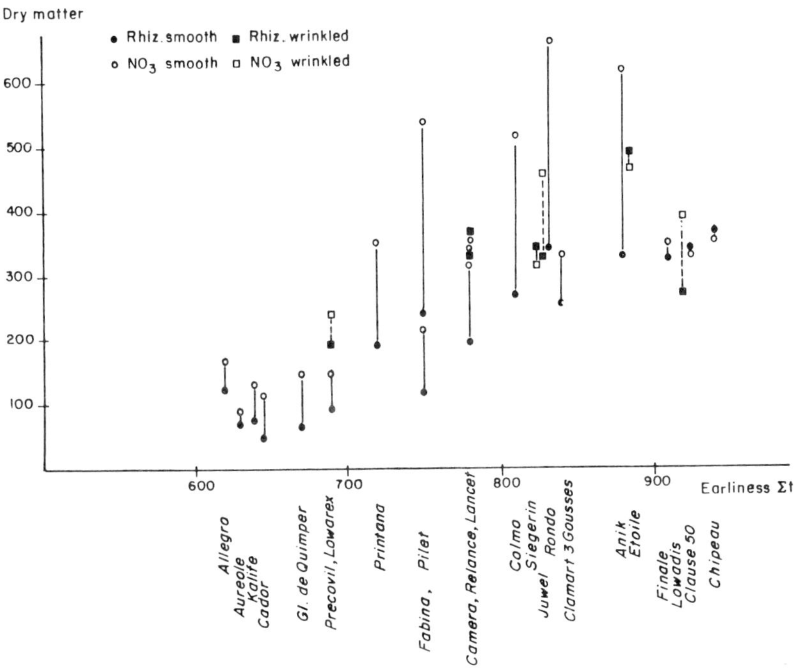

Fig. 17. Productivity of some varieties in hydroponic solutions:
a) Without nitrogen fertiliser (atmospheric nitrogen fixation by Rhizobium) ● Rhizobium smooth seeded varieties
■ Rhizobium wrinkled seeded varieties
b) With nitrogen fertiliser (nitrogen assimilation only)
○ NO_3 smooth seeded varieties
□ NO_3 wrinkled seeded varieties.

DISCUSSION

V. Haahr *(Denmark)* You said the winter barley could survive night temperatures of -15°C. If this was under artificial conditions, our experience would indicate that it could survive lower temperatures in field conditions, perhaps to as low as -23 or -24°C. This is because soil temperatures in the field are not reduced to the same levels as the air temperature, whereas in artificial conditions air and soil (i.e. root) temperatures are similar.

M. Cousin *(France)* In field conditions the winter survival is more complex than in artificial conditions.

C. Calet *(France)* I thought the last graph (Figure 17) was very important and I would like to have a fuller explanation of it.

J. Picard *(France)* Perhaps because of language difficulties I should try to explain it. Probably the most significant finding represented in Figure 17 is the marked contrast between some cultivars in their response to added nitrogen, especially among the later maturing cultivars. This interaction between genotype and level of *Rhizobium* activity may provide scope for selection for greater efficiency of nitrogen fixation.

R. Davies *(UK)* It is generally believed that the same biomass is obtained when nitrogen is obtained by *Rhizobium* fixation or from applied nitrate and I wonder whether the *Rhizobium* in this experiment was functioning effectively.

J. Picard As I remember it these experiments were done in the greenhouse. The peas were sown on 1st April and grew up to the end of June or beginning of July. There was no stress as far as water is concerned and growth without nitrate indicated that fixation occurred. Now we have diallel crosses between varieties showing a big difference between *Rhizobium* and nitrogen fertilisation on growth and those which show a small difference, to try to determine the genetic situation in these varieties.

M. Frauen *(FRG)* It is surprising to me that the cv. Rondo for example just about doubled its yield when given nitrogen, but there was no effect on maturity.

J. Picard This may be in part explained by the fact that the values for earliness used on the abscissa are based on maturation in the field and not of the actual plants used in this experiment. Certainly during early stages of growth, plant size was greatly favoured by the provision of an inorganic nitrogen source.

POTENTIAL AND ACTUAL YIELDS OF DRIED PEAS IN NORTH WESTERN EUROPE

G. Dantuma
Centre for Agrobiological Research,
Bornsesteeg 65, P.O. Box 14, 6700 AA Wageningen,
The Netherlands.

Growing peas in North West European conditions is associated in many years with considerable difficulties. The main defects under favourable growing conditions, are excessive vegetative growth and susceptibility to lodging, with the result that at a certain stage of development the plants almost invariably collapse. This results in considerable yield loss and a crop that is difficult to harvest. Thus, dried peas are not suited to modern farming practices and we have seen in the last 25 years, a sharp decline in the area of this crop in North Western Europe, with, for example, a decrease of about 90% in the Netherlands.

If the crop is again to play a significant part in modern farming it has to be capable of being cut and gathered into swaths or windrows which can be combined after drying. This means that lodging has to be prevented. Direct combining would be preferable of course, but this would be possible only under ideal harvest conditions which are rare in our climate.

Most crops of peas show signs of lodging well before they are ready to harvest, sometimes at mid-flowering or early pod-filling. Lodging is coupled with a number of side-effects such as:

- unfavourable light penetration into the crop,
- a humid environment within the crop, promoting the development of disease e.g. *Botrytis*, *Peronospora*, *Fusarium* sp., rapid ageing or decay of leaves, stems and pods,
- unfavourable translocation patterns.

Early lodging means that in many years yields fall far short of the potential. Lodging also makes the estimation of yield potential very difficult but it can be approached by providing net supports to prevent lodging. Attack by pests and diseases also needs to be reduced to a minimum.

Another approach is to study the results of a series of yield trials over a number of years and select the yields of the best variety under the best growing conditions. Such a series was made over the last ten years by the Institute for Research on Varieties in the Netherlands. Average yields of a number of varieties in these experiments ranged from 3 300 kg/ha in 1972 to 6 050 kg/ha in 1978. Highest yield of the most productive variety was 7 650 kg/ha in 1978 and 4 450 kg/ha in 1972.

Finally, we can try to calculate the yield potential by means of growth rates, duration of growth and harvest index. The last two methods give a yield of dried peas of about 7.5 to 8 tons/ha.

Pea varieties have been generally divided in two groups, those for soils that produce abundant vegetative growth and those for soils poorly supporting vegetative growth. This division now seems obsolete because the newest varieties, such as Finale and Maxi, are characterised by short and relatively stiff straw and are the most productive under all circumstances.

In some years on both light soils and heavy clay soils the new pea varieties remain too short in a dry period, but this problem can be overcome to some extent by the use of narrower rows e.g. 25 cm instead of 33 cm, and a somewhat higher plant density. Although peas are quite susceptible to drought, insufficient vegetative growth is the exception in most years and in most regions.

The character, however, which is still lacking and needs to be bred into new varieties is improved yield stability under stress conditions and this is far more important than breeding for potential yield. Stress here relates to particular climatic conditions that favour an abundant growth, an excessive leaf area, the susceptibility to lodging and to diseases that are favoured by these conditions. A good example of this is shown by some of the Dutch results of the EC joint trials in 1980, a year with a long cloudy wet period in June and July (Table 1). Lodging occurred at an early stage and as a result total crop yield was reduced. The trial in the NO Polder gave the highest

TABLE 1

TOTAL CROP YIELD (TON/HA 86% d.m.) AND SEED YIELD FOR THREE CULTIVARS AT TWO SITES

Variety	Location	Total crop yield	Seed yield
Finale	Wageningen	10.0	6.0
Finale	NO Polder	10.3	6.0
Amino	Wageningen	9.2	5.2
Amino	NO Polder	10.0	5.0
Columba	Wageningen	8.4	4.4
Columba	NO Polder	9.3	4.0

crop yields, but on the heavy clay soil near Wageningen higher seed yields were obtained.

An important breeding objective is a high and more stable harvest index as this improves yield stability and good examples are the varieties Finale and Maxi. Results with these varieties have shown that, although they are a considerable improvement over older varieties, they still do not meet the requirements of modern farming, neither with regard to yield stability nor the difficulties of harvesting. This means that for the wet regions of North Western Europe suitable varieties are still not available. Promising lines however, are those with short straw and a short flowering period as found with Finale and Maxi, but also with addition of the semi-leafless character. If with these plant types lodging can be delayed for about two or three weeks in wet years a considerable increase in both total plant yield as well as seed yield could be obtained. The more open structure of the canopy also suppresses early attack by some diseases. It is not known yet whether complete collapse of the crop can be prevented during adverse weather conditions in the ripening period, and it is likely that the harvest will remain one of the most critical phases in the production of dried peas.

It should be remembered of course, that the appearance of seed required for animal feed is less important than that for human consumption, and conditions at harvest are therefore also less important in seed intended for animal feedstuffs.

DISCUSSION

Plancquaert *(France)* How do you explain the depressed yield following irrigation when this year (1981) was dry in the Netherlands?

G. Dantuma *(Netherlands)* It was dry this year only for a short period, from the beginning of flowering until the beginning of pod fill, a period of about three weeks. Early irrigation promoted substantial additional vegetative growth. With the advent of adverse weather conditions severe lodging occurred in the irrigated plot, resulting in lower biological and seed yield as well as a lower harvest index compared with that from the non-irrigated plot.

W. Williams *(UK)* Is there evidence from the EEC Joint Trial of a genotype/ environment interaction that would suggest whether a single variety may be suitable for the whole of Europe, or whether regional variation must be taken into account?

G. Dantuma A regional approach is necessary, it would be impossible to breed a variety suitable for the whole of western Europe, either for peas or for field beans.

BREEDING OF DRIED PEAS FOR DANISH CONDITIONS

[1]M.H. Poulsen and [2] Chr. Sebro

[1]Department of Crop Husbandry and Plant Breeding, Research Station Højbakkegård Tåstrup,

[2]Danish Plant Breeding Ltd., Boelshøj, Store Heddinge, Denmark.

ABSTRACT

Dried peas (Pisum sativum) in Denmark are produced from culinary peas and from the marbled pea. The average acreage for the years 1971 - 1980 was 2 749 ha with a mean yield of 2.7 t/ha.

BREEDING OBJECTIVES

The pea crop ripens in August which, in Denmark, is a period with unstable weather conditions. It is the month with the highest average rainfall. Thus harvesting of the crop may be difficult and quantitative as well as qualitative losses are often encountered. Introduction of the more dwarf types of cooking pea has eased harvesting, but further improvement in reliability of performance is needed. Therefore, the current breeding work places a strong emphasis on producing cultivars which are easier to harvest; and suffer reduced losses of seed.

Early and uniformly ripening genotypes with short and stiff stems and pods clear of the ground are favoured. In this work it would be useful to include also the non-shattering pod-type now available in the USSR.

A further change of canopy has been introduced by two genes for leafless habit. The semi-leafless type appears to be interesting under Danish conditions.

High seed yield is of major importance. Seed yields below those of the currently available cultivars are unacceptable. Recent improvements in seed yield have been associated with increased seed-size but this relationship needs to be broken because a further increase in seed size is undesirable.

Seed quality is taken account of with respect to feeding value and processing value. The main attention is now focused on the culinary pea in order to avoid the tannins present in the

TABLE 1

SUMMARY OF BREEDING OBJECTIVES FOR BREEDING DRIED PEAS FOR DANISH CONDITIONS

	Present	Wanted
A. Earliness		
Early development of canopy		
Early flowering	20 June	10 June
Short flowering period	4 weeks	2 weeks
Early ripening	middle August	Begn. August
Uniform ripening		
B. Seed and protein yield		
Many seeds		
1. Many pods		
2. Many seeds/pod		
Two pods per inflorescence		
Long non-shattering pods		
Intermediate seed weight		200 mg
Large embryo		
High protein content	25%	30%
C. Canopy		
Many short nodes before flowering		
Small leaf-lets (semi-leafless)		
Strong stems		
1. Short		
2. Erect		
3. Stiff		
Higher harvest index		
D. Seed quality		
(Tannin free)		
Uniform seed size		
High volume weight		
Thin testa		
Strong testa		
Good colour		
Good processing characteristics		

marbled pea and which, if fed to monogastric animals may inhibit growth. The main aim is for a dual purpose cultivar combining high protein content with good processing characteristics (short imbibition and cooking time, tender and good colour etc.). Testing for protein content is practised throughout the breeding programme. Although for feeding purposes a high protein content may be required protein yield is considered more important than protein content. Laboratory tests for processing characteristics are limited to screening of potential cultivars late in the breeding programme.

BREEDING METHODS

Exploitation of variation within cultivars and generation of new variability by intercrossing are the two techniques employed in the breeding programme. Examples resulting from the success of the first method are the cultivars Bodil (ex. Birthe) and Salome (ex. Lysima).

A current hybridisation programme includes parents of European and American origin chosen for one or more of the following characteristics: short erect growth habit well adapted to Danish conditions, good seed and processing characters and high yield.

The handling of the breeding material follows a pedigree method. Single plants (I_2 - I_5) are selected on the basis of their growth habit, flowering characteristics and maturity. Yield is evaluated and the seeds are visually checked for colour, shape and size, unwanted types are discarded. Protein content is determined and the progeny are sown for further observation in 1.5 m rows. When segregation appears to have ceased a yield evaluation first in small plots and later in normal yield trials is made.

Finally the progress in both earliness and seed yield obtained by the new cultivar Salome is illustrated by data from the official Danish testing authorities (Table 2).

TABLE 2

IMPROVEMENT WITH THE NEW CULTIVAR SALOME OVER THE PARENT CULTIVAR LYSIMA. THE DATA ARE MEANS OF 14 EXPERIMENTS OVER THREE YEARS.

Name	LYSIMA	SALOME	LSD 95
Breeders designation	64-01	76-02	
Seed production			
Yield, 85% DM, t/ha	3.1	3.2	2.5
Protein content, %	26.5	26.5	0.5
Protein yield, kg/ha	701	725	57
Testa content, %	11.8	11.1	0.4
Weight/l, g	781	787	8
Seed weight, mg	224	237	14
Plant height, cm	150	143	12
Date for			
Beginning of flowering	17/6	12/6	
End of flowering	23/7	12/7	
Maturity	23/8	17/8	
Green harvest			
DM/ha relative	100	95	
Protein, %	17.2	17.4	
Protein/ha	100	97	

DISCUSSION

C. Calet *(France)* Is it possible, as you suggested, to obtain seeds with testas that are both thin and strong?

M.H. Poulsen *(Denmark)* This is possible because the split testa which is quite common is controlled by different genes from those which control the testa thickness.

C. Calet It may be that selection for a thin testa would influence protein content of the seed.

W. Williams *(UK)* It is interesting that in the soyabean the testa is very thin and strong, and this has increased protein level by reducing the non-proteinaceous testa component.

J. Picard *(France)* Dr. Poulsen spoke of a 'good colour' for the seed. What is a good colour?

M.H. Poulsen Good colour is determined by the processing industry according to the use to which the product is put.

A.G. Prendergast *(CEC)* I would like to put a question to the whole meeting. If the present member countries of the EC were to concentrate on one crop for the production of protein for animal supplementary feed, does the meeting think that it should be peas or small seeded *Vicia faba*? It may be that if work is concentrated on one crop then there might be a chance of achieving the same results that the Americans have achieved with the soyabean.

J. Picard I believe it would be dangerous to focus on one species only; perhaps the participants would bear that in mind when they are considering Dr. Prendergast's question.

CRITERIA AND METHODS IN THE SELECTION OF *L. ALBUS*

M. Lenoble
Station D'Amélioration des Plantes, Lusignan, France.

INTRODUCTION

Amongst the species of lupin that we have tried to adapt for grain production in Lusignan, the white lupin proved to be more interesting than the other species. For four or five years we have been trying to define those criteria of selection necessary for the improvement of the yield or the quality of the grain.

For seed quality we have tried to show the wide range of genetic variability for factors considered to be important by nutritionists and biochemists. Taking into consideration economic demands, it was established that in the short term emphasis should be placed upon the improvement of yield, either by improvement of the yield of spring lupins without a decrease in the quality, or by the selection of very productive winter lupins.

The first approach is discussed in this paper.

WHITE SPRING LUPIN

The known sweet varieties of white lupin do not show significant variability in yield. There is compensation between the different yield components such as the number of pods on the main stem, the number of pods on the branches, the number of seeds/pod and the weight of 1 000 seeds.

I have tried to see if, by crossing with bitter types, recombination of the different characters could be obtained and, in this way, increase the variability. The variability of other characters was also measured particularly that concerning yield.

The first crosses were made in 1975 from selected parents having a high protein content. The F2 plants and following generations were cultivated in lines of thirty plants. The F3 and F4 generations were selected from one pod/plant. Simultaneously an individual genealogical selection was conducted for yield, earliness, absence of alkaloids and an *a posteriori* selection for quality was made.

Yield components

Comparison of the F3 families with the F2 mother plant shows that it is impossible to predict the value of the F3 family from the F2 data. Here and there correlations appear between the characters whose variability is very strong in F2. Where the intracross correlations are very weak, on the other hand, those from the intercrosses are very strong (Table 1). There are no differences between the families from bitter plants and those from sweet plants.

TABLE 1

CORRELATION BETWEEN F_2 PLANTS AND F_3 FAMILIES

	Intracross	Intercross
100 seed weight	0.21	0.86
Oil content	0.11	0.72
Protein content	0.09	0.77
Soluble sugar content	0.25	0.86
Seed yield	0.10	0.23

In comparison of F3 and F4 families the correlations between characters are also scattered and non-systematic. Nevertheless, the yield of the F4 lines is well explained by the complex number of pods, fertility* and weight of 1 000 seeds (R from 0.76 to 0.98, depending on the cross).

The individual breeding of F3 plants is very efficient for the improvement of all the characters but principally for the crosses which demonstrate an important variability between the F3 families, from which are observed important recombinants (Tables 2 and 3).

Quality components

We have analysed the oil content, the content of soluble carbohydrates and the nitrogen content. Very weak intracross variability was observed (coefficient variations from 3 - 4%),

* number of grains/pod.

TABLE 2

F_4 TRANSGRESSIONS FOR 2 CROSSES WITHOUT SELECTION.

		P1	P1 x P2	P2	P2 x P3	P3
Total height		115	75 - 110	60	55 - 110	115
Pods number	MS	10	5.4 - 16.6	9.8	7.5 - 15.5	6.0
	RI	11.7	4.9 - 16.6	13.0	5.4 - 19.8	11.0
Seeds per pod	MS	4.5	3.9 - 5.1	3.6	3.9 - 5.1	4.8
1 000 seed weight		420	286 - 432	350	310 - 439	420
Oil content		10.3	9.3 - 11.0	10.3	9.3 - 11.0	10.7
Protein content		46.0	39.6 - 44.7	42.0	41.4 - 44.7	41.7

P = parent

MS= main stem

RI= first branches

TABLE 3

GENEALOGIC SELECTION EFFICIENCY. COMPARISON BETWEEN AN F5 LINE AND ITS PARENTS.

		F5	P1	P2
Flowering time		4/6	30/5	6/6
Pods number in July	MS	13.3	8.9	7.7
	RI	22.3	16.3	11.5
Pods number at harvest	MS	11.1	7.0	6.0
	RI	19.4	12.4	10.0
Maturity date		29/8	1/9	20/9
100 seeds weight		22.2	24.9	25.8
Seeds/pod (MS)		4.5	4.0	4.3
Protein content		6.00	6.24	6.3
Sugar content		9.00	11.3	11.0
Oil content		9.5	9.2	10.3
Testa proportion		24.4	23.2	20.3
Total yield		700	550	515

and there was no link between the yield components and quality. Sometimes there exists a negative correlation between the oil content and the nitrogen content.

As for the yield components, the intracrossing correlations are not significant, but on the other hand, the intercross correlations are very significant, which permits very efficient selection (Table 1).

Individual genealogical selection for protein content is also efficient, but the lines thereby selected proved to be low yielding.

Other criteria for selection

In the species *L. albus* there is significant variability for other useful characters, for example: vigour at emergence, particularly in compact soils, does not depend only upon the size of the seeds; the proportion of the testa of the seed is related to its size (r = 0.70) but certain families or ecotypes are clearly outside the regression; the resistance to *Thielaviopsis basicola*, a disease which occurs in certain soils; the round form of the seed, a recessive character, unfortunately linked to a chlorophyll deficiency.

Conclusion

The results obtained from these first crosses suggested that the variability, and consequently the possibilities for selection, could be considerably increased by using bitter parents. The estimation of the progeny value in relation to that of the parents is a problem to be resolved. We are studying this subject at present.

After crossing, the only selection done in the F2 generation is on the sweet types and sometimes on precocity. It is therefore necessary that the number of plants be sufficiently high. On the other hand, the selection between crosses can be carried out between F2 for quality components and certain yield components. Genealogical selection can begin at F3 and one can carry out yield tests with F5 families. Selection for better fertility, number of pods and number of seeds/pod, has an indirect effect on acceleration of the maturity and diminution of the weight of 1 000 seeds.

It should be pointed out that important 'grand parent' effects are possible in the F2, as seen in Table 4.

TABLE 4

RECIPROCAL EFFECTS IN F2 CROSSES (1981) (A, B, C, D = PARENTS).

		A	AxB	BxA	B	C	CxD	DxC	D
Flowering date		5/6	29/5	1/6	3/6	3/6	5/6	5/6	19/6
Pods number	MS	15.0	6.8	7.4	9.5	13.7	13.3	16.4	8.1
	RI	21.3	12.8	13.4	16.2	17.9	11.5	12.9	16.9
Leaf area		101	52	58	84	81	99	77	125
Stem height		40	28	30	47	30	39	32	60
Stem diameter		8.8	8.6	7.7	10.0	7.7	9.6	6.5	11.0
Length of branches		26.5	25.0	27.1	39.8	20.3	28.6	23.3	38.5

These results were obtained in nurseries at a low planting density. At a normal density the factor modified most by competition is the number of pods/plant. The size of the seed and the number of seeds/pods are the least modified. An improvement of this last character would probably be an efficient way to increase the yield. On the other hand, a selection of lines having a higher number of pods/plant should permit the creation of varieties which could be planted at a low density, a factor important for economic progress considering the cost of seeds.

WHITE WINTER LUPIN

The vernalisation needs of the white lupin are variable and one may define the winter lupins as those requiring vernalisation, for when spring sown they remain in a rosette stage for a long time. There are, however, not really two distinct groups, winter or spring.

The winter lupins in Lusignan have a high yielding potential of about 7 t/ha. Nevertheless, the actual productivity is often lower than that of the spring lupins because of winter damage.

The winter lupins coming from a Mediterranean climate have other disadvantages, namely a high alkaloid content, lateness in maturation (although early flowering) and a relatively low protein content, due partly to a high percentage of seed coat.

Priority in selection is given to the sweet type, which should be early and resistant to winter, rather than to productivity as such.

The absence of alkaloids

We began the crosses between the bitter winter types and the sweet parents in 1976. The sweet parents were of two types:

(a) line 34, selected in Arkansas, which is a sweet winter type and is probably from a cross between *Lupinus graecus* and a spring sweet lupin;

(b) spring sweet lines.

Although the Arkansas line is relatively sensitive to the cold and is slightly dehiscent, we have obtained good sweet lines in the progeny of its crosses, which are now being used as parents. The spring parents are disappointing because they transmit winter sensitivity to their progenies and also, above all, a sensitivity to parasites. Hence, it is necessary to have a large population size in F2 and F3. The spring type was found to be dominant. In all cases, the disjunction for sweetness is simple and monofactorial.

Earliness of maturity

Because of the conditions existing in Lusignan the winter types have a very long growth cycle of almost twelve months. Their root system, being probably deeply rooted, makes them resistant to drought.

To improve the earliness of maturity we began to cross our best lines, sweet or bitter, with *Lupinus graecus*. This subspecies has a rapid maturity rate, although it is late flowering. The maturation of the pods begins before defoliation. This subspecies shows good fertility with respect to number of pods and number of seeds/pod. On the other hand, the pods are dehiscent, the seeds are too small and the plants are relatively sensitive to winter (Table 5). Our first F2 in 1981 seems very promising. The Greek ecotypes also appear to mature early.

Resistance to winter

Resistance to winter is a complex including resistance to cold, to asphyxia and to parasites. It is, at this time, difficult to state an order of importance for these characteristics.

TABLE 5

WINTER WHITE LUPIN CHARACTERISTICS

		GR	Lucky	Arkansas	C9	LA 26	Italiens	*L. graecus*
Flowering date		18-24/4	18/24/4	5-12/5	1-12/5	2/5	15-19/5	12/5
Pleiocheta		S	S	S	AR	AR	R	R
Lodging		TS	S	S	AR	R	R	S
Pods number	MS	6-7	2-7	4-6	6-8	4-8	8-17	7
	R1	6-10	9-16	7-11	12-18	8-18	12-30	18
	R2	6-17	12-20	10-18	9-17	0	0	18
Maturity date		11-20/8	13-30/8	29/8	29/8	3-15/9	1-30/9	15/8
100 seed weight		41-50	46	19-27	24-27	28-30	34	15.0
Seeds/pod		4-5	5.0	4.0	4.4	4.4	4.0	5.8
Protein content		36	36	38	40	37.0	40.0	36
Sugar content		9.8	12.5	8	9	10	9.8	12
Oil content		10.2	12.3	9.5	9.7	12	12.2	6
Testa proportion		17.0	18	24	20	20	18.5	30

TS - very susceptible
S - susceptible
R - resistant
AR - somewhat resistant

MS = main stem
R1 = first branches
R2 = second branches

There are two kinds of sensitivity to cold. First, sensitivity of the apical region which could be destroyed. In this case, the plants remain alive but all development is stopped. Second, there exists sensitivity of the hypocotyl axis, which leads to the destruction of the plants. This sort of sensitivity is often incorrectly attributed to the damping off of diseased seedlings. Temperatures of about 5°C can harden the plants. A study of the resistance to cold is to begin soon. At the moment our best cold-resistant parents come from the Italian ecotypes. We hope to be able to obtain positive results with the Spanish ecotypes of the Salamanca region.

The white lupin suffers badly from excessive winter water, which leads to poor rooting. However it is difficult to calculate the consequences of root asphyxia and no genetic resistance exists.

One of the most important diseases of the winter lupin in Lusignan is *Pleiocheta setosa*, which can lead to a 100% destruction of the crops by the end of winter. This leaf disease appears during the first improvement of the weather and can completely invade the plant. It is not known if this disease is a result of winter frosts. Fortunately, genetic variability exists, and resistance is a recessive factor. All the spring types are sensitive. The winter Italian types are resistant, whereas the Greek, and the majority of the Spanish, ecotypes are sensitive. Selection for resistance to *Pleiocheta* would be facilitated by establishing a method of artificial infection.

The disease is rarely found generally in the nursery. Mortality apart, the disease often decreases the fertility of those plants affected.

Yield components

The most important part of the yield in early winter lupin is first and second branching. The variability for both the number of pods/plant and the number of seeds/pod is high.

Because of the influence of climatic factors and diseases we have not yet been able efficiently to select for yield.

Conclusion

It appears that the selection of high yielding white winter lupins which are resistant to diseases is progressing. Nevertheless, the evaluation of true frost resistance remains to be studied. The adaptation range for this variety type is still unknown. If we consider the programmes now being undertaken in this area, I believe that the answers to these questions will come in two or three years.

FUTURE PROSPECTS

With the present sweet, white lupin varieties such as Kalina and Lublanc, in our area we obtain a yield of 3 - 4 t/ha at the agricultural level. The feeding experiments with animals give good results with all animal species except pigs.

I believe that I have shown in this paper that by using

bitter lupins as parents one can considerably increase the production of the spring lupin. The 4.5 t/ha yield would seem an attainable objective.

At the same time, it should be possible to create winter varieties giving 6 t/ha, at least where winters are mild.

It is now necessary to think about the improvement of quality. It is particularly necessary to increase the oil and protein content, to decrease the carbohydrate content and to diminish the proportion of testa in the seed, thereby increasing the energy value for monogastric animals.

DISCUSSION

R.J. Heringa *(Netherlands)* Can you give figures for the yield of *Lupinus albus* in your area in comparison with other crops such as cereals?

M. Lenoble *(France)* In 1981 we had about 100 ha of cultivated spring lupin in our area of Lusignan with a mean yield of 35 quintals/ha, whilst the mean yield for cereals was about 50 quintals/ha. We have no large areas of winter lupins but in small plots, this year, we obtained about 25 quintals/ha for spring lupins and 70 for winter lupins.

W. Williams *(UK)* You showed that for spring and winter lupin breeding, emphasis is placed on selection in the F2 for sweetness. Does that mean that the first cross is always between sweet and bitter types?

M. Lenoble The original crosses seven years ago were between sweet and bitter, but present ones are between sweet only. Selection is made at F3 rather than F2 in order to select for yield components and quality.

W. Williams Since a number of people may commence lupin breeding programmes as a result of this seminar, it is important to draw attention to the fact that there are at least five different loci involved in the control of low alkaloid levels in *Lupinus albus*. These are now represented by well known cultivars. It would be prudent for breeders to use only one of these uniformly throughout the breeding programme in Europe. If breeders use different complementary loci then there will be a possibility of outcrossing and the development of bitter forms as contaminants if the crop ever becomes developed.

BREEDING RESEARCH ON PEAS *(PISUM SATIVUM* L.) AND *(LUPINUS LUTEUS, ALBUS* AND *MUTABILIS* L.) AS PROTEIN CROPS IN THE NETHERLANDS

R.J. Heringa
Foundation for Agricultural Plant Breeding,
SVP PO Box 117, 6400 AA Wageningen, Netherlands.

PEAS

Introduction

The acreage of dry peas for human consumption and for compound animal feed, has decreased drastically since 1950. Three reasons can be given for this decrease;

A. The crop has to compete with other, more reliable crops such as spring barley. The farming system in the Netherlands is very intensive, resulting in high yields of crops such as potatoes, cereals and beet. Competition from these crops under Dutch conditions is almost always to the detriment of peas.

B. The crop could not be harvested mechanically until recently. In the past, peas were put on hay racks for drying at harvest time, but that is now impractical due to high labour costs. Cutting the crop in swaths followed by combine harvesting gives problems under less favourable weather conditions.

C. The use of dry peas for human consumption, e.g. in soup also became less favoured, with the advent of convenience foods. Finally the use of dry peas for animal feed is always strongly dependent on the price of imported protein rich crops such as soya.

The interest from breeders has, in general, not been great although two private breeding companies have persevered with their activities with considerable success. The Dutch Government has shown interest in the breeding research of this crop in programmes at the agricultural institutes. The Foundation for Agricultural Plant Breeding was active in this field from 1950 - 1970. The work was concentrated mainly on mutation breeding for minor mutants and breeding for resistance to diseases.

Although many mutants were produced none gave a higher yield than existing cultivars. In the normal breeding programme resistance to mildew was found. After 1970 this work was stopped due to lack of funds and reorganisation of the programme but as a consequence of the renewed interest of the EC in protein rich crops, more attention is now being paid to crops such as peas and field beans. In the UK and France especially, growers and animal feed manufacturers have shown considerable interest.

Breeding activities

Since 1940 twelve varieties have been bred by two companies and placed on the list of varieties in the Netherlands. Table 1 gives these varieties and the years in which they were established and withdrawn.

TABLE 1

VARIETIES OF PEAS BRED IN THE NETHERLANDS

Unica	1927 - 1969
Rondo	1943
Servo	1946 - 1965
Vares	1953 - 1961
Pauli	1957 - 1978
Fertila	1958 - 1969
Rovar	1958 - 1973
Hylgro	1961 - 1969
Dik Trom	1962 - -
Allround	1969 - -
Finale	1973 - -
Maxi	1980 - -

Of these varieties eight have disappeared during the last 40 years. For the remaining four varieties the year of inclusion in the list of varieties was 1962, 1969, 1973 and 1980 in the order given in Table 1. This means that on average a new variety is included every six years.

It is not always clear why a variety disappears; low yield or a poor quality or a disease susceptibility are often the cause. The very famous variety Rondo for instance, bred in 1943 was grown for many years and accounted for more than 50% of the acreage. Nowadays, however, there is little interest in this variety.

On the other hand the variety Allround, which was bred in 1970, increased very rapidly in acreage at the expense of the Rondo, but it disappeared by 1975 due mainly to its poor taste and the introduction of Finale. The cultivar Finale is good quality, can be grown on a wide range of soil types and gives high and reliable yields.

Recent trends in the areas grown of five of the more popular cultivars are given in Figure 1.

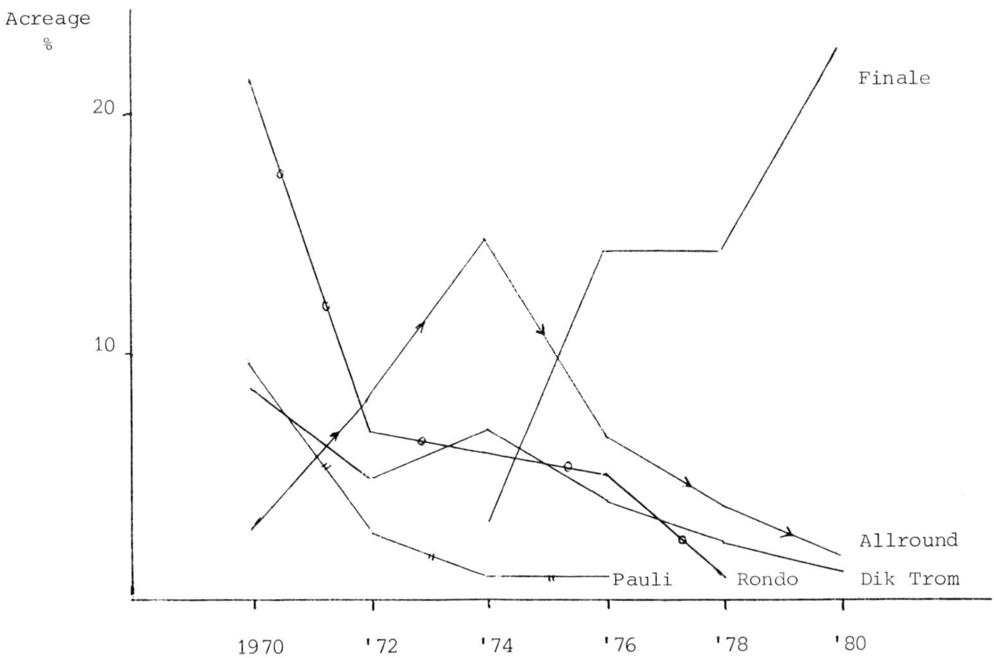

Fig. 1. Areas grown of five dried pea cultivars for the last 10 years.

In most cases, the new varieties are derived from crosses between existing cultivars, therefore the genetic background is somewhat restricted (Table 2).

TABLE 2

PEDIGREE OF SOME COMMON PEA VARIETIES

Rondo	: Unica x (Corona x Viktoria x Maple pea)
Dik Trom	: Pauli x (Rondo x Profusion)
Pauli	: Stijfstro x Mansh. pluk
Allround	: (Cebeco nr x Pauli) x (Rondo x Pois du Chemin long)
Finale	: Dik Trom x Cebeco nr
Maxi	: Allround x Finale

So far characters such as quality, straw length, yield, could be obtained from this material, however it may be difficult to introduce the currently topical character, straw stiffness from this source. Screening of material from gene pools has not offered much hope as the more primitive forms generally have very long and weak straw.

During the past few years more and more attention has been paid to the semi-leafless and leafless types which seem good in this respect and it may be possible to find other material from which stiff straw can be obtained. As has already been mentioned straw stiffness must be improved as well as the height from the first pod to the soil surface. Indeed it may be appropriate for government institutes to contribute in this direction and also towards improved protein content.

Future prospects

There might be a market for peas as a protein rich crop. Peas have some advantages over other crops; the possibility of early sowing in spring leads to effective weed control; the crop requires low inputs of fertiliser, nitrogen fixation reducing the costs considerably; the early harvest means that more time is available for thorough seed bed preparation for the next crop.

For the Netherlands and Western Europe in general, it is necessary to introduce new varieties which can be harvested mechanically in a single pass. The crop must therefore ripen in the field and this may conflict with yield because ripening is associated with lodging. Moreover, the crop must have non-shattering pods, with all pods ripening over a short period.

The development of an ideotype with the above mentioned characters should arouse renewed interest as farmers need a profitable spring-sown break crop. Additionally, farmers in the Netherlands are already acquainted with the cultivation and husbandry of the crop.

Breeding research could also investigate possibilities for export. At the present time several Dutch varieties are grown with great success in France and the UK such as the large blue Finale in France and the white seeded variety Birte in the UK. It is expected that the area grown in France will be about 75 000 ha and for the UK 15 000 ha.

Although seed quality must be good, the demands are not so high for animal feed as for human consumption.

Other types of peas grown in the Netherlands, include white peas, brown marrowfat peas, maple peas but there is only a small demand for these and a large increase is not expected.

LUPINS

In 1950 in an article concerning lupins it was mentioned that the sweet lupin *Lupinus luteus* is now largely used as a fodder crop and that a further increase in production could be expected. The protein content in the seeds is about 40% and could probably be increased by breeding. However, seed production and especially the seed quality are generally poor in the Netherlands due to unsuitable environmental conditions.

From these remarks it is clear that there has been little interest in extending the crop. After World War II it was expected that a new crop may be required, mainly for the lighter sandy soils, where there was a need for improving soil fertility.

The Dutch Government, through the Foundation for Agricultural Plant Breeding, inaugurated a research programme in 1948 with the aim of producing new varieties of sweet lupins for use as a fodder crop. Within ten years two varieties were bred, one sweet and one bitter, with the latter being used as a crop for green manure. Resistance against diseases and rapid early growth was introduced in these cultivars together with a reasonably good yield. Even so there was little interest from the farmer.

Seed production was the main limitation as at harvest time the weather conditions were unfavourable and the risks to the farmer too high. Moreover, nitrogen supplies were no longer a problem. Farmers are, in general, very reluctant to grow crops which cannot compete with other more reliable crops. The use of lupins for green manure on new virgin soils was not needed because no new land was being made available.

Breeding research in the Netherlands was therefore terminated. Productivity is far too low and harvest time too late.

The above remarks relating to *L. luteus* apply even more to *L. albus* and *L. mutabilis*.

DISCUSSION

D.R. Davies *(UK)* In discussing the prospects for peas no-one has discussed the potential value of the haulm in dried peas. There is a substantial weight of haulm per hectare; it is highly digestible and has a protein content of 10 - 12%.

R.J. Heringa *(Netherlands)* Years ago haulm was used for animal feed but at present there is no demand, in the Netherlands.

D.R. Davies But we have no indication of the value of this product in animal feeding. Maybe this is something we should come back to in our "perspectives".

C. Knott *(UK)* Pea haulm has been used in some areas of the UK for animal feed but the pea growing areas tend to be towards the east of England where there aren't many animals.

A.G. Prendergast *(CEC)* Mr. Chairman, I would like to make a short comment arising from your own point about using pea straw. In some parts of Europe investigators are examining whole crop harvesting of some legumes. In our last two programmes we have dealt only with protein crops, and only with some narrow aspects of them because the resources available were very small and it was thought best to use them in this way. The next programme is due to start in 1984 and continue for four years until the end of 1988. There is some possibility of additional resources and we hope that in that programme some attention can be given to the general agronomy of a limited number of crops. If that is so, perhaps then we may be able to arrange co-operation between people who are involved in whole crop harvesting and also those who are involved in the use of plant by-products. Personally, I would like to see some kind of programme regarding cropping systems or even in the broader aspect, farming systems. I should mention that the present protein programme was developed from initial attempts to devise a general programme on plant improvement but with the resources available it was found that we could support only a very narrow area of plant improvement and this was devoted to the proteins.

Plancquaert *(France)* We have tried cereals mixed with peas but I am not sure that this is a satisfactory technique because it gives irregular yield and regularity of production is very important to farmers.

DRIED PEAS: SOME FACTORS INFLUENCING YIELD AND STAINING

R.F. Murphy

The Agricultural Institute, Kinsealy Research Centre,
Malahide Road, Dublin 5, Ireland.

ABSTRACT

In a series of trials, direct combining was assessed as a means of harvesting dried peas. This could take place satisfactorily once the moisture content of the peas (cvs. Maro and Proco) was below 30%. Diquat used as a desiccant increased the rate of drying of the peas by three days. Yields from direct combining were on average 12% lower than from the traditional tripodding system. One of the biggest problems of direct combining was lodging of the pea crop. This led to increased staining and yield reduction when conditions were wet prior to harvesting. Investigations showed that these problems could be alleviated by drilling a nurse crop such as oats (cvs. Mostyn and Astor) at 31 kg/ha with the peas. This increased yields of peas and reduced the staining in the crop. Further reduction in staining was achieved with the use of benomyl applied at or near the flowering stage. Populations in the range of 46 - 62 plants/m² produced the highest yields of peas when grown in association with the oats. Chlormequat proved useful in controlling the vigour of the oats without affecting the peas.

INTRODUCTION

In Ireland, peas are grown either for the dried pea trade or as green peas for freezing and dehydration. Dried peas are harvested at the ripe seed stage with a moisture content of 20% - 25% while green peas are harvested at a much earlier immature stage when the moisture content is approximately 80%. Green peas are much more popular with growers than dried peas because yields are consistently good, averaging over 5.0 t/ha and in most years there are no problems associated with harvesting the crop. Furthermore growers can usually put in a second cash crop after the peas are harvested. With field dried peas however the position is far from satisfactory. Yields vary due mainly to difficulties experienced in harvesting the crop, particularly when wet conditions prevail prior to and during harvesting. These conditions cause lodging of the crop, and the resultant wet microclimate favours disease, losses in yield and staining of the peas. In some years the haulm decomposes causing more difficulty. It is no surprise therefore that yields of dried peas under Irish conditions can vary from 1.2 - 4.0 t/ha with staining ranging from 5% to 40%.

Dried peas reached a peak in Ireland during the 1950s when about 2 000 ha were grown. The crop was cut and put up on tripods to complete the drying process after which they were threshed Yields and quality were generally good from this method of harvesting even when climatic conditions were unfavourable. However, this method was too labour intensive and the shortage of labour forced more and more growers out of growing field dried peas. In the seventies, it became obvious that unless direct combining the crop was possible growers would no longer grow dried peas.

Investigations commenced in 1971 on different aspects of direct combining the crop. Further trials assessed cultivars, plant population and methods of reducing lodging and staining of the crop.

MATERIALS AND METHODS

The trials were conducted at Kinsealy, Co. Dublin on a medium heavy loam. All treatments and cultivars were sown with a Fiona seed drill previously calibrated to give a required plant stand. A mixture of trietazine and simazine applied pre-emergence at 2.2 kg/ha was used for weed control where peas only were grown. Where peas and cereals were grown together dinoseb-amine at 1.68 kg/ha was applied post-emergence when the temperature was above 55°F. Fertiliser was applied at the rate of 376 kg/ha of 0 : 10 : 20 (NPK) compound. Most trials were carried out on Maro (marrowfat), Pauli and Proco (round) and later on on Progreta - a new tare leaved marrowfat cultivar.

All trials were sown in March/April as soon as soil conditions were suitable. Plot size was 18 m x 4 m for each cultivar or treatment. At harvest, plots for tripodding were cut when peas were at approximately 60% moisture content. These were turned in the field a number of times until the foliage was dry and they were then tripodded to complete curing for threshing later on in the autumn. Plots for direct combining were harvested at around 20% - 25% moisture when conditions were opportune usually in August or September. Each plot was separately recorded and 2 kg samples of peas were assessed for staining. The moisture content of each plot sample was also determined and the net yields adjusted to 16% moisture.

RESULTS

Direct combining and tripodding trials

Trials from 1971 - 1972 examined the feasibility of direct combining the dried pea cultivars Maro and Proco. These were previously shown as the highest yielding marrowfat and round seeded cultivars respectively grown in Ireland (Murphy, 1974).

Diquat at 0.84 kg/ha was applied as a desiccant to hasten drying and was applied at 55% moisture stage. Benomyl was used as a seed dressing (1 g benomyl/kg seed) and as a spray to control staining applied at 1.0 kg/ha two weeks prior to harvest. The results of these trials are shown in Tables 1 and 2.

TABLE 1

COMPARISON OF DIRECT COMBINING WITH TRIPODDING USING THE cv. Maro, 1971

Treatment	Gross yield peas t/ha	Yield clean peas t/ha	% staining
1) Control - tripodding	4.3	3.7	13.0
2) Direct combining only	3.4	2.8	16.7
3) Direct combining + dessication + benomyl seed dressing	3.4	2.9	14.5
4) Direct combining + dessication	3.3	2.8	15.3
5) Direct combining + benomyl seed dressing	3.8	3.3	12.7
6) Direct combining + dessication + benomyl applied 2 weeks prior to harvesting	3.1	2.5	20.4
F Test	*	*	
SE of mean (d_f = 10)	0.23	0.29	

In the first trial (Table 1) in 1971 gross yield of clean peas from tripodding averaged 21% greater than the direct combining treatments. All the latter treatments except five which included benomyl seed dressing had a higher percentage staining than the tripodding control (13.0%). Overall the use of diquat

TABLE 2

COMPARISON OF DIRECT COMBINING WITH TRIPODDING USING THE cv. Maro, 1972

Treatment	Gross yield peas t/ha	Yield clean peas t/ha	% staining
1) Control - tripodding	5.5	5.4	1.8
2) Direct combining only	5.3	5.1	3.8
3) Direct combining + dessication	5.3	5.0	5.7
4) Direct combining + dessication + benomyl seed dressing	4.7	4.5	4.2
5) Direct combining + dessication + benomyl applied 2 weeks prior to harvesting	5.0	4.8	4.0
F Test	NS	NS	
SE (d_f = 15)	0.4	0.39	

or benomyl as a spray did not improve yields or significantly reduce staining of the peas.

In 1972, tripodding again produced the highest yield of clean peas (5.5 t/ha) but this was only marginally higher than the direct combining treatments where yields of 4.7 to 5.3 t/ha were obtained (Table 2). Desiccation did not improve yields but advanced maturity by about 3 days. Weather conditions during the harvesting of this trial were excellent. Staining was not a problem and benomyl used either as a seed dressing or as a spray applied prior to harvesting failed to have any effect. Under such favourable harvesting conditions, direct combining with or without desiccation produced very satisfactory yields of peas.

The mean of the two years results of direct combining versus tripodding is given in Table 3. July and August 1971 had a high rainfall while 1972 was much drier for this period. Tripodding produced the highest gross yields of 4.9 t/ha but this was only 12% higher than direct combining alone or direct combining in combination with desiccation. Yield of clean peas was also higher with tripodding due to a lower percentage staining

TABLE 3

THE MEAN OF TWO YEARS (1971 - 1972) RESULTS ON DIRECT COMBINING VERSUS TRIPODDING

Treatment	Gross yield peas t/ha	Yield clean peas t/ha	% staining
1) Control - tripodding	4.9	4.5	7.4
2) Direct combining only	4.3	4.0	10.2
3) Direct combining + dessication	4.3	3.9	10.5
4) Direct combining + dessication + benomyl seed dressing	4.0	3.7	9.3
5) Direct combining + dessication + benomyl spray applied 2 weeks prior to harvesting	4.0	3.6	12.2

(7.4%). Overall, however, direct combining produced acceptable yields and without the high labour input associated with the tripodding method. Benomyl either as a seed dressing or applied two weeks prior to harvesting did not reduce staining to any degree. The tripodded peas had a superior green colour to any of the direct combined treatments in each year.

More detailed work on staining of dried peas was carried out in 1971 - 1973 and results are summarised in Table 4. Benom[illegible] was applied separately six times at weekly intervals on two cultivars Maro and Pauli. Desiccation was also applied to one half of each treatment at the standard rate.

Neither seed treatment with benomyl nor desiccation with diquat had a significant effect on staining. Benomyl sprays however had a significant effect on staining when applications were made between the beginning of flowering and one week after flowering, giving a reduction in staining of approximately 50%. Application made two weeks after flowering or at the time of desiccation gave very poor control of staining as in previous trials. The levels of staining were very much greater in Pauli than in Maro but the relative effects of the fungicidal treatments were practically the same in both cultivars.

TABLE 4

EFFECT OF DESICCATION AND BENOMYL SPRAYS ON PERCENTAGE OF STAINING IN DRIED PEA cvs. Maro and Pauli

Date of benomyl spray*	Maro			Pauli		
	Desiccated	Non desiccated	Mean	Desiccated	Non desiccated	Mean
July 2	4.4	4.2	4.3	18.6	25.7	22.2
July 8	2.6	4.1	3.4	20.4	11.5	16.0
July 16	4.1	3.9	4.0	20.3	8.8	14.6
July 23	5.2	4.9	5.1	12.5	25.6	19.1
July 30	6.2	7.4	6.8	32.7	29.1	30.9
August 17	7.0	11.5	9.3	35.8	37.4	36.6
Control	6.6	10.4	8.5	48.6	44.0	46.3

* Flowering period: June 25 to July 16

Nurse crops

From the experiences gained over the period 1971 - 73 it became obvious that there were disadvantages with direct combining of peas under inclement weather conditions. The main problem was lodging of the crop which caused staining and difficulty in lifting the haulm with the pick-up reel of the harvester. The latter operation, where lodging was severe, caused further shedding of peas during combining. Rainfall figures in Ireland for July and August can often be as high as 60 mm or greater for each of these months (see Appendix 1). It seemed imperative that some means be found of supporting the pea crop to keep it off the ground until harvesting. With this objective in mind a number of nurse crops were grown with peas in 1974. These included wheat, fodder radish, oats, crambe and sunflower and were assessed in observational plots.

Only oats and crambe showed sufficient promise and these were assessed in more detail in replicated trials in 1975. The results of the first of these trials is given in Table 5 where oats cv. Astor was sown at 31, 94 and 124 kg/ha along with the peas at 196 kg/ha. Chlormequat at 1.6 kg/ha was also used to reduce vigour of the oats.

The results indicated that the lowest seeding rate of oats was the least competitive with the peas, gave the highest yield

TABLE 5

YIELD OF MARO (DIRECT COMBINED) GROWN IN ASSOCIATION WITH OATS AT THREE SEEDING RATES, 1975

Treatment	Yield peas t/ha	Yield oats t/ha
1) Oats at 31 kg/ha + peas 196 kg/ha	2.6	1.6
2) Oats at 94 kg/ha + peas 196 kg/ha	1.3	2.2
3) Oats at 125 kg/ha + peas 196 kg/ha	1.5	1.8
4) Oats at 31 kg/ha + peas 196 kg/ha + chlormequat at 1.6 kg/ha	2.4	1.9
5) Control Peas 196 kg/ha	1.2	O
F Test	NS	
SE (d_f = 5)	0.34	

of peas and offered a reasonable measure of support to the pea crop. The higher seeding rates of oats proved too competitive for the peas. With the low rainfall occurring in July and August staining in this equipment was not a problem (Appendix 1). Chlormequat reduced straw length of the oats by 10% but had no effect on the vine length or yield of peas.

Trials continued in 1976 using the two oat cultivars - Astor, previously used in 1975 and Mostyn which was reputed to be more resistant to lodging. Crambe which was drilled at two stages, firstly, three weeks before the peas were sown and secondly at the same time as the peas, was also included. The more vigorous Maro was used in the first trial and the less vigorous Proco in the second trial. Results are given in Tables 6 and 7.

In the first trial (Table 6) Maro again competed satisfactorily when grown with the oat cultivars at the low seeding rate of 31 kg/ha. The two oat cultivars satisfactorily supported the pea crop and matured together at the same time as the Maro and were harvested together. Crambe was also successful as a nurse

TABLE 6

YIELD OF MARO GROWN IN ASSOCIATION WITH TWO CULTIVARS OF OATS AND CRAMBE, 1976

Treatment	Yield peas t/ha	Yield support crop t/ha
1) Peas 196 kg/ha + crambe drilled 3 weeks before at 2.2 kg/ha	2.5	0.01
2) Peas 196 kg/ha + crambe drilled same time as peas at 2.2 kg/ha	2.8	0.01
3) Peas 196 kg/ha + oats (cv. Mostyn) at 31 kg/ha	2.1	1.0
4) Peas 196 kg/ha + oats (cv. Astor) at 31 kg/ha	2.7	1.5
5) Control Peas 196 kg/ha	1.8	0
F Test	NS	
SE (d_f = 5)	0.36	

TABLE 7

YIELD OF PROCO GROWN IN ASSOCIATION WITH OATS AND CRAMBE AS SUPPORT NURSE CROPS, 1976

Treatment	Yield peas t/ha	Yield support crop t/ha
1) Proco 224 kg/ha + crambe seeded 3 weeks earlier than peas	0.6	0.08
2) Proco 224 kg/ha + crambe seeded same time as peas	1.0	0.02
3) Proco 224 kg/ha + oats (cv. Mostyn) 31 kg/ha	1.6	1.6
4) Proco 224 kg/ha + oats (cv. Astor) 31 kg/ha	1.7	1.2
5) Control	0.4	0
F Test	***	
SE (d_f = 5)	.06	

crop. The yields of peas from plots grown with the two oat cvs. or crambe were all higher than the control.

In the second trial (Table 7) a similar pattern emerged but yields of peas from the cultivar Proco were lower and although this cultivar did not compete as successfully with the oats as Maro, the highest yields of peas were still obtained from the plots grown in association with the oats.

In 1978 and 1979 it was decided to determine the optimum seeding rate of Maro and Proco in conjunction with the low seeding rate (31 kg/ha) of oats cv. Mostyn and also further to assess crambe as a nurse crop; the results of these trials are given in Tables 8 to 10. Unfortunately the crambe proved unsuitable for early sowing (beginning March) as it was not hardy enough.

TABLE 8

YIELDS FROM THREE SOWING RATES OF MARO GROWN IN ASSOCIATION WITH OATS, 1978

Treatment	Plants /m^2	Gross yield peas t/ha	% stained peas	Yield oats t/ha
1) Peas 140 kg/ha + oats 31 kg/ha	36	0.82	12	0.54
2) Peas 186 kg/ha + oats 31 kg/ha	46	0.92	6	0.56
3) Peas 248 kg/ha + oats 31 kg/ha	64	0.90	10	0.55
4) Control Peas 217 kg/ha	56	0.81	14	0
F Test		NS		
SE (d_f = 10)		0.04		

In 1978 with the cultivar Maro grown in conjunction with the oats, differences in yield were non-significant between the lowest seeding rate (140 kg/ha) and the highest rate (248 kg/ha). In the following year differences in yields between the lowest rate (155 kg/ha) and the highest rate of peas (279 kg/ha) were even smaller.

TABLE 9

EFFECT OF FOUR SEEDING RATES OF MARO USING OATS AS A SUPPORT CROP, 1979

Treatment	Plants /m^2	Gross yield peas t/ha	% stained peas	Yield oats t/ha
1) Peas 155 kg/ha + oats 31 kg/ha	40	0.66	6	0.10
2) Peas 186 kg/ha + oats 31 kg/ha	48	0.67	5	0.08
3) Peas 242 kg/ha + oats 31 kg/ha	62	0.73	5	0.11
4) Peas 279 kg/ha + oats 31 kg/ha	71	0.65	5	0.07
5) Control Peas 217 kg/ha	56	0.32	13	0
F Test		*		
SE (d_f = 9)		0.06		

TABLE 10

EFFECT OF THREE SEEDING RATES OF PROCO IN CONJUNCTION WITH OATS AS A SUPPORT CROP, 1978

Treatment	Plants /m^2	Gross yield peas t/ha	% stained peas	Yield oats t/ha
1) Proco 93 kg/ha + oats 31 kg/ha	41	0.55	7	2.4
2) Proco 124 kg/ha + oats 31 kg/ha	54	0.67	2	1.8
3) Proco 155 kg/ha + oats 31 kg/ha	68	0.85	4	2.1
4) Control Proco 124 kg/ha	55	0.32	12	0
F Test		NS		
SE (d_f = 4)		0.10		

With Proco which is the smaller seeded cultivar there were small but insignificant yield increases from the lowest seeding rate of 93 kg/ha to the highest of 155 kg/ha. As in the trials with Maro, oats reduced staining at all seeding rates compared to the control, and yields of peas were higher. Pea yields were generally very low in 1978 and 1979 because of extremely wet weather in August of each year and harvesting was delayed till late September.

In 1980, Progreta, a new tare leaved erect-growing marrowfat cultivar having high yield potential (Anon, 1975) and Filby a leafless cultivar were grown to compare with Maro and Proco and again seeded with oats at 31 kg/ha. With the extremely wet conditions prevailing from June to October it was not possible to harvest this trial. Observations, however, indicated that Progreta was supported much better by the oats than Maro.

The crude protein content of three pea cultivars was determined at Oakpark, 1980 and the results are given in Table 11 below.

TABLE 11

CRUDE PROTEIN CONTENTS OF THREE PEA cvs., 1980

Cultivar	% crude protein
Finale (round)	28
Progreta (marrowfat)	30
Proco (round)	27

Growth regulators

Because of the vigorous vegetative growth that peas make under the cool damp Irish weather conditions the effect of the growth regulator aminozie (N - dimethyl - amino succinamic acid) on yield and growth was studied. Observations in 1973 indicated that the compound could reduce vine length. In 1974, two replicated trials were carried out in which aminozide was applied at three rates and at four stages of plant growth (Table 12). However, under the much cooler conditions prevailing none of the aminozide treatments increased yield or reduced the weight of vines significantly ($P < 0.5$).

TABLE 12 (Trial 1)

THE EFFECT OF AMINOZIDE ON DARK SKINNED PERFECTION, 1974

Treatment	Weight vine t/ha	Yield peas t/ha
3rd true leaf stage		
Aminozide 0.5 kg/ha	42	2.6
Aminozide 1.1 kg/ha	43	3.1
5th true leaf stage		
Aminozide 0.5 kg/ha	48	3.1
Aminozide 1.1 kg/ha	45	3.0
7th true leaf stage		
Aminozide 0.5 kg/ha	40	2.7
Aminozide 1.1 kg/ha	43	2.4
10th true leaf stage		
Aminozide 0.5 kg/ha	43	3.1
Aminozide 1.1 kg/ha	42	2.5
Control	43	3.2
F Test	NS	NS
SE (d_f = 27)	2.5	0.6

TABLE 12 (Trial 2)

EFFECT OF AMINOZIDE ON DARK SKINNED PERFECTION, 1974

Treatment	Weight vine t/ha	Yield peas t/ha
3rd leaf stage		
Aminozide 1.5 kg/ha	50	6.3
5th leaf stage		
Aminozide 1.5 kg/ha	60	7.2
7th leaf stage		
Aminozide 1.5 kg/ha	50	5.3
10th leaf stage		
Aminozide 1.5 kg/ha	56	6.7
Control	53	6.3
F Test	NS	NS
SE (d_f = 4)		

CONCLUSIONS

Trials carried out on both wrinkled and round cultivars of dried peas showed that they could be successfully direct combined when they reached a moisture content of 30% or below. The quality of peas from the direct combined plots was excellent when the conditions were dry before harvesting especially during the three weeks immediately prior to harvest. If conditions were wet during this period then lodging was a problem, staining was accentuated and yields suffered. Nevertheless direct combining in 1971 - 1972 produced yields only 12% lower than the traditiona tripodding method of harvesting with its high labour input. The losses in yield were mainly due to shedding of the peas from the pods prior to harvesting.

Lodging was the biggest obstacle to direct combining because of the difficulty experienced in lifting the crop with the combine pick-up reel, but also of significance were the losses incurred due to staining and shedding of the peas. Various nurse crops were grown to provide support for the pea crops and prevent lodging. Of the many crops tested, oats was found the most promising and compatible with the peas. Both crops can be sown at th same time and drilling can take place early in February or March. They also ripen at the same time so harvesting of both at the sam time presents no problem. Satisfactory weed control can be achie ed with post emergence application of dinosebamine without any significant damage to either crop. In addition chlormequat can be used to control oat vigour without any affect on the pea crop.

In subsequent trials on drilling rates, a low seeding rate of 31 kg/ha of oats was found to allow for an optimum yield of peas. Higher rates proved too competitive and reduced the yields of peas considerably. Both cvs. Astor and Mostyn proved suitable for providing support. In some years the oat stems bent with the weight of the peas but they never allowed the peas to lodge completely to the ground. In the wet years of 1978 to 1979 the oat nurse crop considerably reduced staining in Maro and Proco compared with the unsupported control. Further reductions in staining were achieved with benomyl but only when applied during or near flowering of the peas, when a reduction of about 50% in the degree of staining was achieved. This is in close agreement wit work at PGRO (Anon, 1969). An important consideration was that in eacy year when peas were grown with oats the total yield of peas was higher than for the unsupported crop. This was attributed to three main factors (a) less lodging occurred which allowed more of the pea crop to be recovered for threshing by the combine harvester, (b) less of the pea crop was diseased and less shedding of seed occurred, (c) reduced attack by pigeons which always favoured plots without oats.

From the limited range of pea plant populations grown in conjunction with oats, results showed that for the cv. Maro low populations of 46 - 62 plants/m^2 gave the highest yields. Increasing plant populations much above 60 plants/m^2 was accompanied by a reduction in pea yield. These results agree with Nagy (1966), PGRO (Anon, 1968) and King (1966; 1967) who also found relatively low plant density requirements for dried peas. Proco which is a non-tillering cultivar would appear to require a higher plant population than Maro but more work is required on this aspect.

The other nurse crop crambe showed considerable promise for supporting the peas in 1976 but proved non-hardy when sown early in March in 1978 and was therefore not satisfactory. Furthermore a selective weed control programme was not available.

The growth regulator aminozide did not influence either weight of vine, vine length or yield of peas when applied under cool wet conditions favourable for vegetative growth.

In all these trials the wrinkled pea cv. Maro produced higher yields than the round cv. Proco when grown in conjunction with oats. Because wrinkled peas have a higher protein content than round peas as shown by Richter (1976) this system of growing wrinkled peas with a nurse crop of oats is a possible method of producing high yields of crude protein in Ireland where relatively cool, damp weather conditions normally prevail during the growing and harvesting season. Using oats as a support crop also makes for easier direct combining. Even higher protein yields may be obtainable by growing the recently introduced tare leaved cultivar Progreta.

REFERENCES

Anon., 1967-9. Plant Population - Dried Peas, Pea Growing Research Orgn., Peterborough, Annual Report 1967, p. 23, 1968, p. 24, 1969 p. 31-33.

Anon., 1971. Pea Growing Research Orgn., Peterborough, Annual Reports, 1969 p. 38-39, 1971 p. 38.

Anon., 1977 and 1979. Processors and Growers Research Organ., Peterborough, England, Annual Reports, 1977 p. 23; 1979 p. 25.

King, J.M., 1966. Misc. Publ., no. 18, Pea Growing Res. Orgn., Peterborough.

King, J.M., 1967. Agriculture, Lond. 74: 167.

Murphy, R.F., 1974. Evaluation of dried pea cultivars, Ir. J. Agric. Res. 13: 301-306.

Nagy, J., 1966. Novenytermeles, 15: 43.

Richter, E., 1976. Differences in the characteristics of smooth and wrinkled peas, 1. Content of mature seeds in Primary constituents, Gartenbauwinenschaft, 42(2): S. 72-78.

APPENDIX 1

METEOROLOGICAL RECORDS (Kinsealy, 1972 - 1980)

Year	Rainfall (mm)		
	July	August	September
1971	60.6	51.2	36.8
1972	34.8	22.1	4.2
1973	60.4	93.3	82.0
1974	95.4	23.2	129.6
1975	27.8	26.7	117.8
1976	72.3	3.8	131.5
1977	23.1	64.4	27.4
1978	54.3	86.9	37.5
1979	55.6	85.0	27.6
1980	29.8	93.3	99.4

EFFECT OF N-FERTILISER ON YIELD, PROTEIN CONTENT, AND SYMBIOTIC N-FIXATION IN *PISUM SATIVUM* L. GROWN IN PURE STAND AND MIXTURES WITH BARLEY

A.J. Andersen, V. Haahr, E.S. Jensen and J. Sandfaer.

Agricultural Research Department,
Risø National Laboratory,
DK-4000 Roskilde, Denmark.

ABSTRACT

Effects of N-fertiliser on yield, protein content and symbiotic N-fixation in pea (Pisum sativum L.) were studied in a pot experiment and two field experiments in 1980. Barley (Hordeum vulgare L.) was included in the field experiments in pure stands and in mixtures with pea.

The pot experiment clearly demonstrated that fertilisation with varying amounts of ^{15}N-labelled calcium nitrate and times of supply reduces the N-fixation. Levels of reduction between 33 and 98%, depending on amount and time of supply, were observed. Even though the nitrogen content in some treatments was increased by N-fertilisation the yield of protein generally was reduced due to a reduction in dry matter. However, the experiment indicated that fertiliser-N, supplied at an appropriate time, may increase protein production.

In a field experiment in which harvesting took place five times during the growing period of pea and barley, no difference in dry matter production and N-uptake in pea was seen following the addition of 0 and 50 kg/ha of nitrogen fertiliser. Comparing the dry matter production and N-uptake in pea and barley a faster early development of barley than pea was indicated. In a mixture of pea and barley this effect resulted in pea being the weaker competitor. Barley profited from this by having access to more than its proportional share of the soil and fertiliser nitrogen. The total yields of seed dry matter and protein were increased in the mixtures compared with barley in pure stand, especially without application of fertiliser-N.

In a second field experiment with pea and barley sown in different proportions at three N-levels (0, 40 and 80 kg N/ha) it was also shown, by calculating the relative yields (RY) and relative yield total (RYT), that pea was the weaker competitor in mixtures of the two species. Comparison of the RYT-values indicated an improved utilisation of growth factors, especially when the N-content in the soil was low. RYT-values were generally higher than unity but decreased with increasing N-supply.

In mixed cropping total yields of seed dry matter were equal to or higher than that of the respective components in pure stands. The yield of seed protein increased with increasing proportions of peas in the mixtures. The highest protein yield was obtained in the mixtures of 20% barley and 80% pea.

INTRODUCTION

It is well known that high levels of soil and fertiliser nitrogen (N) can reduce or even inhibit nodulation and symbiotic N-fixation in legumes. Small additions of fertiliser-N may stimulate the N-fixation in some instances (Rennie, 1981). The magnitude of these effects depends on the conditions of growth, the form of N-fertiliser, and the time of N-application. The physiological mechanisms behind these effects are not known in detail but both the reduction of NO_3^- to NH_4^+ and the metabolic incorporation of NH_4^+ into amino acids require photosynthetic energy. Thus, the assimilation of mineral N from soil and fertiliser will compete with nodules for photosynthetic energy and its influence may depend on the photosynthetic capacity of the crops (Houwaard, 1979). The supply of mineral N-fertiliser may increase the yield and protein production in legumes in spite of its negative influence on symbiotic N-fixation.

Non-legumes grown in mixtures with legumes may profit from the association with legumes by having access to a more than proportionate share of the soil- and fertiliser-N; the non-legume also may obtain a fraction of the symbiotically fixed N. The reduction of the nitrogen level in the soil caused by the non-legume does not necessarily reduce the yield of the legume because it is able to compensate for the reduced soil-N by increasing the symbiotic N-fixation.

The present investigations were carried out to measure the effects of N-fertilisation on symbiotic N-fixation and the yield of pea grown in pure stands and that of pea and barley grown in mixtures. The proportions of N derived from the soil and fertiliser and from symbiotic N-fixation were estimated by use of ^{15}N-labelled N-fertilisers.

MATERIALS AND METHODS

A pot experiment (Experiment I) was carried out in plastic pots containing 22 kg of a loamy sand with a surface area of 500 cm^2 and a depth of 40 cm. Appropriate amounts of essential plant nutrients except N were carefully mixed into the entire volume of the soil. The ^{15}N-labelled N-fertiliser was applied as a calcium nitrate solution after two-thirds of the soil was transferred to the pots. Fourteen pea plants (cv. 'Bodil') and twenty-five barley plants (cv. 'Nery') were grown in each pot. Barley was used as the non-fixing reference crop in both the pot and field investigations to estimate symbiotic N-fixation. The crops were harvested at maturity and the yield of roots, vegetative parts of above-ground plant material, and seeds were determined and analysed for total N and ^{15}N content.

Different amounts and methods of N-fertilisation were investigated, these were: a) 0, 1.5, 3, and 4.5 g N/pot at seeding time, b) 3 g/pot at one of four different times during the growth period, and c) split application of 0.75 g N/pot on each of four occasions; b) and c) were at seeding time, five weeks later, early blooming, and finally, at the early pod-filling stage of development.

Field experiments (Experiments II and III) were carried out at the experimental field at Risø on a sandy loam soil containing approximately 2% organic matter and 100 kg/ha of mineral N (NH_4^+ + NO_3^-). The NH_4^+ and NO_3^- were determined after two weeks incubation in the laboratory. A randomised block design with three replicates was used.

In Experiment II the accumulated yield of total dry matter and N-assimilation in pea (cv. 'Bodil') and barley (cv. 'Nery') were determined five times during the growing season. The crops from the second (24th June) and fifth (8th August, mature crops) harvests were supplied with ^{15}N-labelled calcium nitrate in order to estimate the amounts of N deriving from soil and fertiliser and from symbiotic N-fixation. The pea and barley crops were grown at two N-levels (0 and 50 kg N/ha) in pure stand and in a 1 : 1 mixture obtained by sowing 50% of the seeds used in the pure stands.

Experiment III was carried out at three N-levels (0, 40 and 80 kg N/ha) without ^{15}N-labellings. Pea (cv. 'Bodil') and barley (cv. 'Nery') were grown to maturity in pure stand and in mixtures of different composition obtained by replacing the seeding rate of one of the components with 0, 20, 50, 80 and 100% of the other.

The N content in plant material was determined by the Kjeldahl procedure. Aliquots of the digest were prepared by a modified Dumas procedure (Fiedler and Proksch, 1975) for the mass-spectrometric determination of the $^{15}N/^{14}N$ ratio, which was performed at the Chemistry Department, Risø National Laboratory.

The amounts of soil and fertiliser nitrogen in the non-fixing crop, barley, were determined by the well-known isotopic dilution method, and the amount of symbiotically fixed nitrogen in the pea crops was calculated by the procedure presented by Fried and Middelboe (1977):

$$\text{N fixed} = \left[1 - \frac{\text{atom \% } ^{15}\text{N-excess, pea}}{\text{atom \% } ^{15}\text{N-excess, barley}}\right] \text{ x total N in pea}$$

In these calculations we assume that the proportions of N-uptake from soil and fertiliser are the same for both the pea and barley crops. This may not be completely justified because of possible differences in their rooting patterns or in the time and mode of their N-uptake.

An evaluation of the effectiveness with which growth factors were utilised in pure stands and mixtures was obtained by calculating the Relative Yield Total (RYT) (de Wit, 1960) which is defined as the sum of the Relative Yields (RY) of pea and barley:

$$RYT_{\text{pea-barley}} = RY_{\text{pea}} + RY_{\text{barley}}$$ where

$$RY_{\text{pea}} = \frac{\text{mixture yield of pea}}{\text{pure stand yield of pea}}$$ and

$$RY_{\text{barley}} = \frac{\text{mixture yield of barley}}{\text{pure stand yield of barley}}$$

RESULTS AND DISCUSSION

Experiment 1

Results from the pot experiment clearly demonstrated that large amounts of fertiliser nitrogen reduced both the yield and symbiotic N-fixation in the pea crops (Tables 1 and 2). The highest dose of N almost completely inhibited N-fixation. However, the smallest dose (1.5 g N/pot) tended to increase the

TABLE 1

CROP YIELDS AND N CONTENT OF PEA SUPPLIED WITH INCREASING AMOUNTS OF N AT SEEDING (EXPERIMENT 1)

N-supply g/pot	Dry matter, g/pot Root	Straw + pod	Seed	Harvest index DM	N content, % Root	Straw + pod	Seed	Seed* protein g/pot
0	3.5	39.4	60.7	61	2.51	1.53	3.79	14.4
1.5	3.7	44.6	62.8	58	2.38	1.19	3.46	13.6
3.0	3.2	35.6	43.5	55	2.31	1.43	4.02	10.9
4.5	3.1	33.9	39.5	54	2.61	1.72	4.24	10.5

* Total N x 6.25

TABLE 2

NITROGEN (mg N/pot) DERIVED FROM THE FERTILISER AND SOIL AND FROM SYMBIOTIC N-FIXATION IN PEA CROPS SUPPLIED WITH INCREASING AMOUNT OF N AT SEEDING (EXPERIMENT 1)

N-supply g/pot	Total N-uptake	N derived from Fertiliser	Soil	Fixation	Fixation % of total N	Recovered % fert. N
0	2 989	-	562*	2 427	81	-
1.5	2 792	1 312	563	917	33	87
3.0	2 330	1 665	358	307	15	55
4.5	2 337	1 906	273	44	2	42

* Calculated by multiplying available soil N (A_N) as estimated with barley with the percentage recovered fertiliser-N from treatment 1.5 g N/pot.

yield of seed dry matter, but the yield of protein was slightly reduced. Higher N-supply (3.0 and 4.5 g N/pot) decreased the seed yield considerably, and even though the protein contents were increased the yield of seed protein was also reduced.

The effects of N-fertilisation depended greatly on the time of its supply (Tables 3 and 4). Delaying the N-supply (3 g N/pot) until the crop was well established (9th June) or even to early flowering (26th June) and early pod-fill (10th July) increased both seed yield and protein production. Dividing the N in four applications (4 x 0.75 g N/pot) at the above-mentioned dates gave the highest seed yield and protein production obtained in this experiment. Also the influence on symbiotic N-fixation depended on the time of N-supply (Table 4). The most

TABLE 3

CROP YIELDS AND N CONTENT OF PEA SUPPLIED WITH 3 g N/POT AT DIFFERENT TIMES DURING THE GROWTH PERIOD (EXPERIMENT 1)

Time of N-supply (1980)	Dry matter, g/pot			Harvest index DM	N content, %			Seed* protein g/pot
	Root	Straw + pod	Seed		Root	Straw + pod	Seed	
8 May (seeding)	3.2	35.6	43.5	55	2.31	1.43	4.02	10.9
9 June	3.4	41.4	63.9	61	2.38	1.24	3.71	14.8
26 June	3.7	43.1	66.3	61	2.59	1.33	3.85	16.0
10 July	3.7	43.9	66.3	60	2.68	1.45	3.95	16.4
4 x 0.75 g N	4.5	48.8	74.4	60	2.36	1.19	3.69	17.2

* Total N x 6.25

TABLE 4

NITROGEN (mg N/POT) DERIVED FROM THE FERTILISER AND SOIL AND FROM SYMBIOTIC N-FIXATION IN PEA CROPS SUPPLIED WITH 3 g N/POT AT DIFFERENT TIMES DURING THE GROWTH PERIOD (EXPERIMENT 1)

Time of N-supply (1980)	Total N-uptake	N derived from			Fixation % of total N	Recovered % fert. N
		Fertiliser	Soil*	Fixation		
8 May (seeding)	2 330	1 665	358	307	15	55
9 June	2 968	2 229	(478)	(261)	(9)	74
26 June	3 219	2 000	(426)	(793)	(25)	66
10 July	3 355	1 285	(271)	(1 799)	(54)	42
4 x 0.75 g N	3 433	2 481	534	418	12	83

* Approximate estimate since it was not possible to determine accurately the amount of N deriving from soil.

sensitive period apparently was the early vegetative growth (9th June) when only 9% of total plant-N was derived from symbiotic N-fixation. Later applications did not reduce N-fixation to the same extent but the recovery of applied fertiliser-N was very low compared to early or split applications.

The present results indicate that N-fertilisation may increase the seed yield of pea crops even if it reduces symbiotic N-fixation. The amounts of fertiliser-N used in this investigation were probably too high to reveal the potential response. It may be that lower doses given at an appropriate time, e.g., at early pod-fill, would give an improved response both in seed yield and seed protein.

Experiment 2

In this experiment the accumulated yields of dry matter and N-assimilation were measured five times during the growing period. The total yield of dry matter in above-ground parts of the pea crops was almost equal at the two N-levels, but a small negative effect of N-fertilisation was detected in the final harvest at maturity (Tables 5 and 6). The total yield of dry matter in barley was considerably increased by N-supply, especially during the later part of development (Figure 1). It was clearly demonstrated by the growth curves that the very early development of barley crops is much faster than that of pea crops, even if the two species matured at the same time.

TABLE 5

CROP YIELDS (g/m^2) OF PEA AND BARLEY GROWN IN PURE STAND AND IN MIXTURES AT 2 N-LEVELS IN A FIELD EXPERIMENT 1980 (EXPERIMENT 2)

kg N/ha	Pea				Barley			
	Root	Straw	Pod	Seed	Root	Straw	Glume	Seed
0, pure stand	18	290	123	456	63	263	81	412
50, pure stand	16	312	105	358	70	402	112	550
0, mixture	3.4	52	33	140	52	260	85	339
50, mixture	3.2	34	17	96	68	394	106	462

TABLE 6

YIELDS OF NITROGEN (g/m^2) IN PEA AND BARLEY GROWN IN PURE STAND AND IN MIXTURE AT 2 N-LEVELS IN A FIELD EXPERIMENT 1980 (EXPERIMENT 2)

kg N/ha	Pea				Barley			
	Root	Straw	Pod	Seed	Root	Straw	Glume	Seed
0, pure stand	0.4	6.2	1.8	18.0	0.5	1.3	0.4	6.5
50, pure stand	0.3	7.8	1.9	14.5	0.6	2.3	0.7	9.5
0, mixture	0.1	0.8	0.3	5.3	0.4	1.5	0.7	6.0
50, mixture	0.1	0.6	0.2	3.1	0.6	2.5	0.8	8.6

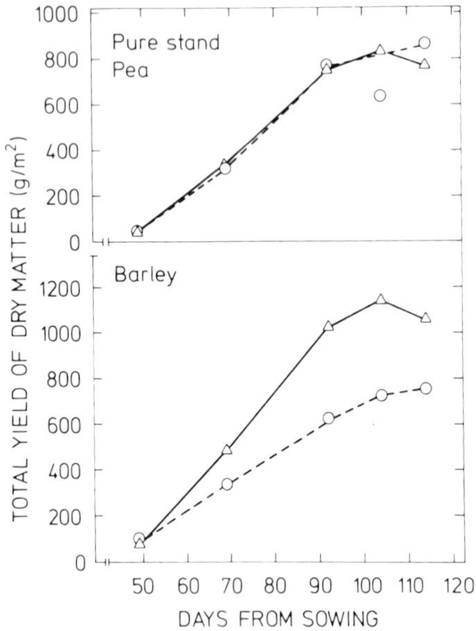

Fig. 1. Accumulation of dry matter in above-ground plant parts of pea (above) and barley (below) grown in pure stands. Without N-supply (broken lines); with 50 kg N/ha (solid lines).

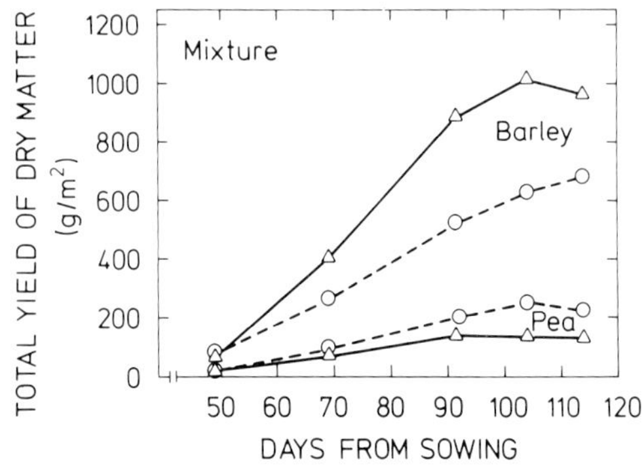

Fig. 2. Accumulation of dry matter in above-ground plant parts of pea and barley grown in mixtures. Without N-supply (broken lines); with 50 kg N/ha (solid lines).

A clear negative effect of N-fertilisation was observed in the dry matter production of the pea crops when grown in a mixture with barley (Table 5 and Figure 2). This may not be a direct effect of the N-supply since that was not seen in the pure stands. It is probable that the increased competition through the early vigorous growth of barley caused a corresponding depression in the growth of pea.

Differences in growth pattern were also demonstrated in the rate of N-assimilation during the growth period. At the first harvest (4th June) almost one-third of the maximum N-uptake was found in barley, whereas only one-tenth of the maximum uptake was found in the pea crops at that early stage of development. More than 80% of the maximum N-uptake in barley was found in the second harvest (24th June), while the pea crops continued to assimilate N at almost a constant rate during the entire growth period (Figure 3).

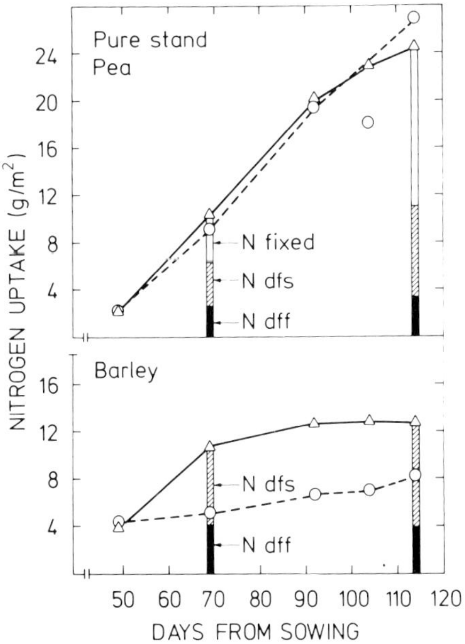

Fig. 3. Accumulation of N in above-ground plant parts of pea (above) and barley (below) grown in pure stands. Without N-supply (broken lines); with 50 kg N/ha (solid lines). Columns indicate the amounts of N derived from each of the three sources: fertiliser (N dff), soil (N dfs), and biological N-fixation (N-fixed) at early bloom/anthesis and maturity, respectively.

It is interesting to note that the maximum amount of N derived from the fertiliser (Figure 3, N dff) was found in the barley crops harvested on 24th June, and some of the N dff was lost before the harvest at maturity on 8th August (Figure 3). This agrees with the wellknown fact that some of the N in cereals is lost during the later part of ripening and is presumably caused by leaching under wet conditions or loss of leaves. However, it was found that the N derived from soil (N dfs) increased in barley during that period. This may have been due to a continued liberation of mineral N from the decomposition of organic matter in the soil. Also, N assimilated during that period in barley receiving no N-fertiliser is presumed to have originated from mineralisation of soil organic matter.

The N dff in pea continued to increase during the entire growth period (Figure 3) in contrast to that for barley. The increase in N dfs in pea, however, was more directly related to the continuing provision of mineralised soil-N, than was that for barley. These differences between barley and pea in the pattern of N-uptake will, to some extent, invalidate the use of barley as a reference crop for estimating symbiotic N-fixation in peas. The accuracy of the estimates may be influenced by soil type, growing conditions, and length of the growing period.

However, in standardised experiments and as a relative measure this technique can give useful information on symbiotic N-fixation.

The pea crop was shown to be a very weak competitor compared with barley in relation to the uptake of soil- and fertiliser-N (Figure 4). The amount of N dff in pea grown in mixture was very low but in barley almost as high as in pure stand. This again may be due to the difference in rate of development during the early part of the growing period.

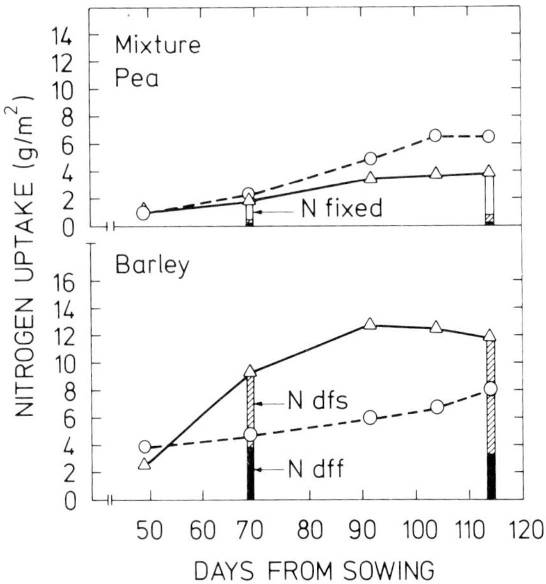

Fig. 4. Accumulation of N in above-ground plant parts of pea (above) and barley (below) grown in mixtures. Without N-supply (broken lines); with 50 kg N/ha (solid lines). Columns indicate the amounts of N derived from each of the three sources: fertiliser (N dff), soil (N dfs), and biological N-fixation (N-fixed) at early bloom/anthesis and maturity, respectively.

The final yield of seed dry matter and seed protein is summarised in Table 7. The addition of 50 kg N/ha at seeding time reduced both the seed and protein yield of pea in pure stand by about 20%, whereas the yield of barley was increased by 33% and 46%, respectively, for grain and grain protein. Comparison of the relative yield (RY) showed that pea was the weaker competitor when growing in mixture with barley. The

TABLE 7

YIELD OF SEED DRY MATTER AND SEED PROTEIN (N x 6.25) IN PEA AND BARLEY GROWN IN PURE STAND AND IN MIXTURE (EXPERIMENT 2)

Crop	kg N/ha	Seed g/m^2	Protein %	Protein g/m^2	Harvest Index Seed	Harvest Index Protein
Pea	0	456	24.6	112.3	53	69
	50	358	25.4	90.8	46	60
Barley	0	412	9.9	40.6	55	79
	50	550	10.8	59.6	52	76
Pea + Barley*	0	479	14.7	70.6	53	78
	50	558	13.0	72.7	50	74

*Total yield and calculated protein content of the crop mixture.

relative yield total (RYT) for seed and seed protein was greater than unity at both N-levels, which indicates an improved utilisation of resources when the two species were grown in mixtures compared with that in pure stands.

Experiment 3

Mixed cropping of pea and barley was further evaluated in this experiment where the two species were sown in different proportions at three N-levels (0, 40 and 80 kg N/ha). The N-fertiliser was applied at seeding time. The yield of seed dry matter and seed protein was measured and compared in replacement diagrams (Figure 5), and the results are summarised in Table 8 on the basis of the relative yield data. Again it was shown that pea was the weaker competitor in these mixtures. The high RY-pea values obtained in the treatment without N-supply were caused by the surprisingly low yield of pea in pure stand. RY-pea values from the other treatments were consistently lower than would have been suggested by the proportional difference in the amount of seed sown, especially at the highest N-level. As mentioned above, this effect of N-fertilisation was probably not caused by an influence of fertiliser-N on symbiotic N-fixation, although it could not be directly detected in the present experiment. It may be explained by the more vigorous development of the barley crop, at the expense of pea growth.

TABLE 8

RELATIVE YIELD (RY) AND RELATIVE YIELD TOTAL (RYT) OF SEED DRY MATTER AND SEED PROTEIN FROM PEA AND BARLEY GROWN IN PURE STANDS AND IN MIXTURES OF VARIOUS COMPOSITIONS AT 3 N-LEVELS IN A FIELD EXPERIMENT (EXPERIMENT 3)

Mixture barley/pea	kg N/ha	Relative yield of seed DM RY_s-barley	RY_s-pea	RYT_s	Relative yield of seed protein RY_p-barley	RY_p-pea	RYT_p	Total yield of seed, g/m^2 Dry matter	Protein*
100/0	0	1.00	-	-	1.00	-	-	404	38.6
80/20	0	0.90	0.27	1.17	0.89	0.25	1.14	416	46.9
50/50	0	0.86	0.67	1.53	0.89	0.62	1.51	479	65.3
20/80	0	0.64	1.19	1.83	0.70	1.09	1.79	495	82.0
0/100	0	-	1.00	-	-	1.00	-	199	50.4
100/0	40	1.00	-	-	1.00	-	-	546	53.1
80/20	40	0.95	0.10	1.05	1.04	0.09	1.13	554	63.3
50/50	40	0.84	0.27	1.11	0.90	0.26	1.16	555	69.7
20/80	40	0.58	0.64	1.22	0.69	0.62	1.31	538	88.5
0/100	40	-	1.00	-	-	1.00	-	351	84.3
100/0	80	1.00	-	-	1.00	-	-	578	66.2
80/20	80	0.90	0.07	0.97	0.92	0.07	0.99	546	66.6
50/50	80	0.83	0.22	1.05	0.84	0.22	1.06	553	73.5
20/80	80	0.59	0.59	1.18	0.63	0.57	1.20	536	88.3
0/100	80	-	1.00	-	-	1.00	-	330	81.4

*Total N x 6.25

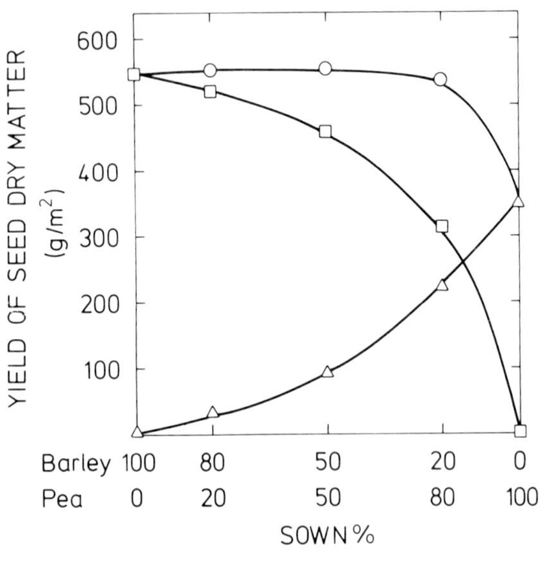

Fig. 5. Seed yields of barley and pea in relation to their proportion when sown in a mixture supplied with 40 kg N/ha. Barley (squares); pea (triangles); total seed yield (circles).

The relative yield total (RYT) depended both on N-fertilisation and proportion of seed sown. RYT-values were generally higher than unity, but decreased with increasing N-supply.

Apparently the most favourable mixture at all N-levels was 20% barley and 80% pea. The greatest advantage of mixed cropping of pea and barley was obtained when the N content in the soil was low.

REFERENCES

Fiedler, R. and Proksch, G., 1975. The determination of nitrogen-15 by emission and mass-spectrometry in biochemical analysis: A Review. Analytica Chimica Acta 78: 1-62.

Fried, M. and Middelboe, B., 1977. Measurement of amount of nitrogen fixed by a legume crop. Plant and Soil 47: 713-715.

Houwaard, F., 1979. Effect of combined nitrogen on symbiotic nitrogen fixation in pea plants. Diss. Wageningen, Holland, 1979.

Rennie, R.J., 1981. Potential use of induced mutations to improve symbioses of crop plants with N_2-fixing bacteria. Induced Mutation as a Tool for Crop Plant Improvement, IAEA, Vienna, 9-13 March 1981. (In press).

Wit, C.T. de, 1960. On competition. Landbouwk. Onderzoek. 66: 1-81.

DISCUSSION

G. Dardenne *(Belgium)* Do farmers generally provide nitrogen for peas?

V. Haahr *(Denmark)* In our loamy sands there is about 50 kg/ha nitrogen naturally available in the spring and, therefore, there is no need for nitrogen fertilisers for pea. The same applies to *Vicia*.

Plancquaert *(France)* Were nodules present on the pea roots?

V. Haahr Estimates of activity were made by the acetylene reduction technique, taking roots up until early flowering. There was no nodulation of roots given a high application of nitrogen fertilisers.

R.C. Hardwick *(UK)* The theory of Sinclair and de Wit is that nitrogen is abstracted from the rest of the plant for ovule growth, resulting in an early cessation of photosynthesis. Was there a rapid reduction in nitrogen in the rest of the plant corresponding to the increase in nitrogen in the seed?

V. Haahr We are aware of this problem but I really cannot answer your question. We can achieve high yields, for example this year we obtained 6 t/ha. It seems that there is a surplus of carbohydrates in the pea crop and, therefore, may not be an important determinant of production.

R. Thompson *(UK)* In one of your pot experiments you applied nitrogen on several occasions. The later you applied it, the more effective it appeared to be both in terms of total dry matter production and grain yield. Is there any evidence that the increasing demand for carbohydrate by the developing ovules depleted that required for efficient nodule activity, and late application of supplementary nitrogen compensated for reduced nodular nitrogen?

V. Haahr One of the reasons we did the experiment was to study this in more detail but the levels of nitrogen used were too high and disrupted fixation too much. We were using 3 g/pot and smaller amounts would have been more appropriate.

M. Frauen *(FRG)* Dr. Haahr can you put a figure on the amount of residual nitrogen in the soil following a pea crop and which is available for use by the next crop.

V. Haahr The residual effect of the haulm when ploughed in on subsequent cropping has not been determined. However, there is no doubt that there is a large amount of nitrogen left in the soil, perhaps up to 100 kg/ha nitrogen.

R.C. Hardwick *(UK)* The validity of the hypothesis of Sinclair and de Wit is extremely important as an explanation for the generally low yields of legumes. The implications behind the enormous residual effect of nitrogen must throw very strong doubt on the validity of the Sinclair and de Wit hypothesis.

C. Knott *(UK)* A major problem with peas is harvesting at the correct time otherwise unacceptable losses can occur. Is it not compounding the problem of time of harvesting by mixing a legume with any other crops?

V. Haahr Synchronous maturity of the pea and cereal can be obtained by selection of appropriate cultivars.

R. Thompson The problem does not arise of course for crops of mixed species intended for forage purposes.

SESSION III

PROTEIN QUALITY ASPECTS

Chairmen: D.R. Davies and D. Boulter

REGULATION OF STORAGE PROTEIN SYNTHESIS AND DEPOSITION IN DEVELOPING PEA SEEDS

D. Boulter
Department of Botany, University of Durham,
Durham, DH1 3LE, UK.

INTRODUCTION

It is a wellknown fact that peas, a high protein crop, are nutritionally first limiting in their protein sulphur amino acids. This is because the storage proteins, the major seed proteins, are low in sulphur amino acids. The object of this paper is to give an account of our present understanding of the regulation of the synthesis and deposition of these storage proteins in developing pea seeds and to relate this, wherever possible, to the feasibility for crop improvement.

GENERAL DESCRIPTION

Peas contain three types of storage protein, legumin (Croy et al., 1979; Casey, 1979); convicilin, (Croy et al., 1980c); and vicilin, (Gatehouse et al., 1981). Legumin, although relatively low in sulphur amino acid content, has a better protein profile in this respect than convicilin, which in turn is better than vicilin which contains very small amounts of sulphur amino acids.

Immunological techniques have shown these proteins are present in very small amounts in the very early stages of pea seed development (Domoney et al., 1980), but that at about a third of the way through the course of seed development, their rate of synthesis is greatly increased. This increased rate of synthesis continues for most of the middle seed development period and then stops as the seed starts to dry out and mature. In the mature seed, as harvested, the storage proteins constitute as much as 80% (the exact percentage varies in different varieties) of the total seed protein. The storage proteins have been shown to be synthesised exclusively on the polysomes of the rough endoplasmic reticulum (RER) (Evans et al., unpublished observations) and are eventually deposited in membrane-bound protein bodies, the origin of which has not been clearly established (Pernollet, 1978). A generally held opinion, however, is that protein bodies are of dual origin, formed from sub-divided vacuoles or from vesicles derived either from the endoplasmic reticulum or from the golgi.

REGULATION

Investigations using *in vitro* protein synthesising systems and short-term pulse chase *in vivo* experiments show that considerable co- and post-translational modifications of the storage protein polypeptides take place after their initial synthesis on the polysomes of the RER. All these proteins have additional sequences of about 1 000 mol. wt. which are subsequently removed by enzymes in the RER (Croy et al., 1980a, b; Gatehouse et al., 1981; Higgins and Spencer, 1981; and unpublished observations).

Furthermore, legumin is made as a subunit of about M_r 60 000 which is subsequently cleaved to the M_r 40 000 and 20 000 subunits (Croy et al., 1980a). Vicilin polypeptides are synthesised initially as M_r 50 000 and 47 000 subunits which are in part subsequently cleaved to the M_r 33 000, 19 000, 13 000 and 12 500 subunits found in isolated vicilins (Croy et al., 1980b; Gatehouse et al., 1981; Higgins and Spencer, 1981). Some artefactual proteolytic cleavage may occur during the very initial stages of protein isolation, although the extent of this is very hard to establish, but it is clear that some vicilin polypeptides are 'nicked' *in vivo*. The intracellular locations of the enzymes responsible for these several changes have not yet been established, but their effects are not observed in pulse chase labelling experiments until several hours after protein synthesis.

Some of the vicilin polypeptides, but not those of the other two storage proteins, are glycosylated and carbohydrate is found mainly, if not exclusively, in the 12 000 - 13 000 M_r polypeptides of isolated vicilin (Gatehouse et al., 1981).

In a typical seed development period of about 24 days from flowering to maturity (depending on variety), vicilin accumulates initially faster than legumin, although the synthesis of the latter persists longer, to 14 and 20 days respectively. There is little turnover of the protein (Boulter et al., unpublished observations) and the changed protein levels are reflected in a changed mRNA population. Both vicilin and legumin mRNAs are present in increasing amounts in the cotyledon cells from 8 - 14 days as demonstrated by using cloned cDNA of the respective messages. The demonstration of a partially α-amanitin-sensitive burst of transcriptional activity in nuclei isolated from cotyledons 8 - 10 days old accounts therefore for this increase *in vivo*. This burst of transcriptional activity leads to the build-up of mRNA which continues to accumulate for at least 14 days i.e. even after the transcriptional activity is once again reduced to very low levels, because the mRNAs are relatively stable with half-lives of about 1½ days (Boulter et al., unpublished observations). Investigation of pea genomic DNA with cDNA probes using Southern blotting shows that legumin genes are present as a few single copy genes; about 10 vicilin genes also occur in one or a few copies (Croy et al., 1981).

CONCLUSIONS

The results presented in this paper suggest that an overall increased protein content of pea seeds might be achieved if the transcriptional regulation event described herein could be manipulated so that transcription either started earlier or continued longer.

Since there are relatively few storage protein genes, it may also be possible to improve the storage protein amino acid profile by using recombinant DNA technology. Copy DNA has already been used in our laboratories to isolate some storage protein genomic genes from restriction digests of pea DNA and it may be possible, in future, to re-introduce suitable mutagenised genes into adult pea plants, if and when suitable vectors become available. Apart, however, from the many techincal problems to be overcome, there is no evidence at present that it is the template, rather than the supply of sulphur amino acids, etc., that is the major constraint.

Sulphur feeding can change the protein profile of pea storage proteins, i.e. increase the ratio of legumin to vicilin and hence increase the sulphur amino acid of the seed, but not at levels which could be of importance agriculturally, i.e. the cut-off point is below the levels of sulphur found in most agricultural soils (Evans et al., unpublished data). In any case, there is no guarantee that sulphur feeding is reflected in increased sulphur amino acids in the seeds, which is the important consideration. Another avenue for improvement could be to increase the albumin content, especially of the few albumin proteins, which occur in larger amounts (unpublished data) and which possess a good sulphur amino acid profile.

ACKNOWLEDGEMENTS

I wish to thank Drs. Croy, Evans, Lycett and J. Gatehouse for the use of their unpublished data.

REFERENCES

Casey, R., 1979. Immunoaffinity chromatography as a means of purifying legumin from *Pisum* (Pea) seeds. Biochem. J. 177, 509-520.

Croy, R.R.D., Derbyshire, E., Krishna, T.G. and Boulter, D. (1979). Legumin of *Pisum sativum* and *Vicia faba*. New Phytol. 83, 29-35.

Croy, R.R.D., Gatehouse, J.A., Evans, I.M. and Boulter, D., 1980a. Characterisation of the storage protein subunits synthesised *in vitro* by polyribosomes and RNA from developing pea (*Pisum sativum* L.). I. Legumin. Planta 148, 49-56.

Croy, R.R.D., Gatehouse, J.A., Evans, I.M. and Boulter, D., 1980b. Characterisation of the storage protein subunits synthesised *in vitro* by polyribosomes and RNA from developing pea (*Pisum sativum* L.). II. Vicilin. Planta 148, 57-63.

Croy, R.R.D., Gatehouse, J.A., Tyler, M.L. and Boulter, D., 1980. The purification and characterisation of a third storage protein (convicilin) from the seeds of pea (*Pisum sativum* L.) Biochem. J. 191, 509-516.

Croy, R.R.D., Lycett, G.W., Gatehouse, J.A., Yarwood, J.N. and Boulter, D., 1981. Cloning, characterisation and sequence analysis of cDNAs encoding *Pisum sativum* L. (Pea) storage protein precursors. Nature (in press).

Domoney, D., Davies, D.R. and Casey, R., 1980. The initiation of legumin synthesis in immature embryos of *Pisum sativum* L. grown *in vivo* and *in vitro*. Planta 149, 454-460.

Gatehouse, J.A., Croy, R.R.D., Morton, H., Tyler, M. and Boulter, D., 1981. Characterisation and subunit structures of the vicilin storage proteins of Pea (*Pisum sativum* L.). Eur. J. Biochem. 118, 627-633.

Higgins, T.J.V. and Spencer, D., 1981. Precursor forms of pea vicilin subunits. Plant Physiol. 67, 205-211.

Pernollet, J.C., 1978. Protein bodies of seeds: ultrastructure, biochemistry, biosynthesis and degradation. Phytochemistry 17, 1473-1480.

DISCUSSION

C.L. Hedley *(UK)* You said that you found increased synthesis of protein after eight days. Would you describe the seed at this stage of development?

D. Boulter *(UK)* The seeds are very small, endosperm is present and cell division has virtually stopped; the tissues have formed and the enlargement phase has not yet started.

C.L. Hedley Have you sought variants in relation to seed development?

D. Boulter No, the problem is that work on variants really requires isogenic lines.

R. Casey *(UK)* Do you have a gene copy number for convicilin?

D. Boulter No. We have not been able to find a convicilin clone. An important point is that the vicilin clones do not react. The 50 000 vicilin clones do not cross-react with the 47 000 vicilins, so these are separate genes.

P. van Oostveldt *(Belgium)* You said that peas are favourable crops for doing this microbiological work. Our experience is that lupins are the legumes with the highest amount of satellite DNA, about 40%. This could be a method for reducing the total amount of DNA into two separate parts and may allow isolation of genes from such a plant.

D. Boulter Yes, but the most important requirement eventually will be to mesh the molecular map. On the molecular basis we work in the level of thousands of base pairs but in classical genetics much bigger cross-over units are used. Eventually we must use the information from classical genetics. We have already found this essential in our work with peas, but it is not available for lupins. It is possible to find small technical reasons of this nature which will suggest work on one plant rather than another but the basic point, genetically and cytologically, is that there is an excellent genome map of pea. There is a vast experience of pea genetics and we have very good lines which have been constructed and which are stable. Without these we should only make limited progress in molecular biology.

P. Guldager *(Denmark)* You indicate that with the legumin gene you found only four fragments. Does this mean that you don't have a lot of duplicated genes in the native DNA?

D. Boulter This is because we titrate the copy number. We know the approximate size of the pea genome and the approximate size of the gene for legumin. We know the size of our pBR 322 genome, the amount of DNA, we know in that we only have single copy and we know the size. In this way we can run out and standardise our gel by taking into account these two ratios. When we do that we find a single copy of these particular genomic fragments. The same is true for vicilin. It could be argued that with our probe we have not hybridised every legumin sequence. However, by analogy with other systems this could only make a small difference. It could also be argued that we are looking in a rather subjective way at the amount of radioactivity on a gel and that this is not highly quantitative, but it is quantitative enough to say whether you have one, two or three copies. We do not have many copies. We believe we can say that we have between one and two copies.

W. Williams *(UK)* In view of the beautiful, detailed work that you have described to us I wonder if I dare ask whether you have any evidence that peas differ in some of the characters you described. For example, you said that one method of increasing protein would be to find genotypes in which synthesis went on longer. Is there any evidence that these genotypes do exist among the wealth of genetic material available to you for this type of work?

D. Boulter There is a fair amount of variation in rates of synthesis, timing of the synthesis, and so on, although I don't know precisely how that is related to the overall seed cycle. However, there is evidence of variation, both quantitative and qualitative, for most of these characters.

W. Williams From your experience would you identify this as an important area that might be rewarding in the fairly near future from the point of view of increasing protein content?

D. Boulter I don't think I can answer that question. I think it is unlikely in the very near future because we have to learn how to transfer genetic information stably in a controlled way, and that is a big question. I believe that in European agriculture we need to have these techniques available in the future and know how to use them. As they are not likely to be of immediate impact, we cannot expect the farmer, or even the research institutes which are concerned with everyday improvement, to support them. It is the responsibility of governments to be far-sighted in these matters because they are the only people who can identify the long range programmes. In Europe we have tended not to do this and here is a golden opportunity which the EC should not let slip in this particular area of expertise where Europe is very well represented having made some of the initial discoveries. I believe we have a better reservoir of know-how in Europe even than the Americans.

MOLECULAR PROPERTIES AND THE COMPOSITION OF THE GLOBULINS FROM LUPIN SEEDS

M. Duranti and P. Cerletti

Department of General Biochemistry, University of Milan, Via Celoria, 2 - I-20133 Milano, Italy.

ABSTRACT

Extraction and separation of proteins from seeds of Lupinus albus is discussed. The molecular properties of the purified globulins are examined. From the types and relative amounts of subunits isolated from the legumins (i.e., proteins 8 and 9a) we propose a molecular composition for these oligomeric proteins, based on a tetramer-octamer equilibrium of polypeptides of molecular weights 64.4 and 61.5 x 10^{-3} respectively, part of which have been 'nicked' by proteolytic enzymes.

A similar situation probably also exists for the vicilins (i.e., proteins 4, 5, 6 and 7), but the larger number of subunits does not allow us to establish the size of the parent polypeptide(s) and the number of components in the oligomeric molecules.

Some microheterogeneity in subunit molecular weight and charge is observed in both vicilins and legumins.

The protomer compositions of globulin 1 and globulin 9b, two major components distinct from vicilins and legumins, are also described; globulin 1 shows consistent subunit charge heterogeneity.

Globulins are glycoproteins; the amount and type of bound sugar differs for the various groups of proteins. Sugar heterogeneity may be one cause of subunit microheterogeneity.

INTRODUCTION

The increasing demand of plant proteins, particularly those from legume seeds, and the need to find optimal conditions for their production and use, has recently stimulated worldwide studies aimed at assessing the composition and molecular properties of such proteins, their nutritional and physico-chemical characteristics, while increasing attempts are being made to relate the latter to the functional properties of the proteins.

Lupin has been increasingly studied, since its protein content is one of the highest among legume seeds (from 31 to 44% of whole dry seed, depending on the species). Its chemical composition is similar to that of soybean, with a reduced oil content (from 3 to 14%), but also with very small amounts of antinutritional factors, mainly in sweet varieties. Lupins grow under soil and climate conditions unfavourable to soybean culture.

Lupin seed proteins have been investigated using many approaches; molecular types and their distribution in related varieties, amino acid composition, amino acid sequence and the influence of nutrients on the protein and amino acid composition of the seed have all been studied (Gerritsen, 1956; Joubert, 1956, 1957; Blagrove and Gillespie, 1975, 1978; Gillespie and Blagrove, 1975; Blagrove et al., 1976; Herbert and Hill, 1976; Ellman, 1977; Cerletti et al., 1978; Gillespie et al., 1978; Duranti and Cerletti, 1979); nevertheless, knowledge of lupin proteins is not as advanced as that of the proteins from soybean, pea and other legumes.

The molecular properties of the major proteins from different legume seeds are similar in many respects. This suggests a common genetic origin; lupin, however, seems to be distinct in several respects. These aspects are discussed in the present report, which collates the data obtained in our laboratory on the seed proteins of *Lupinus albus*.

RESULTS AND DISCUSSION

Protein extraction and identification

Protein extraction is generally performed on 10 g samples of defatted flour. The full procedure, summarised in Scheme 1, requires two days in addition to the time needed for dialysis and freeze-drying. Globulins in the first extract are only 61% of total but display on cellulose acetate electrophoresis the same composition as the product of three extractions. Therefore we usually perform only one extraction, in order to reduce time. With this practice, from 50 to 200 mg pure protein are recovered in each fraction; this rather low yield results from the single extraction and from the amounts discarded after ion exchange and gel filtration so as to improve the resolution of peaks.

Proteins isolated as described in Scheme 1 appear homogeneous on gel filtration, ion exchange chromatography on DEAE cellulose, cellulose acetate electrophoresis and have a characteristic amino acid composition. This indicates that they are individual components of total globulins. Other preparative

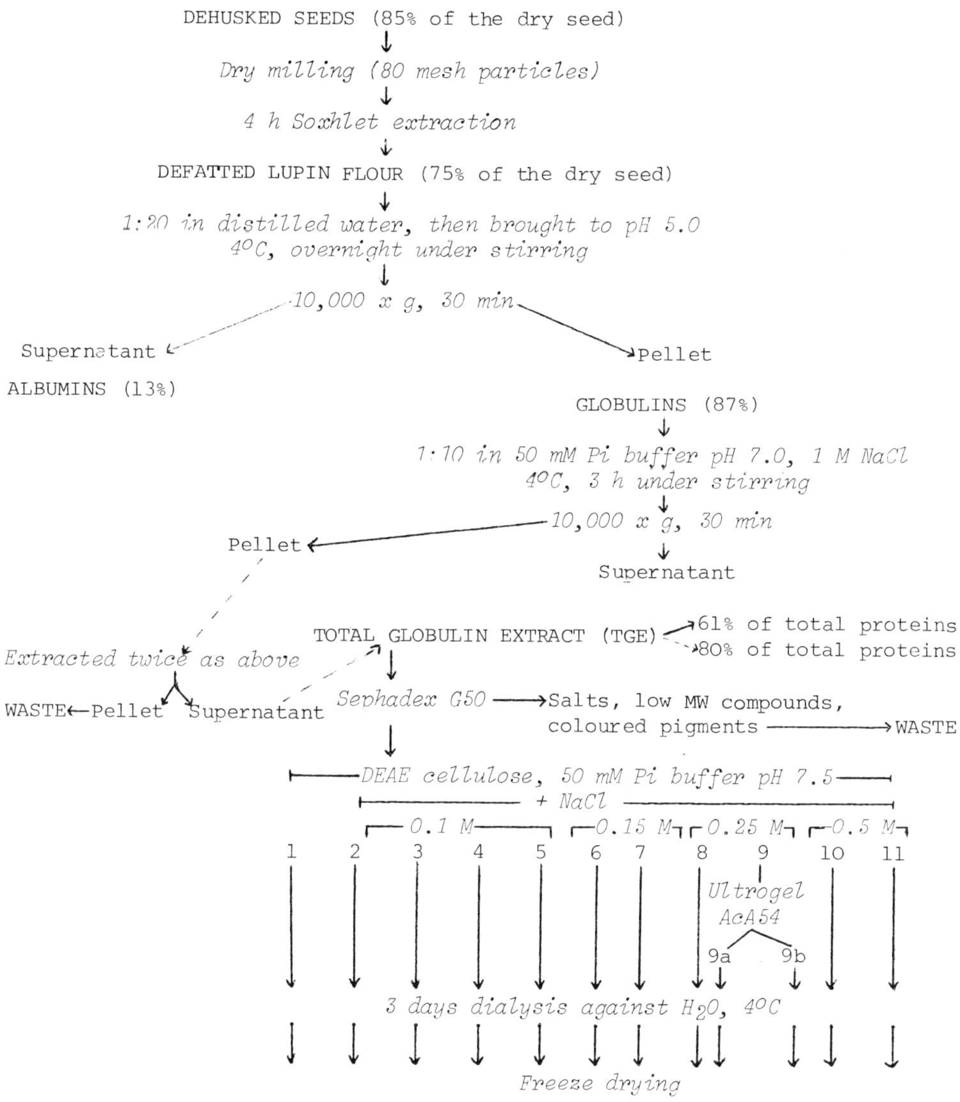

Scheme 1. Extraction and purification of proteins from *L. albus* seeds.

procedures reported in the literature for lupin, or other legumes, may appear more rapid but they either discard part of the proteins; or these are not satisfactorily separated; or else they undergo treatments which may dissociate or denature them, as for example happens when alkaline extractions are used.

Table 1 gives some of the molecular properties of the purified globulins. Cellulose acetate electrophoresis does not

TABLE 1

MOLECULAR WEIGHT AND DISTRIBUTION OF PROTEIN AND NEUTRAL SUGAR IN THE GLOBULINS OF *Lupinus Albus*.

Globulin		Electrophoretic component	MW (KD)	Protein % of total globulins	Neutral sugar mg/100 mg peptide
	1	γ	199	6.0	4.3
	2	-	n.d.	0.5	36.9
	3	-	123	0.7	15.5
vicilins	4	β	270	10.0	2.4
vicilins	5	β	275	2.1	3.7
vicilins	6	β	335	30.2	2.1
vicilins	7	β	143	1.9	3.4
legumins	8	α	240-470	21.2	1.5
legumins	9a	α	240-470	12.0	0.8
legumins	9b	δ	44	12.5	0.7
	10	-	226	2.6	1.5
	11	-	n.d.	0.4	21.5

resolve proteins 4, 5, 6 and 7 which all move as band β, nor proteins 8 and 9a which move together as band α. This behaviour explains why previous workers, who used this technique, recognised only one component for each group, which they called conglutin β and conglutin α respectively. They identified also conglutin γ, which corresponds to globulin 1 but did not note band δ. We identified this band as being due to protein 9b, a component previously recognised in sedimentation studies as a 2S protein. Globulins 4, 5, 6 and 7; and 8 and 9a are similar in many of their properties and we have assigned them, by analogy with other legume seeds, to two homogeneous groups, the 'vicilins' and the 'legumins'.

The two proteins, which are eluted together as fraction 9 from DEAE cellulose and are separated by gel filtration into globulins 9a and 9b, differ with respect to many properties. Separation on gel filtration depends on their widely different molecular weight; the interaction which makes them move together on the ion exchanger, either before or after purification

(Duranti et al., 1981), is probably electrostatic in nature and is related to their different acidity. Indeed the associated proteins have an intermediate elution volume on the ion exchanger as compared to the two separated molecules. The molar ratio in the interaction of purified 9b and 9a is 5 : 1 and corresponds to the molar ratio of these proteins in the total globulin extract; this suggests that they may also be associated in the seed. The other legumin, globulin 8, does not show any affinity for globulin 9b (Duranti et al., 1981) and this is one of the features which distinguish it from globulin 9a.

Legumins

Both globulins 8 and 9a, the two legumins, display on gel filtration a monomer-dimer association equilibrium between components of molecular weight 240 x 103 and 470 x 103, with identical subunit composition.

Two dimensional SDS-PAGE under non reducing conditions resolves a number of subunits from both globulins in the first run, some of which appear to contain more than one polypeptide when subjected to a second run in the presence of a reducing agent. This behaviour indicates the presence of interchain disulfide bonds. The subunit and protomer composition differs in the two globulins. This is shown in Figures 1 and 2, which give the molecular weights and the molar ratios for each of the major subunits and for the protomers forming the subunit. Subunits differ considerably in molar ratios; although each legumin is homogeneous in many respects. Therefore the overall composition of the molecules is similar although they contain different subunits. We propose that in each legumin the observed subunits derive from the breakdown of a single parent polypeptide, as shown in Figures 3 and 4. The fragments of proteolysis all remain present in the molecule, forming the oligomeric structure; on SDS-PAGE they liberate the observed subunits. Under reducing conditions, intersubunit disulfides are split and the characteristic pattern of protomers is formed. The large disulfide bonded peptides are more acidic in character than that one of molecular weight 19.0×10^{-3}; this may result from a high content of

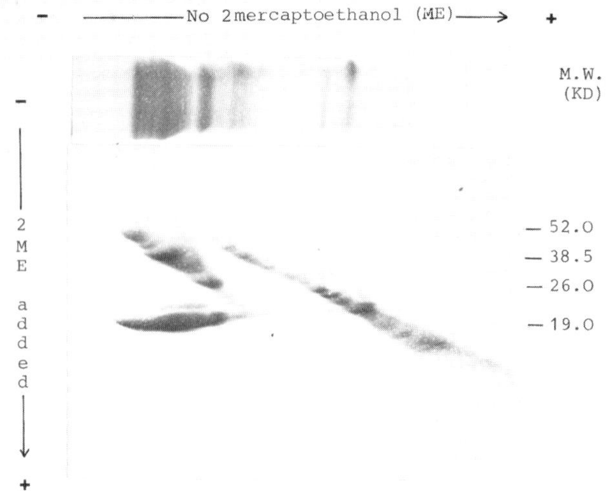

M.W. (KD) and *molar ratios* of the major subunits and protomers

64.5 *13*	55.5 *64*	44.0 *17*	21.0 *6*
52.0 *10*			
	38.5 *64*		
		26.0 *20*	
19.0 *13*[+]	19.0 *64*[+]	19.0 *17*[+]	21.0 *6*[+]

[+]*Values calculated from determined total amount.*

Fig. 1. Subunit and protomer composition of globulin 8; two dimensional SDS polyacrylamide gel electrophoresis.

glutamate and aspartate, which together represent 35% of total amino acids in the protein (Duranti et al., 1981).

In view of the identity in molecular weight of the two legumins and of their great similarity in amino acid composition, we propose that they both may derive from a same original polypeptide 'nicked' by proteases in different positions.

The proteolysis advocated here most likely occurs in the seed. Significant changes during the experimental procedures are unlikely and the peptide distribution was unchanged in samples kept for convenient times under the experimental conditions (data not shown).

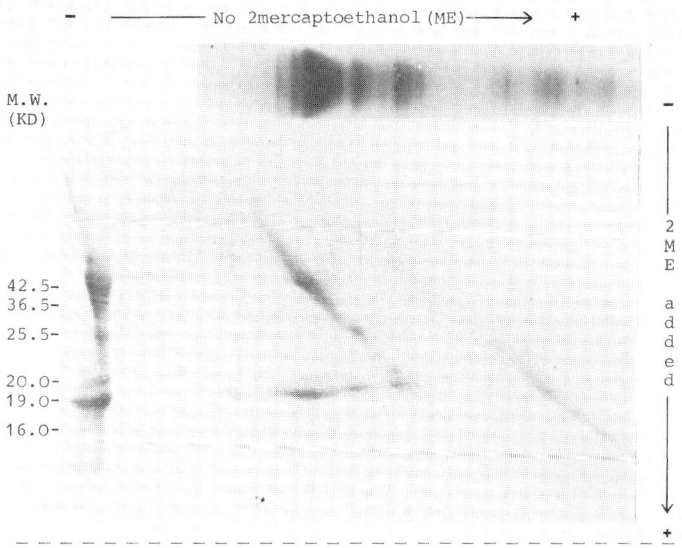

M.W. (KD) and *molar ratios* of the major subunits and protomers

61.5 *8*	55.5 *32*	45.0 *15*	36.5 *19*	21.0 *10*	16.0 *16*
42.5 *4*					
	36.5 *33*				
		25.5 *17*			
19.0 8^{+}	19.0 31^{+}	19.0 15^{+}	19.0 19^{+} 19^{+}	21.0 10^{+}	16.0 *17*

$^{+}$*Values calculated from determined total amount.*

Fig. 2. Subunit composition of globulin 9a; two dimensional SDS polyacrylamide gel electrophoresis.

If now we come to the quaternary structure, in globulin 8 the monomer appears to contain four component units with molecular weight 64.5 x 10^{-3} and the dimer eight such units. For globulin 9a the same equilibrium is observed, with a unit of molecular weight 61.5 x 10^{-3}. The molecular weights and molar ratios of subunits and their protomers allow us to propose the compositions and relative amounts of component units as shown in Table 2. Nothing specific can be said about the specificity of association of units within each molecule. Hydrophobic interactions are important binding forces, as shown by the dissociating effect of detergents and urea. Values of molecular weights determined for the major protomers practically coincide

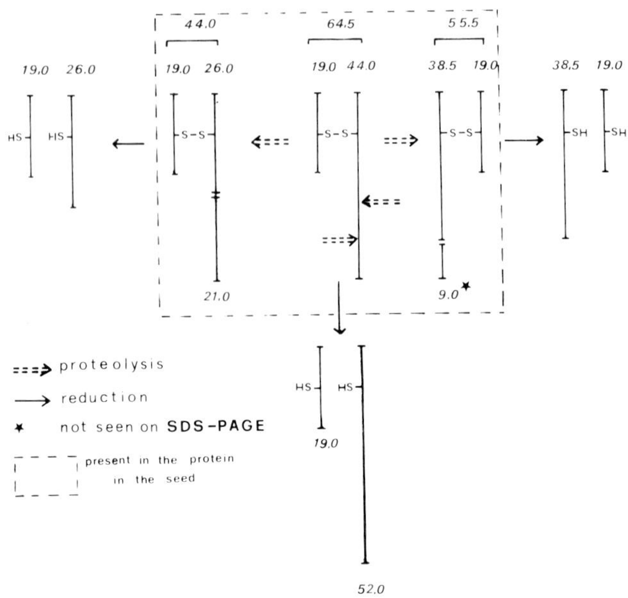

Fig. 3. Subunit and protomer formation by proteolysis and reduction; globulin 8.

with those found by other authors for corresponding proteins of other legumes. However, in our hands the whole proteins appear to yield a tetramer-octamer equilibrium, whereas in other legumes legumins are described as hexamers.

Legumin subunits are microheterogeneous both for charge and for molecular weight. If IEF is followed by SDS-PAGE in the two-dimensional technique (both under reducing conditions), many of the bands separated by IEF yield, on SDS-PAGE, components having the same molecular weight, thus indicating charge heterogeneity (Figure 5). This could be attributed to different degrees of amidation of the glutamate and aspartate residues. Heterogeneity may also result from minor differences in amino acid composition, as has been observed for storage proteins in *Vicia faba*, *Pisum sativum*, etc. Heterogeneity in molecular weight is revealed by gel filtration in 8 M urea followed by SDS-PAGE; by the latter technique several similar protomers are resolved in the peaks from gel filtration, having molecular weights close to that of the peak itself (Duranti et al., 1981). They do not derive from breakdown of a larger unit but very likely were

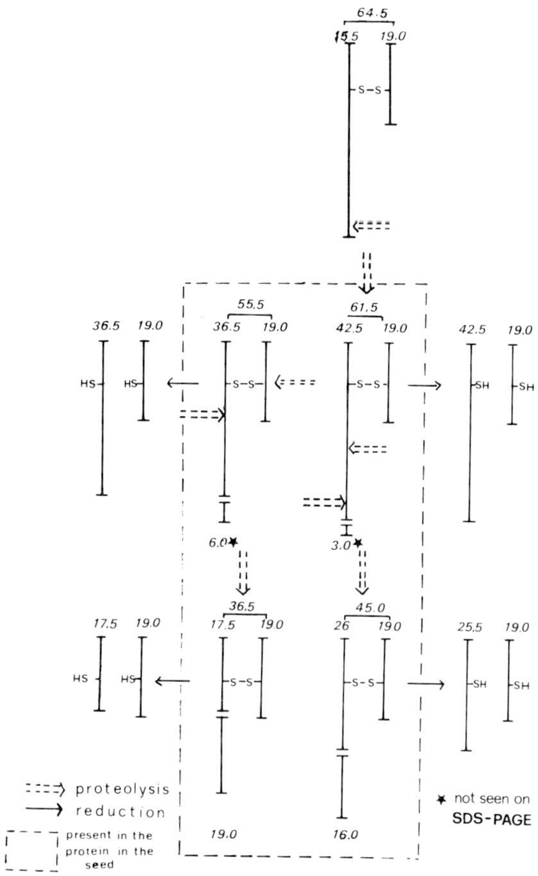

Fig. 4. Subunit and protomer formation by proteolysis and reduction; globulin 9A.

TABLE 2

SUBUNIT STRUCTURE OF LUPIN LEGUMINS

	MW (KD) of the native protein	MW (KD) of the protomers (S-S bound)	% of total	Number of subunits
		(45.0 + 19.0)	14	
Globulin 8	240 ⇌ 470	9.0* + (38.5 + 19.0)	68	4 ⇌ 8
		21.0 + (26.0 + 19.0)	18	
		(45.0 + 19.0)	5	
Globulin 9a	240 ⇌ 470	6.0* + (36.5 + 19.0)	46	4 ⇌ 8
		6.0* + 21.0 + (17.5 + 19.0)	25	
		16.5 + (25.5 + 19.0)	24	

*not seen in SDS-PAGE

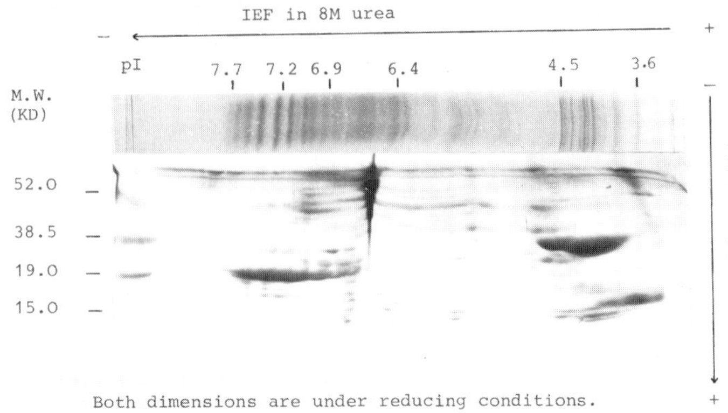

Fig. 5. Two-dimensional IEF-SDS-PAGE of globulin 8.

originally present in the molecules. This molecular weight heterogeneity appears also when single bands in IEF yield similar, but separate bands on SDS-PAGE. Heterogeneity is an intrinsic characteristic of lupin storage proteins and is present when proteins from a single seed are analysed. The multiplicity of bands increases slightly in the flour from a genetically selected cultivar and is significant in seeds of commercial origin.

As shown in Table 3, covalently bound sugar in legumins is qualitatively typical of this group of globulins and the relative amounts of bound hexoses are characteristic for each of them. Bound carbohydrate is, however, little and appears to be associated with subunits which represent only a minor fraction of the total. Thus the major molecular components of legumins may not contain bound carbohydrate and its presence or absence may be one of the causes of heterogeneity.

Vicilins

In vicilins all subunits contain only one peptide; there are no interchain disulfide bonds because of the lack of cysteine in these molecules.

As shown in Figure 6, all vicilins display a pattern on SDS-PAGE which is qualitatively identical and shows only minor differences in the quantitative distribution of subunits. Subunits with corresponding molecular weight in separate proteins

TABLE 3

MONOSACCHARIDE COMPOSITION OF CARBOHYDRATE IN GLOBULINS, %

Sugar		Vicilins			Legumins		
	1	4	5	6	8	9a	9b
Xylose	3.9	-	-	-	tr	tr	3.3
Arabinose	20.8	-	-	-	tr	tr	7.5
Galactose	61.2	-	tr	tr	2.9	20.1	31.2
Mannose	10.7	96.3	72.0	94.2	79.9	68.9	41.5
Glucose	2.8		22.9	0	6.3	1.3	13.1
N-acetil glucosamine	0.6	3.7	5.1	5.8	10.9	9.7	3.4

are likely to have similar amino acid compositions, in view of the overall amino acid similarity of the vicilins. The relative amounts of the various subunits in each molecule do not indicate a definite structure which might include them all. It is therefore possible that the peptides originate, as discussed for the legumins, by different fragmentation of the same ancestor polypeptide so that their associations yield the same overall amino acid composition and molecular weight. However, the large number of peptides resolved and their relative molar ratios do not allow as straightforward an interpretation of their associations as for the legumins.

Some of the major subunits are accompanied by a number of minor polypeptides of similar molecular weight. This heterogeneity is a further cause of difficulty in formulating a structure for these globulins. Vicilins isolated from a single seed also appear heterogeneous. There is no evidence for subunit fragmentation during purification.

The same subunits are likely to be associated in multiple ratios, as indicated by the molecular weights of the major vicilins (namely globulins 4 and 6; and also globulins 5 and 7) and this is probably the basis of the molecular identity of the vicilins. Further differences may result from minor differences in amino acid composition and in heterogeneity of bound carbohydrate. Vicilins contain significant amounts of covalently bound sugar, characteristic of each (Table 3). Again,

M.W. (KD)	Moles,% 4	5	6	7
64.0	15	15	18	18
43.0	9	5	9	8
34.0	11	11	10	12
28.0	11	9	11	8
21.0	14	14	12	19
16.5	27	25	20	17
13.5	13	21	20	18

Fig. 6. Subunit SDS-PAGE pattern of vicilins from *Lupinus albus*.

carbohydrate is associated only with minor subunits which appear to correspond in molecular weight in the different vicilins. The sugar has a role in characterising the molecule, especially in the case of protein 5, which is the only vicilin containing glucose.

Other globulins

Two other important globulins are protein 1 and protein 9b, which are characteristic in terms of their amino acid composition and which will be described in another report (Cerletti et al., 1982). Here we will discuss their molecular features.

Protein 1 displays only one type of subunit (molecular weight 45.0 x 10^{-3}) which contains two different, disulfide-bonded peptides of molecular weights 16.0 and 28.0 x 10^{-3}; the smaller is more acidic in character; while the larger is less acidic and has a significant amount of bound sugar with a variety of hexose and pentose components. Molecular weights and molar ratios indicate the native protein to be association of four subunits of molecular weight 45.0 x 10^{-3}. No heterogeneity of the subunit or of the constituent peptides appears on SDS-PAGE, but heterogeneity is observed when SDS-PAGE follows IEF, indicating charge, but not molecular weight, heterogeneity; the isopeptides of the heavier protomer have isoelectric points which range from 6.2 to 6.8, whereas those of the lighter one have pIs in the range from 5.6 to 5.9 (Restani et al., 1981).

Protein 9b has the most acidic character and the smallest molecular weight of all the globulins. It does not contain bound sugar in significant amounts. Its structure has been described (Restani et al., 1981). Proteins 2, 3, 10 and 11 are very minor globulins. All, except protein 10, are proteoglycans, containing much bound carbohydrate. Their amount varies considerably, indicating that they are not constant components, unlike vicilins and legumins (Restani et al., 1981).

ACKNOWLEDGEMENT

This research was supported in part by grants from the Italian National Research Council (CNR), programme: New Protein Food.

REFERENCES

Blagrove, R.J. and Gillespie, J.M., 1975. Aust. J. Plant Physiol. 2, 13.
Blagrove, R.J., Gillespie, J.M. and Randall, P.J., 1976. Aust. J. Plant Physiol. 3, 173.
Blagrove, R.J. and Gillespie, J.M., 1978. Aust. J. Plant Physiol. 5, 651.
Cerletti, P., Fumagalli, A. and Venturin, D., 1978. J. Food Sci. 43, 1409.
Cerletti, P., Restani, P., Duranti, M., Mariani, A. and Semino, A. *In vitro* nutritional properties and the effect of bound carbohydrate and of conformation on the digestibility of globulins from lupine seeds. This volume.
Duranti, M. and Cerletti, P., 1979. J. Agric. Food Chem. 27, 977.
Duranti, M., Restani, P., Poniatowska, M. and Cerletti, P., 1981. Phytochemistry 20, n. 9, 2071-2075.
Ellman, T.C., 1977. Aust. J. Biol. Sci. 30, 33.
Gerritsen, T., 1956. Biochim. Biophys. Acta 22, 269.
Gillespie, J.M. and Blagrove, R.J., 1975. Aust. J. Plant Physiol. 2, 29.
Gillespie, J.M., Blagrove, R.J. and Randall, P.J., 1978. Aust. J. Plant Physiol. 5, 641.
Herbert, S.J. and Hill, G.D., 1976. Proc. Agric. Soc. NZ 6, 65.
Joubert, F.J., 1956. Biochim. Biophys. Acta 19, 172.
Joubert, F.J., 1957. J.S. Afr. Chem. Inst. 10, 21.
Restani, P., Duranti, M., Cerletti, P. and Simonetti, P., 1981. Phytochemistry 20, n. 9, 2077-2083.

DISCUSSION

D. Boulter *(UK)* What is the molecular weight of legumin, or 11S protein?

M. Duranti *(Italy)* We found an equilibrium between a monomer and a dimer between 240 and 470 kilodaltons. In our first experiment we found a molecular weight close to 350 000 but we were able to improve the separation. With a longer column and a different resin, we obtain two peaks.

D. Boulter Do you plan to measure in the ultracentrifuge? This is important because this is the only basis for suggesting that there are 4 60 000 sub-units, which is quite atypical for this type of protein.

M. Duranti We hope to measure in the ultracentrifuge.

D.R. Davies *(UK)* You did say that your legumin is glycosylated.

M. Duranti Yes. Actually if we stained them with Periodic Acid - Schiffs Reagent (PAS) for sugar we found that sugar is covalently bound on very minor peptide components - protein components. So we cannot dismiss the possibility that there may be some artefacts or contamination. Thus probably legumin is not a glycoprotein.

G. Dardenne *(Belgium)* What separation technique do you use in gel electrophoresis?

M. Duranti We use gradient polyacrylamide gel electrophoresis.

QUANTITATIVE AND QUALITATIVE MICROSPECTROPHOTOMETRIC TECHNIQUES FOR THE ANALYSIS OF SEED VARIETIES AT DIFFERENT STAGES OF DEVELOPMENT

P. van Oostveldt

State University of Gent,
Faculty of Agricultural Science,
Coupure 533 B9000 Gent, Belgium.

The commercial availability of reliable and easy to use microdensitometers or microspectrophotometers has resulted in the application of cytophotometric techniques to a broad range of objects and products. Although most cytometric techniques have been tested on selected objects of botanical origin, the overall application of these techniques still remains limited in plant physiological research.

In this paper we will discuss

1. The main difficulties that are encountered during cytometric research on plant material;
2. The potential of cytometric techniques as a tool for the study of plant development.

An evaluation of the reliability and the specificity of the different cytochemical staining methods that are now available have been considered in other papers or in articles in specialised journals (Bahr, 1979; Van Duijn, 1976).

The difficulties encountered during cytometry on plant tissues depend on the method used. In most cases there is a choice between cytometric evaluation by absorption or by fluorescence measurements.

For absorption measurements the main difficulties are the heterogeneous distribution of the stained material, the so-called distributional error, and glare or non-specific light loss caused by differences in the refractive index of the plant tissue, for example, starch granules or cell wall.

The distributional error can be reduced to less than 2% by using scanning microdensitometry or the two-wavelength method of Patau (1952) and Ornstein (1952). However, some plants, for example lupins, have such an extreme heterogeneous distribution of their chromatin in the nucleus that this will inevitably result in serious distributional errors (Van Oostveldt and Boeken, 1976). When using two-wavelength cytometry, the high transmittance of the measuring spot will result in an increased error of the estimation of the transmittance at the wavelength of choice. When using the scanning microdensitometer the results will increasingly be influenced by residual distributional error which is proportional to the ratio of the measuring spot size and the object size (Duyndam et al., 1979). The object size cannot be altered and the measuring spot size is limited by the light intensity which is still measurable without too much noise.

In order to eliminate or correct for non-specific light loss we can use a two-wavelength scanning method as already applied for plant tissues by Miksche et al. (1979), but such a correction is not possible in the two-wavelength method of Patau and Ornstein (1952). Two-wavelength scanning can only be done on the assumption that the absorption spectrum is constant and the integrated absorbance of a blank at the two-wavelengths represents the same non-specific light loss as the scanned nucleus or any other object.

A computer-directed scanner is much more efficient in correcting for different absorption errors because the light intensity of every individual image point is stored on a storage medium. The corrections applicable in this case are limited by computer memory or available software but are always time consuming when used for routine analysis.

When measuring chromophore amounts by fluorometry, the problem of distributional error is totally eliminated. Furthermore the non-specific light loss is constant and, in contrast to absorbance measurements, its effect upon the fluorochrome estimate is independent of the concentration of the dye.

The main disadvantages of fluorometry are the non-specific background fluorescence, the fast photofading of the fluorochrome and the narrow linear relation between fluorescence and dye concentration.

The problem of photofading is controlled by commercially available shutters, which can limit excitation time to less than 0.25 seconds.

Background fluorescence can have its origin in non-specific

staining or be caused by endogenous fluorescent products such as chlorophyll for example. Endogenous fluorescence or autofluorescence can sometimes be reduced by changing the fixative or applying some preliminary extraction procedures. In other cases an appropriate selection of the dye in combination with narrow band excitation and/or emission filters is possible.

Once these problems are solved it must be ensured that the absorbance of the dye never exceeds 0.1 in the measuring object. The reduction of the dye concentration in the staining bath can help to keep absorbance low.

Fluorometric determinations are very rapid and extremely accurate. The price of a commercially available microfluorometer is much less than a scanning microdensitometer and hence fluorometry is ideally suited for application in routine and applied research.

REFERENCES

Bahr, G.F., 1979. Frontiers of Quantitative Cytochemistry. Anal. Quant. Cytology. 1: 1.

Duijndam, W.A.L., Smeulders, A.W.M., van Duijn, P. and Verwey, A.C., 1980. Optical Errors in scanning stage absorbance cytophotometry. J. Histochem. and Cytochem. 28: 388.

Miksche, J.P., Dhillon, S.S., Berlyn, G.P. and Landauer, K.J., 1979. Non-specific light loss and intrinsic DNA variation problems associated with Feulgen DNA cytophotometry. J. histochem. Cytochem. 27: 1377.

Ornstein, L., 1952. Distributional error in microspectrophotometry. Lab. Invest. 1: 250.

Patau, K., 1952. Absorption microspectrophotometry of irregularly shaped objects. Chromosoma 5: 341.

van Duijn, P., 1976. Prospects for microscopical cytochemistry. Histochem. J. 8: 653.

van Oostveldt, P. and Boeken, G., 1976. Absorption Cytophotometry: Evaluation of three methods for the determination of DNA in Feulgen stained nuclei Histochemistry 50: 147.

GENETIC VARIATION OF THE PEA STORAGE PROTEINS STUDIED BY IMMUNOELECTROPHORETIC METHODS

P. Guldager

Department of Genetics,
Royal Veterinary and Agricultural University,
Bülowsvej 13, DK-1870 Copenhagen V, Denmark.

ABSTRACT

Two hundred lines of pea, including a spectrum of distinct varieties, were investigated by crossed immunoelectrophoresis. Significant differences were detected in the content of major storage protein legumin. High legumin lines showed a precipitation arc area three times larger than low legumin lines. The content of the other major storage protein, vicilin, varied little between the lines. Vicilin from one line had a higher electrophoretic mobility than normal.

Albumins varied considerably among the lines. In some lines only six proteins could be detected, while in other lines 14 distinct proteins were present.

For lectin, only moderate differences were observed. One specific line showed an altered precipitation pattern due probably to the deviant binding specificity which this lectin was later shown to possess.

Crossed immunoelectrophoresis was found to be a reliable method for detecting quantitative and qualitative differences. The method is, however, too time-consuming for use in large scale screening.

RESULTS AND DISCUSSION

Multi-cross immunoelectrophoresis (MCI) is a method suitable for screening differences in seed protein content (1) between lines, (2) in mutagen-treated material and (3) during analysis of segregation in second generation offspring from controlled crossings. Examples of the use of the method are given.

The major storage proteins, vicilin and legumin, interfere with the detection of many of the minor proteins present in a total seed extract. This interference particularly influences the results obtained with multi-cross immunoelectrophoresis, where the vicilin precipitation arc covers minor protein arcs, but the problem can be solved by use of the intermediate gel technique. Specific antibodies against vicilin and legumin are placed in the intermediate gel to precipitate vicilin (or legumin) and consequently the final precipitation pattern obtained in the gel containing total antiserum is free of the vicilin arc.

The intermediate gel technique was used to examine four distinct varieties of pea.

Tandem-crossed immunoelectrophoresis and crossed-line immunoelectrophoresis are suitable methods for the comparison of diverging precipitation patterns of different varieties. In cases where precipitation arcs of a protein from two varieties exhibit perfect fusion, identical protein is present in the two lines, or the molecular structure of the antigenic sites are identical in the two proteins. Partially fusing arcs indicate that the proteins are similar, whilst clear-cut crossing (no fusing at all) indicates that the proteins are distinctly different, without common antigenic sites.

A major advantage of the immunoelectrophoretic methods is the ability to detect differences in concentration of a given protein. Under suitable conditions there is a linear relationship between the arc precipitation area and the concentration of proteins. A protein extracted from different varieties might, however, show differences in the arc area/protein concentration regression.

In pea, the legumin molecule might vary in structure between lines and consequently in structure or number of available antigenic sites that the applied antiserum can react with. This type of variant legumin molecule is likely to show too high a precipitation arc area compared with the legumin used to raise the antiserum. By crossed-line immunoelectrophoresis it is possible to elucidate the existence of significant differences in antigenic sites between molecules of a given protein originating from different varieties.

Where antigenic differences are not found among varieties, crossed immunoelectrophoresis and MCI are probably two of the most reliable methods for quantitative analysis of total seed extracts. The methods are hardly, if at all, affected by incomplete separation of the proteins to be analysed and the formation of a specific precipitation arc is rarely influenced by the presence of antigenetically different proteins.

REFERENCES

Axelsen, N.H., Krøll, J. and Weeke, B., 1973. A manual of quantitative immunoelectrophoresis, methods and applications. Oslo: Universitetsforlaget. pp. 169.

Casey, R., Sharman, J., Wright, D., Bacon, J. and Guldager, P., 1981. Quantitative variability in *Pisum* seed globulins, its assessmant and significance. Qualitas Plantarum (In press).

Guldager, P., 1978. Immunoelectrophoretic analysis of seed proteins from *Pisum sativum*. L. Theor. Appl. Genet. 53, 241-250.

Guldager, P., 1981. Screening of the protein content from *Pisum sativum* seeds by multi-cross immunoelectrophoresis. (In press).

NEAR INFRA-RED SPECTROSCOPY AS A RAPID AND ACCURATE TECHNIQUE FOR ESTIMATING *LUPINUS* SEED PROTEIN AND *VICIA FABA* SEED AMINO ACIDS AND PROTEIN

R. Biston[1], G. Dardenne[2], M. Sindic[2], P. Dardenne[1], B. Wathelet[2] and B. De Sousa[3]

[1] Station de Haute Belgique, CRA Gx., 48, rue de Serpont, (6600) Libramont-Chevigny, Belgium.

[2] Faculté des Sciences Agronomiques, Chimie organique et biologique, (5800) Gembloux, Belgium.

[3] Instituto Superior de Agronomia, Lisbon, Portugal.

ABSTRACT

The technique of near infra-red reflectance spectroscopy (NIRS) has been applied to the determination of protein in Lupinus seeds and of amino acids and protein in Vicia faba seeds. The interlaboratory and intra-laboratory reproducibility of analysis by Kjeldahl and the Amino Acid Auto-analyser indicate very satisfactory accuracy in the estimation of protein and amino acids by NIRS. More than 200 analyses a day may be performed. The results clearly show that the near infra-red technique predicts protein amino acids independently of protein.

INTRODUCTION

The development of new laboratory techniques is an essential prerequisite for the increasing activities of plant breeders, of manufacturers interested in quality improvement and of farmers to control production.

Recently, new physical measurements, as already reported in some reviews, have begun gradually to replace old chemical and physico-chemical methods (Kjeldahl, Biuret, DBC, ion ex-change resins, etc.). The superiority of such techniques is due to the fact that they are less destructive, for example they leave the material intact for further different analyses. It is now possible routinely to keep the plant sample intact for further cultivation.

However, it is interesting to consider an important fact which is sometimes forgotten or misunderstood, namely that with this kind of technique, it is absolutely necessary to calibrate the instrument with values obtained from classical methods. The calibration has also to be established on each cultivated plant: wheat, rye, oats, lupin, beans, etc.

Near infra-red reflectance spectroscopy (NIRS) has been recently developed for the rapid prediction of the composition of cereals, legumes, forage grasses, silage, etc. The technique was originally developed for the measurement of moisture content but has now been expanded to measure protein, starch, oil, sugar, cellulose, lignin, digestibility and also protein amino acids (Hymowitz et al., 1974; Pomeranz and Moore, 1975; Williams, 1975; Watson et al., 1976; Norris et al., 1976; Hunt et al., 1977; Miller et al., 1978; Biston and Clamot, 1981). Near infra-red light is an electromagnetic radiation of a specific wavelength ranging between 1.1 μm (1 100 nm) to 2.5 μm (2 500 nm). The ability to scan a spectrum rapidly, coupled with the ability to measure extremely low light levels, means that large amounts of data are generated each fraction of a second. Converting this vast amount of data into quantitative results requires the use of high speed computation circuitry. Fortunately, with the advent of the micro-processor, almost any kind of data manipulation becomes possible at reasonable cost. The most complex algorithms (first derivative, second derivative, etc.) provide greater protection from unwanted optical interference effects caused by other constituents.

The NIRS method is now simple, rapid (a few minutes, including grinding) and economical. Protein, amino acids, oil, cellulose, etc. can be measured simultaneously. The NIRS method requires no weighing and no calculations, as results are given in direct readout. The NIRS is not destructive. After analysis, it will be possible to sow the seeds. The greater source of variability results from sample preparation (difficulty in obtaining homogeneous samples of uniform small particles), from the choice and number of species and cultivars and from the difficulty in obtaining accurate calibration methods. (Rinne et al., 1975; Graham, 1975; Watson et al., 1976; Shenk, 1979).

The minimum quantity of material is about 3 - 4 g. In many cases, only 1 g or less is supplied, especially in breeding research projects. With a modified sample cell and cap, it is now possible to work with a quantity of flour as small as 0.9 - 1 g and such samples are successfully processed (Williams et al., 1978).

Near infra-red technology has opened up vast horizons for its application in the food and feed industry. Neotec recently introduced the model RCA 6350 which is very sophisticated and allows the user to perform fundamental research. It is, perhaps, costly but it will allow new users to take full advantage of the benefits of NIRSs.

NIRS can accurately determine protein levels in seeds. Correlation coefficients have been calculated between Kjeldahl and NIRS protein levels for corn, wheat, oats and barley and they are in general better than 0.95. Very little work has been done with NIRS and legume seeds or with NIRS and amino acids from cereals or legumes. Protein amino acids have only been determined on wheat; lysine alone was analysed in oats and barley (Rubenthaler and Bruisma, 1978).

The present work compares the efficiency of the NIRS technique with the Kjeldahl method and the Amino Acid Autoanalyser. We analysed 105 varieties of *Vicia faba* for protein content and 68 of these samples for amino acid content. We also studied 146 varieties of *Lupinus* seeds for protein content.

MATERIALS AND METHODS

Plant material

The *Vicia faba* collection of the Plant Breeding Station, CRA, Gembloux, containing 105 accessions (sub-species minor, equina, major, 1973 - 1980), was investigated for protein content. Sixty eight samples were analysed for 9 amino acids. *Lupinus* seeds were obtained from different countries throughout the world. The 146 varieties corresponded to different species and all samples were phenotypically different.

As the NIRS method does not measure directly the protein or amino acid contents, it must be calibrated using standards. For the development of standard curves, 65 samples of *V. faba* and 74 samples of *Lupinus* seeds were selected to represent virtually the entire range of Kjeldahl protein estimation. Forty samples of *V. faba* seeds were chosen to calibrate the NIRS instrument, based on the results obtained from the LKB 4400 Amino Acid Autoanalyser.

Methods of analysis

Kjeldahl

The equipment used in this study was the Digestion System DS 20 and the Kjeltec Distilling Unit II (Tecator Inc., Boulder, Colorado). Sample size was 0.3 - 1 g and protein content was calculated by multiplying the nitrogen value by 6.25.

Amino Acid Autoanalyser

Amino acids were analysed, after 24 h hydrolysis in 6 M HCl at 105°, by chromatography using the LKB 4400 Autoanalyser (LKB Biochrom Ltd., Cambridge, England) with sodium citrate buffers. This new generation of instruments used a narrow bore column (4 mm id x 27 cm) packed with 8 μm cation exchange resin. Sixteen analyses were performed every 24 h.

Near infra-red reflectance spectroscopy

A Neotec 'Research Composition Analyser' model RCA 6350 was used for all NIRS analysis. The basic operating principles and procedures have been described by Trevis (1974) and the detailed instructions are provided in the operator's manuals (Neotec Instruments, Inc. Silver Spring, MD, USA).

As only 1 - 2 g of seeds were available in many cases, the RCA sample cup was modified by the introduction of a ceramic disc between the sample and the plastic back plate of the cell (Williams et al., 1978). In this manner, samples as small as 1 g were successfully analysed.

RESULTS AND DISCUSSION

Table 1 presents the means and range of protein and amino acid contents of the standard (used to develop the calibration set) and independent (used to develop the prediction set) samples, obtained through the Kjeldahl method and the ion exchange chromatography technique. Fifty percent of *V. faba* samples have a protein content ranging from 30 - 32% db, while only one sample has a protein value of 22%. For *Lupinus*, 46 varieties have a protein content higher than 40% with 5 samples above 45% db. Seven samples have a protein percentage lower than 25%.

TABLE 1

MEANS AND RANGES OF PROTEIN AND AMINO ACID CONTENT OF STANDARD AND INDEPENDENT SAMPLES OBTAINED BY KJELDAHL

Samples	Constituents	Calibration set n	Calibration set Mean	Calibration set Range	Prediction set n	Prediction set Mean	Prediction set Range
Vicia faba seeds	Protein	65	30.9	22.4 - 36.0	40	31.4	28.9 - 34.4
	Amino acids	40			28		
	Lysine		1.91	1.59 2.25		1.90	1.71 2.12
	Arginine		2.91	1.65 4.08		2.83	2.35 3.35
	Threonine		1.10	0.83 1.34		1.13	0.97 1.29
	Valine		1.31	1.00 1.61		1.30	1.11 1.54
	Leucine		2.20	1.71 2.77		2.22	1.92 2.54
	Isoleucine		1.13	0.90 1.44		1.15	0.98 1.37
	Histidine		0.79	0.56 1.03		0.82	0.71 0.95
	Phenylalanine		1.29	0.99 1.55		1.29	1.11 1.52
	Glutamic acid		4.60	3.52 5.60		4.66	4.26 5.14
Lupinus seeds	Protein	74	33.9	19.5 - 44.0	72	37.2	28.0 - 43.0

n = number of samples; proteins are in g/100 g db for *Vicia faba* and g/100 g (92% db) for *Lupinus*, Amino acids are in g/100 g seeds db.

Figure 1 presents graphically the relationship between the NIRS and the Kjeldahl values for calibration samples of *V. faba* and *Lupinus* protein and for predicted sample values of *Lupinus* protein. We also consider calibration curves (Figure 2) between NIRS and LKB Autoanalyser values for 3 amino acids: threonine, lysine and leucine.

A statistical evaluation of the performances of the NIRS method for the standard and independent samples is shown in Table 2.

The multicorrelation coefficient (MC) and the residual standard deviation, expressed either as absolute value (SE) or as the percentage of the mean ($CV_{y.x}$), make it possible to measure the accuracy of the NIRS method in relation to the reference Kjeldahl or ion exchange chromatography methods. The observed values indicate in general that the determinations made on the calibration set are more accurate than those made on the prediction set. Exceptions to this may be due to the

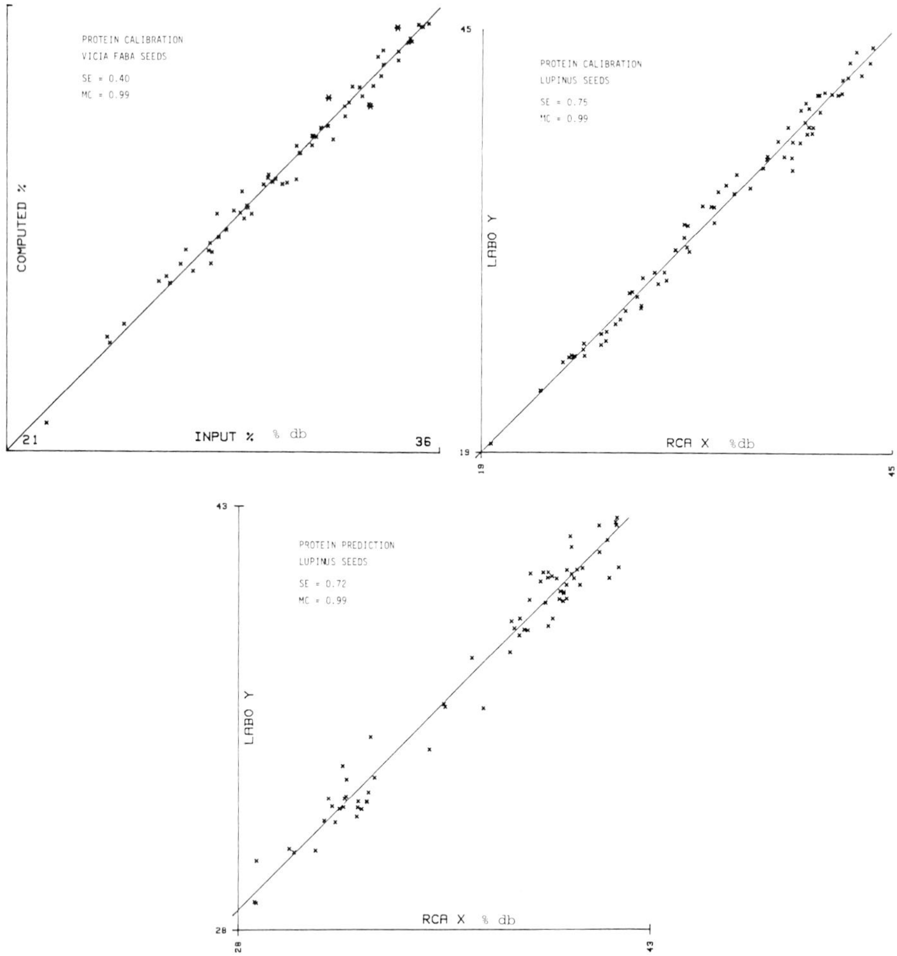

Fig. 1. Relation between NIRS and Kjeldahl methods for 40 and 74 standard samples of *Vicia faba* and *Lupinus* seeds, respectively, and for 72 independent samples of *Lupinus* seeds.

fact that, for lupin, it was very difficult to prepare the sample cups (quantity of seeds lower than 1g).

The results of this study indicate that the near infrared technique is satisfactory for estimating the protein and amino acid content of legume seeds in a breeding programme. The NIRS method is simple, rapid, relatively inexpensive and requires a minimum number of steps.

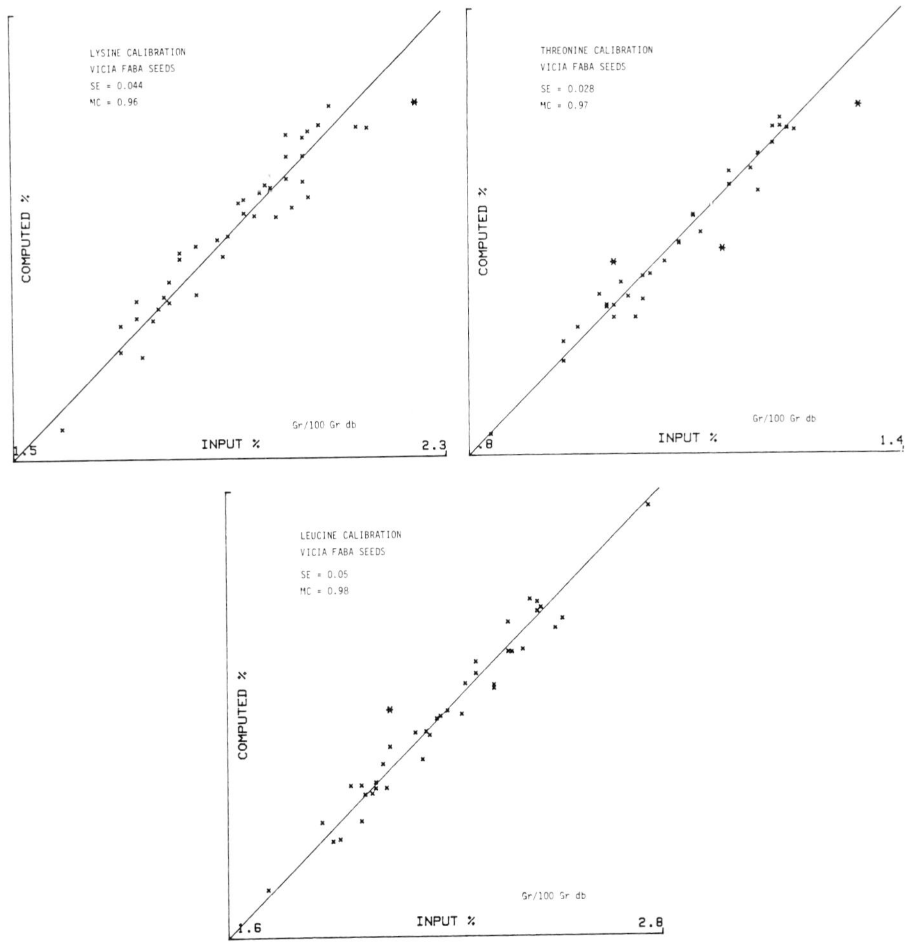

Fig. 2. Relation between NIRS and LKB Autoanalyser methods for 40 standard samples of *Vicia faba* seeds.

ACKNOWLEDGEMENTS

The authors wish to express their gratitude to the NI Vavilov All Union Institute of Plant Industry, Leningrad, USSR and to The Zentralinstitut für Genetik und Kulturpflanzenforschung der Akademie der Wissenschaften der DDR, Gatersleben for samples of *Lupinus* seeds. They are grateful to Dr. Derenne, Station d'Amélioration des Plantes, CRA Gembloux for the gift of *Vicia faba* seeds. Thanks are due to Mrs. L. Francois, Mrs. Mouteau and M.R. Loriau for their competent technical assistance.

TABLE 2

RELATIONSHIP BETWEEN NIRS AND KJELDAHL DETERMINATION OF PROTEIN AND BETWEEN NIRS AND ION EXCHANGE CHROMATOGRAPHY (AUTOANALYSER) DETERMINATION OF AMINO ACIDS

Samples	Constituents	Math. forms	Calibration set			Prediction set		
			MC	SE	CV_{yx}	MC	SE	CV_{yx}
Vicia faba seeds	Protein	D_2OD 10/20	0.99	0.48	1.6	0.91	0.68	2.2
	Amino acids	D_2OD 8/20						
	Lysine		0.96	0.04	2.1	0.90	0.08	2.2
	Arginine		0.98	0.12	4.1	0.84	0.14	4.9
	Threonine		0.97	0.03	2.7	0.91	0.03	2.7
	Valine		0.96	0.04	3.1	0.85	0.06	4.6
	Leucine		0.98	0.05	2.3	0.92	0.06	2.7
	Isoleucine		0.96	0.04	3.5	0.83	0.06	5.2
	Histidine		0.96	0.03	3.8	0.91	0.03	3.7
	Phenylalanine		0.95	0.05	3.9	0.90	0.04	3.1
	Glutamic acid		0.93	0.17	3.7	0.84	0.22	4.7
Lupinus seeds	Protein	D_2OD 8/20	0.99	0.75	2.2	0.99	0.72	1.9

C = mult' rrelation coefficient; SE = residual standard deviation in relati ll the observations; CV_{yx} = residual coefficient of variation in re. to the general mean.

REFERENCES

Biston, R. and Clamot, G., 1981. Use of near-infrared reflectance spectroscopy and dye-binding techniques for estimating protein in oat groats. Cereal Chem, in press.

Graham, T.F., 1975. Analysis of samples with the Infra Analyser unit. TIS News Infra Letter.

Hunt, W.H., Fulk, D.W., Elder, B. and Norris, K., 1977. Collaborative study of infrared reflectance devices for determination of protein in hard red winter wheat and for protein and oil in soybeans. Cereal Food World 22, 534.

Hymowitz, T., Dudley, J.W., Collins, F.I. and Brown, C.M., 1974. Estimations of protein and oil concentration in corn, soybean and oat seed by near-infrared light reflectance. Crop Sci. 14, 713.

Miller, B.S., Pomeranz, Y., Thompson, W.O., Nolan, T.W., Hugues, J.N., Davis, G., Jackson, N.G. and Fulk, D.W., 1978. Interlaboratory and intralaboratory reproducibility of protein determination in hard red winter wheat by Kjeldahl and near-infrared procedures. Cereal Food World 23, 198.

Norris, K.H., Barnes, R.F., Moore, J.E. and Shenk, J.S., 1976. Predicting forage quality by infrared reflectance spectroscopy. J. Animal Sci. 43, 889.

Pomeranz, Y. and Moore, R.B., 1975. Reliability of several methods for protein determination in wheat. The Bakers Digest 44.

Rinne, R.W., Gibbons, S., Bradley, J., Seif, R. and Brim, C.A., 1975. Soybean protein and oil percentages determined by infrared analysis. USDA Publication, ARS-NC, 26.

Rubenthaler, G.L. and Bruisma, B.L., 1978. Lysine estimation in cereals by near-infrared reflectance. Crop Science 18, 1039.

Shenk, J.S., Westerhaus, M.O. and Hoover, M.R., 1979. Analysis of forage by infrared reflectance. Dairy Science 62, 807.

Watson, C.A., Carville, D., Dikeman, E., Daigger, G. and Booth, G.D., 1976. Evaluation of two infrared instruments for determining protein content of hard red winter wheat. Cereal Chem. 53, 214.

Williams, P.C., 1975. Application of near-infrared to analysis of cereal grains and oilseeds. Cereal Chem. 52, 561.

Williams, P.C., Stevenson, S.G., Starkey, P.M. and Hawtin, G.C., 1978. The application of near infrared spectroscopy to protein testing in pulse breeding programmes. J. Sci. Food Agric. 29, 285.

DISCUSSION

C. Calet *(France)* A correlation has been established by a colleague (Mossé) between the amount of protein and the various amino acids, firstly in cereal and subsequently for peas and *Vicia faba*. The method is simple, what is your opinion of it?

G. Dardenne *(Belgium)* We have done similar work with a number of varieties of *Vicia faba* with similar results to those of Mossé.

R. Casey *(UK)* I understood you to say that the technique could be non-destructive. By that do you mean that you take a portion of the seed and grind it?

G. Dardenne The technique is non-destructive because it is possible to analyse the seed directly and then sow it. Within the next year or two it will be possible to analyse seeds individually before they are sown, for starch, protein, amino acid and so on, all at the same time.

Of course, the research instrument is expensive at $120 000, but it is very sophisticated. The routine instrument costs $60 000 but it is very important to realise that with a centralised research instrument results can be transmitted automatically to regional routine instruments.

J. Picard *(France)* In relation to Professor Calet's comments on estimation of amino acid content from total protein, it is important to recognise that the method is not suited to use on variants where amino acid ratios differ.

STORAGE PROTEINS OF ROUND AND WRINKLED PEAS *IN VIVO* AND *IN VITRO*

D.R. Davies and C. Domoney
John Innes Institute, Colney Lane, Norwich NR4 7UH, UK.

ABSTRACT

Round and wrinkled peas are shown to differ in the proportion of storage proteins which they accumulate in their seed, wrinkled forms on average having less legumin than round forms. The wrinkled types are homozygous recessive for the r_a allele, and the known effect of this mutation on carbohydrate synthesis together with the possibility that it also affects the accumulation of one particular storage protein suggests an analogy with some of the maize and barley mutants which affect both seed-protein and carbohydrate synthesis. When cultured under defined conditions in vitro, the round and wrinkled forms do not differ in the proportion of legumin in their protein; further studies of the influence of environmental conditions on seed protein content and composition have included analyses of the effect of sulphur deprivation in the culture medium on round seed and of the addition of abscisic acid to the medium used for culturing wrinkled seed. No effect of sulphur deprivation was observed, but abscisic acid enhanced the protein content of the seed.

INTRODUCTION

An important objective in the production of peas is to increase both the amount of protein and the sulphur amino acid content of the seed. There are two ways in which these objectives can be attained - by changing the genotype of the plants and by changing the environment in which they are grown. There is now much evidence to suggest that considerable variation exists in the structural genes for particular storage proteins (see Casey, 1982); manipulation of this variation could well result in a significant improvement in protein quality, since the various polymorphic forms differ in sulphur amino acid content (Casey and Short, 1981). It is known also that genetic variation exists for the amount and proportions of the main globulins - legumin, vicilin and convicilin - (Casey et al., 1982; Davies, 1976; Guldager, 1978; Millerd et al., 1978; Thomson et al., 1978, 1979; Gatehouse et al., 1980) and this could also be manipulated for practical purposes. Recently we have shown that round and wrinkled seeded peas can differ markedly in the proportion of legumin which they synthesise (Davies, 1980). The round and wrinkled forms used were homozygous for the R_a and r_a

alleles respectively; at present we cannot show whether it is the r_a locus itself or a closely linked one which influences legumin accumulation, but it is worth noting that the main effect of the r_a locus is on carbohydrate synthesis in the seed. As such, this mutation may well be comparable to many in maize, barley and sorghum which not only influence the synthesis of carbohydrates but also the synthesis of some of the storage proteins - usually the prolamins (Nelson, 1981).

The second way in which we can alter the amount and kind of protein in pea seeds is by changing the environmental conditions. The importance of the latter is well recognised; for example the α-amino nitrogen and available methionine in the seed of any one of six pea cultivars grown in replicated trials in four different regions of eastern England, and sown at two different times, was found to differ markedly (Matthews et al., 1975). The variety Greenshaft grown in different environments had an α-amino nitrogen content of from 1.91 to 3.44% of the dry meal; significant variation was also recognised between plants of a genotype within a single plot.

Slinkard (1981) came to an essentially similar conclusion on the role of environmental conditions from his studies of peas grown in different conditions within Canada. The variation observed in such studies has not been related to any particular component of the environment, but its magnitude is such as to constitute an impediment to rapid progress in breeding by conventional methods for improved protein and sulphur amino acid content. The only environmental factors known to influence the protein composition of pea seed are the levels of sulphur, potassium and phosphorus in the nutrient supply; a deficiency of sulphur decreases the proportion of legumin, while deficiencies of potassium or phosphorus enhance the proportion of that same protein (Randall et al., 1979).

We have developed a technique for culturing the pea embryo and its cotyledons under defined conditions (Stafford and Davies, 1979) and have used this as an experimental system in which to study the effects of different environmental conditions and components on the content and quality of protein synthesised in the seed. In the present report we first compare the synthesis of legumin in round and wrinkled seed cultured *in vitro*, then examine the effect on legumin synthesis in a round variety of depriving the nutrient medium of sulphur. Finally we have studied the effect, on a wrinkled seeded variety, of incorporating the plant growth hormone abscisic acid (ABA) in the medium. Sussex and Dale (1979) and Sussex et al. (1980) have shown that embryos of *Phaseolus vulgaris* cultured in the presence of ABA synthesise more of the Glycoprotein II storage protein.

LEGUMIN CONTENT OF ROUND AND WRINKLED PEAS *IN VIVO*

Round and wrinkled phenotypes are distinguishable on the appearance of the seed, in terms of starch and sugar levels and of starch composition (Kooistra, 1962; Richter, 1975). Two unlinked loci are involved in the control of these factors - r_a and r_b. $R_a\ R_b$ genotypes are round whereas $R_a\ r_b$, $r_a\ R_b$ and $r_a\ r_b$ are wrinkled. Some of the other attributes can be summarised as follows in an adaptation of the data of Kooistra (1962).

Genotype	Phenotype	Starch grains	Starch level	Sugar level	Amylose in starch
$R_a\ R_b$	Round	Simple	High	Low	Medium
$R_a\ r_b$	Wrinkled	Simple	Medium	Medium	Low
$r_a\ R_b$	Wrinkled	Compound	Medium	Medium	High
$r_a\ r_b$	Wrinkled	Compound	Medium	Medium	Medium

The two genotypes we used were $R_aR_a\ R_bR_b$ and $r_ar_a\ R_bR_b$. The amount of α-amino nitrogen was measured and the proportion of legumin versus vicilins plus convicilin assayed by cellulose acetate electrophoresis (Davies, 1976). A survey of 171 genotypes from the John Innes pea germplasm collection showed that the percentage of legumin in the globulin of these seeds could vary markedly, but the mean value for the wrinkled genotypes (18.6%) was markedly lower than that of the round (37.7%). An analysis of two F2 populations derived from crossing round and wrinkled peas gave a similar result - the wrinkled always had a lower mean value than the round seed (Davies, 1980). A third comparison made use of a set of seven near-isogenic lines for round/wrinkled seed, generated by Dr. A. Slinkard of the

University of Saskatchewan, Canada. The mean dry weight of the seed, its percentage of α-amino nitrogen, and percentage of legumin in the globulin are given in Table 1. There is no consistent difference in the mean dry weight of round and wrinkled seed; the percentage of α-amino nitrogen is consistently lower in round forms, confirming the results of Shia and Slinkard (1977) and possibly reflecting the higher amount of starch in round seed.

TABLE 1

MEAN DRY WEIGHTS, PERCENTAGE α-AMINO N CONTENTS AND PERCENTAGE LEGUMIN IN TOTAL PROTEIN, OF ROUND (R) AND WRINKLED (r) FORMS OF NEAR-ISOGENIC LINES (Davies, 1980)

	Mean weight/seed (mg)	% α-amino N in meal	% legumin in total protein
9R	244 + 9.2	2.027	24.4
9r	299 + 5.9	2.486	10.7
10R	255 + 15.4	2.011	28.7
10r	198 + 9.1	2.442	6.1
11R	250 + 7.8	1.899	29.3
11r	212 + 7.6	2.424	15.2
25R	170 + 7.9	2.334	33.3
25r	183 + 5.9	2.982	30.3
26R	166 + 6.3	2.468	39.0
26r	171 + 4.3	2.519	18.9
43R	246 + 12.4	2.327	27.9
43r	211 + 9.1	2.523	8.0
44R	239 + 7.8	2.208	30.7
44r	218 + 5.2	2.471	9.8

As round forms have higher yields than wrinkled (Shia and Slinkard, 1977) then the total yield of protein per unit area is likely to be similar in both. The proportion of legumin, calculated as a percentage of the total protein, or of the globulin was consistently higher in round seed.

These results imply that the r_a or a closely linked locus influences the pattern of storage protein synthesis in the seed.

The magnitude of the difference between the two phenotypes depended on the background genotype but in some instances wrinkled seed had less than one third as much legumin as their round isogenic equivalent; this was not always accompanied by a change in the absolute amount of α-amino nitrogen content. The suggestion from these data of an interaction between control of carbohydrate and of protein synthesis in these two phenotypes, indicates a broad similarity to many of the cereal mutants (Nelson, 1981). In many of those mutants also, as starch synthesis is reduced so is the synthesis of one class of proteins - the prolamins. The nature of the primary lesion in the wrinkled mutant is currently being studied.

In practical terms, these results favour the round forms of peas as the phenotype of choice; as legumin contains more S amino acids than vicilins, the other main group of storage proteins (Derbyshire et al., 1976), increasing the former is likely to be beneficial. Round seeded forms have a higher yield and are easier to handle in milling.

COMPARISON OF LEGUMIN CONTENT OF ROUND AND WRINKLED PEAS CULTURED *IN VITRO* WITH THOSE GROWN *IN VIVO*

The r_a locus influences storage product biosynthesis through an effect on the embryo and its cotyledons, rather than by changing the metabolism of the plant on which the seeds are borne. This is shown by the fact that the two phenotypes can be seen within a single pod, when the plant is heterozygous for alleles at the r_a locus. Tsai et al. (1978) have suggested that the effects on starch and protein synthesis seen in one of the maize mutants, *bt2o2*, is due to the high sucrose concentration of its cells. We have therefore examined whether the synthesis of legumin and of protein would be comparable in round and wrinkled embryos and cotyledons, if they were cultured *in vitro* in a medium with a high osmoticum. Two of the isogenic lines for round (*10R*) and wrinkled (*10r*) were cultured in a standard embryo culture medium with 18% sucrose (Stafford and Davies, 1979). Amounts of protein (estimated by a dye binding method; Bradford, 1976), of legumin (estimated by rocket immunoelectrophoresis (Weeke, 1973) using a monospecific antibody raised against legumin) and fresh and dry weights were determined. The fresh weight of the embryos cultured ranged from 15 to 200 mg; legumin synthesis can be first detected when the embryos are approximately 2 mg (Domoney et al., 1981). The amounts of legumin and of total protein in peas of a range of developmental stages from the two genotypes *in vivo* are shown in Figures 1a and 2a. There is a similar amount of protein in both, but the amount of legumin is higher in the round peas, particularly in those of higher fresh weight.

After embryos of a range of initial fresh weights had been grown *in vitro* for a seven day period at 20°C, there was no detectable difference between round and wrinkled forms either in amount of protein or of legumin (Figures 1b and 2b). There was a considerable amount of scatter in the data and we will discuss its basis later. If we plot the amount of legumin against protein, then it becomes apparent that the round seeds have more legumin than the wrinkled *in vivo* (Figure 3a) but the two forms are similar *in vitro* (Figure 3b). The proportion of dry weight in the total fresh weight is similar in round and wrinkled, and in those grown *in vivo* and those cultured *in vitro* (Figure 4).

We referred earlier to the scatter observed in the data which related amount of protein or of legumin to the final fresh weight of the embryos cultured *in vitro* (Figures 1b and 2b); it is worth noting that this scatter does not occur in our plot of legumin vs. protein or of dry vs. fresh weight of embryos *in vitro*. If, instead of plotting the amount of protein or of legumin against the final fresh weight attained *in vitro*, we plot these two parameters against the initial fresh weight of the embryo at the beginning of the culture period, then a much clearer relationship is seen (Figure 5). (Note that different scales are used in Figures 1 and 2 vs. 5). This means that an embryo of a given initial size has the potential to synthesise a given amount of total protein or of legumin; the scatter that is seen in Figures 1b and 2b therefore indicates that some of the embryos have not accumulated the expected quantity of starch, which is the other main component contributing to the fresh weight. We have noted this phenomenon in other experiments;

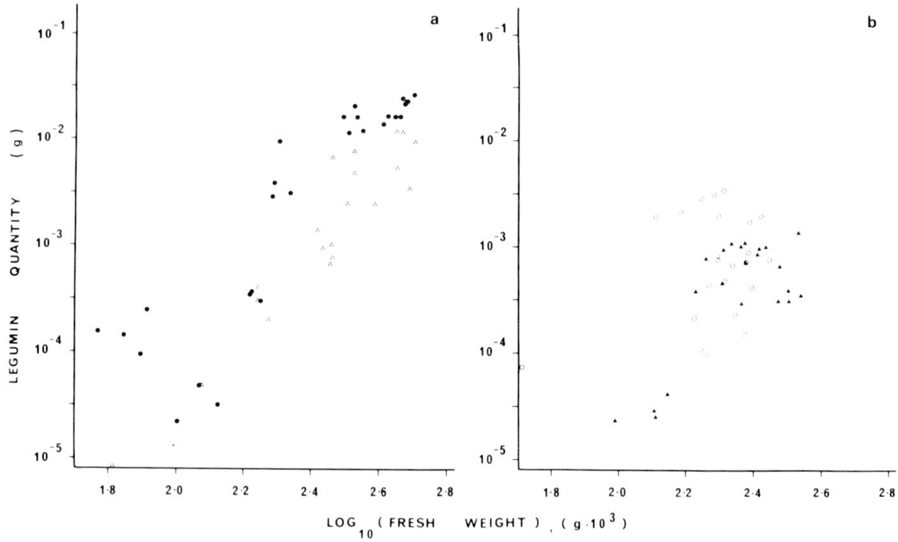

Fig. 1. Legumin amounts in *10R* (●, o) and *10r* (Δ, ▲) embryos developing *in vivo* (a) and *in vitro* (b), in logarithmic relation to total embryo fresh weight (embryo fresh weight after culture in (b)).

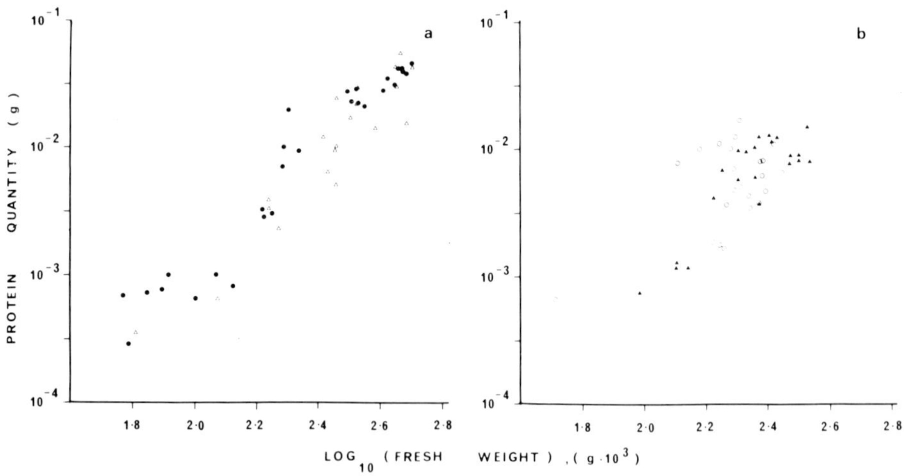

Fig. 2. Total protein amounts in *10R* (●, o) and *10r* (Δ, ▲) embryos developing *in vivo* (a) and *in vitro* (b), in logarithmic relation to total embryo fresh weight (embryo fresh weight after culture in (b)).

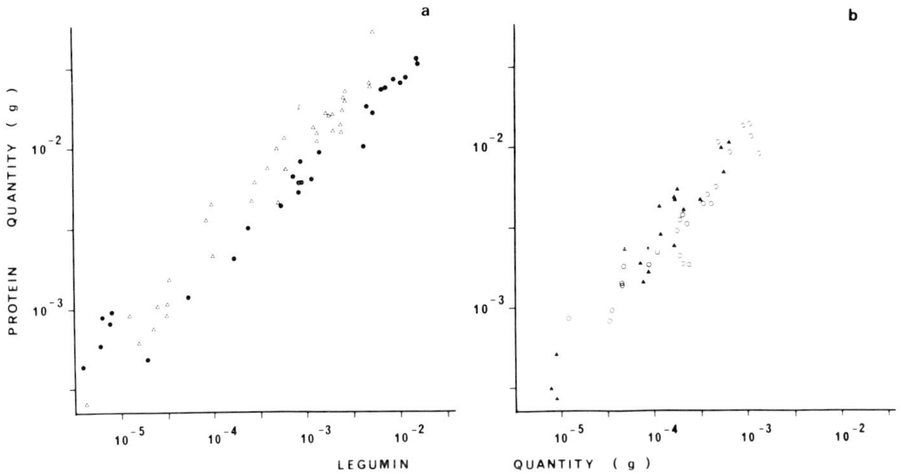

Fig. 3. The amount of total protein present in one cotyledon from each *10R* (●, o) and *10r* (Δ, ▲) embryo grown *in vivo* (a) and *in vitro* (b), in logarithmic relation to the amount of legumin present in the same cotyledon.

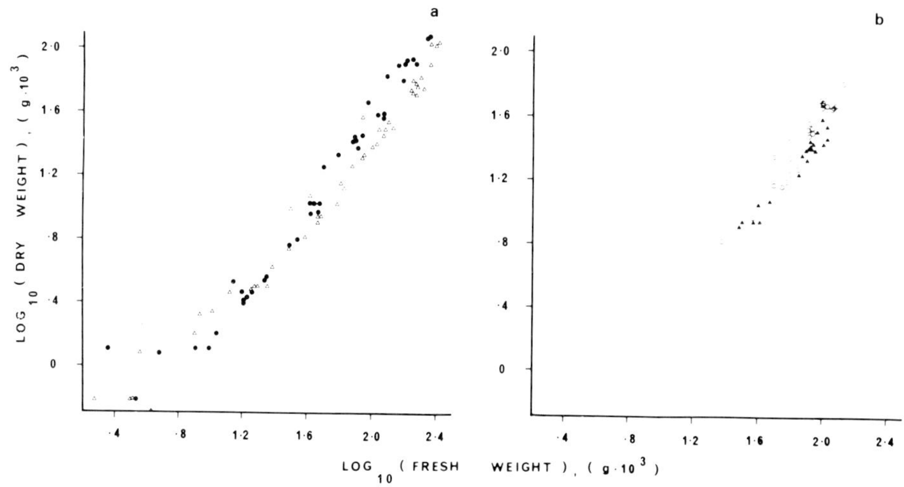

Fig. 4. The dry weight of one cotyledon from each *10R* (●, o) and *10r* (Δ, ▲) embryo grown *in vivo* (a) and *in vitro* (b), in logarithmic relation to the fresh weight of the same cotyledon. Cotyledons were dried at 80°C for 48 h.

embryos which have attained less weight than expected can still have the expected amount of protein, i.e., starch synthesis is more easily perturbed.

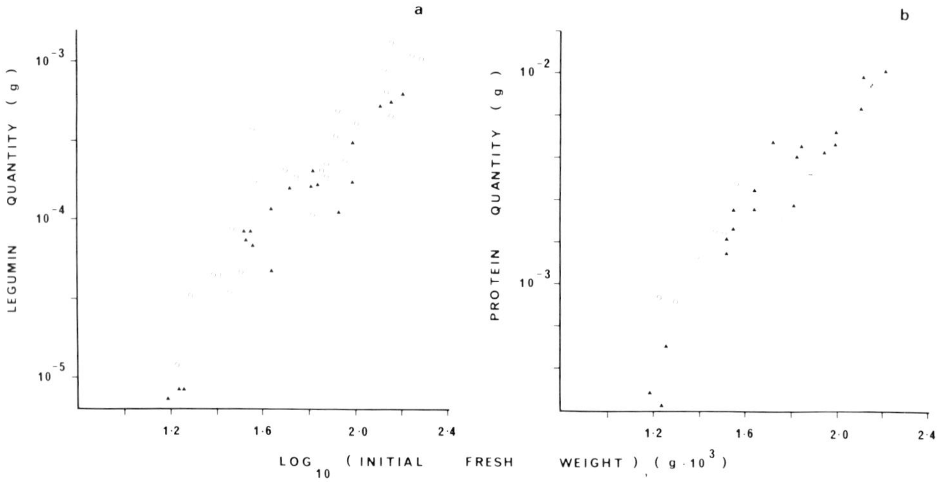

Fig. 5. The amounts of legumin (a) and total protein (b) present in one cotyledon from each *10R* (o) and *10r* (▲) embryo after the culture period, in logarithmic relation to initial embryo fresh weight.

In summary, round and wrinkled embryos and cotyledons had similar proportions of legumin when cultured in defined conditions at high sucrose concentrations; these conditions allowed good growth, and comparable proportions of dry weight accumulation as *in vivo*.

THE EFFECT OF SULPHUR DEPRIVATION ON ROUND PEAS CULTURED *IN VITRO*

A round seeded variety, JI 1068, was used for this experiment as it has a high legumin content, and thus any effect of sulphur deprivation in lowering the legumin amount would be more easily detectable.

The embryos and cotyledons of JI 1068 were cultured in one of three levels of sulphur; the control had a 1.73 mM inorganic sulphate (50 ppm) and the two reduced levels had 0.23 mM (7 ppm) and 0.13 mM (4 ppm) respectively. Embryos of a range of developmental stages from those of less than 5 mg to those of nearly 200 mg initial fresh weight were cultured for 7 days at 20°C. The growth of the control embryos was consistently better than those at the two low sulphur levels, in terms of both fresh and dry weights (Figure 6). The amounts of protein and of legumin per unit dry weight were, however, higher when the embryos were deprived of sulphur; in other words, the smaller embryos produced under the latter conditions still had the same amount of

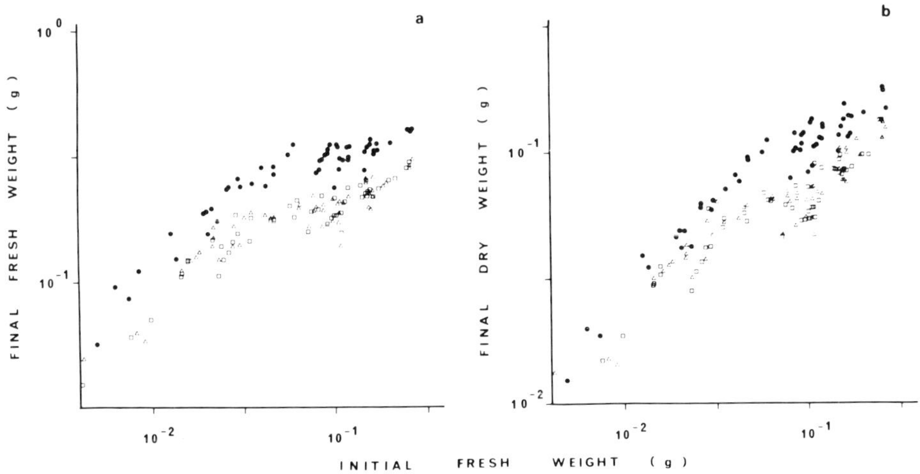

Fig. 6. The growth of JI 1068 embryos *in vitro*, in medium containing 1.73 mM (●), 0.23 mM (□) and 0.13 mM (Δ) sulphur. Embryos were harvested from seeds at a range of developmental stages and cultured for 7 days at 20°C with one embryo and 5 ml medium per petri dish. Figure 6a shows the final fresh weight of the embryos, while Figure 6b shows the final dry weight of one cotyledon plus the equivalent dry weight of the second cotyledon from each embryo, in logarithmic relation to initial embryo fresh weight. One cotyledon from each embryo was dried at 80°C for 48 h.

protein or of legumin as the larger embryos in the control medium. The proportions of legumin in the protein remained unchanged at all the sulphur levels (Figure 7).

These results differ from those of Randall et al. (1979); they grew their control pea plants in 40 ppm sulphur and for the sulphur deprivation, at levels of 4, 0.2 and 0.02 ppm. The effect of reduced sulphur levels was seen at 4 ppm and less. The basis of the difference in the two experiments could be due to: (a) the level of sulphur deprivation in the present experiments being insufficient to influence legumin synthesis although growth was affected; (b) the embryos having had a sufficient pool of sulphur at the time of excision from the seed; (c) the effect of sulphur deprivation possibly becoming apparent only in the final stages of embryo growth, i.e., in larger embryos than those cultured, when the greatest absolute rate of legumin synthesis occurs.

Again it is worth noting that with sulphur deprivation, the levels of protein synthesised were comparable to those in

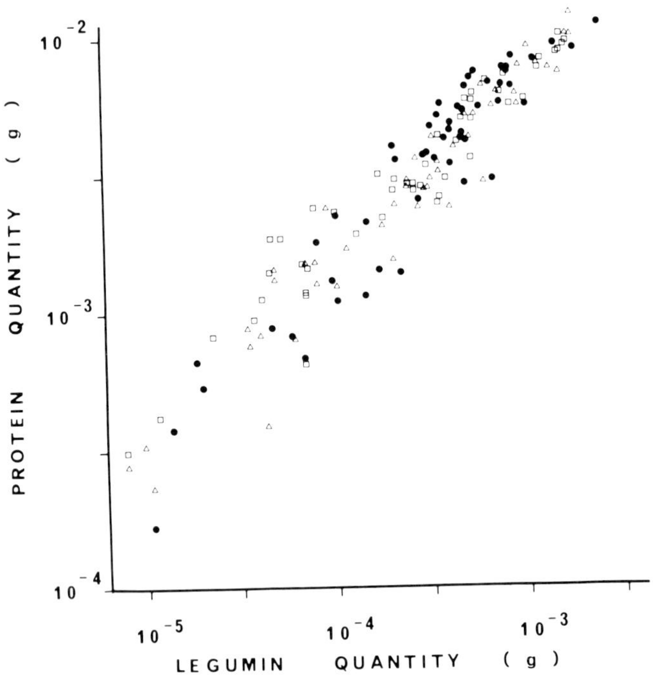

Fig. 7. The amount of total protein present in one cotyledon from each JI 1068 embryo grown in CM (●), LS1 (□) and LS2 (△) (complete medium and two media of lower sulphur contents, respectively), in logarithmic relation to the amount of legumin present in the same cotyledon.

the control, and the lower fresh and dry weights were probably indicative of a reduced starch synthesis. This supports our conclusion from the previous experiment that starch synthesis is more easily perturbed by stress.

THE EFFECT OF ABSCISIC ACID ON WRINKLED PEAS CULTURED *IN VITRO*

Sussex and Dale (1979) cultured embryos of *Phaseolus vulgaris* in a nutrient medium containing 3% sucrose and in the presence of concentrations of ABA from 10^{-8} to 5×10^{-4}M. Concentrations of 10^{-6} to 10^{-4}M enhanced the amount of glycoprotein II per embryo by a factor of up to 2.5. Total protein amounts were not quoted; the embryos, which had an initial fresh weight of approximately 1 mg, grew to 3 mg in the absence of ABA in six days, and to 2 and 1 mg respectively in 10^{-6} and 10^{-4}M ABA.

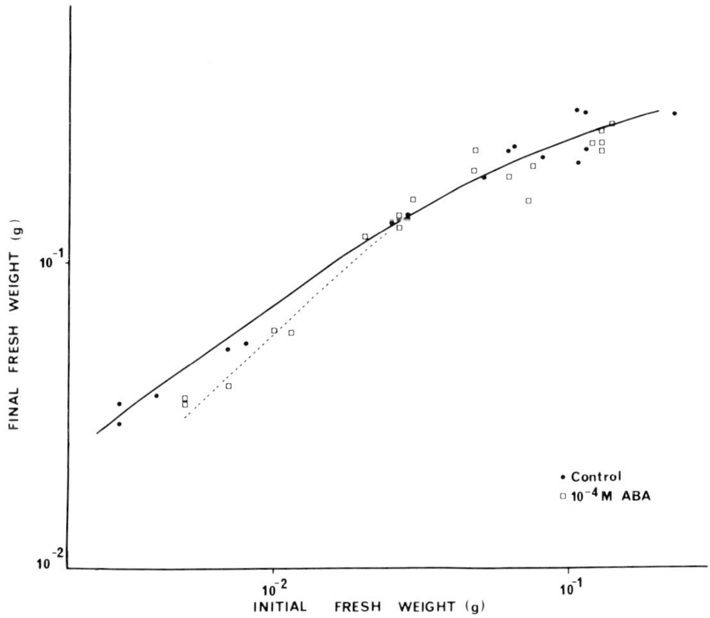

Fig. 8. The fresh weight attained by embryos of DSP *in vitro*, in complete medium and in complete medium supplemented with with 10^{-4} M ABA. The culture conditions were as described for Figure 6.

In our experiments, embryos of Dark Skinned Perfection having an initial fresh weight from 3 to 230 mg were cultured for seven days at 20°C and with 18% sucrose in the medium to inhibit precocious germination (Stafford and Davies, 1979); the final fresh weight ranged from 30 to 300 mg (Figure 8). Concentrations of 10^{-6} and 10^{-4}M filter sterilised ABA were used, but as the results were similar, only those from the higher concentration are quoted. The ABA slightly depressed the growth of smaller embryos, but the amounts of total protein and of legumin were higher in embryos cultured in the presence of ABA (Figure 9). The proportion of legumin in the total protein was the same irrespective of whether ABA was present or absent. Thus, in contrast to the results of Sussex and Dale (1979), we could find no evidence for a selective overproduction of one storage protein but rather for an increase in total protein. We can only speculate as to the basis of the increase in protein; by using a very high osmoticum in the medium (18% sucrose) we eliminate any visible sign of germination, but it is possible that some

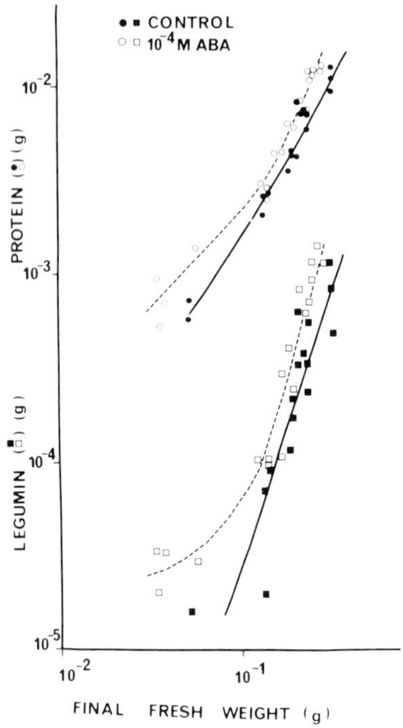

Fig. 9. The amounts of legumin and total protein in embryos of DSP grown in complete medium and in complete medium supplemented with 10^{-4} M ABA, in logarithmic relation to total embryo fresh weight after the culture period.

hydrolysis of the storage proteins is still occurring during culture. ABA may curtail this incipient germination. The endogenous levels of ABA in the seed of the variety used in this experiment have been measured by Browning (1980); over the range of developmental stages cultured, the ABA per unit weight of embryo continues to increase with increasing embryo weight. There is therefore no hint in these measurements of the basis of the different responses observed in the presence of ABA (Figure 8).

CONCLUSION

The round and wrinkled phenotypes of peas may provide the first example in a legume of a locus which acts as a regulator of the synthesis of one particular storage protein. Elucidating the molecular basis of the action of r_a or a closely linked locus will be a matter of considerable interest, in indicating how we can make progress in altering the protein as well as the carbohydrate composition of the seed. Clearly there is in the cereals a distinct likelihood that these two biosynthetic processes are interlinked. The evidence obtained from culturing pea embryos under defined conditions shows that some of these processes can be manipulated by changing environmental conditions; detailed investigations of this kind may indicate why the protein contents of crops of legume seed grown in different environments can differ so markedly and in such an apparently unpredictable manner.

REFERENCES

Bradford, M.M., 1976. A rapid and sensitive method for the quantitation of microgram quantities of protein utilizing the principle of protein-dye binding. Anal. Biochem. 72, 248-254.

Browning, G., 1980. Endogenous cis-trans abscisic acid and pea seed development: evidence for a role in seed growth from changes induced by temperature. J. Exp. Bot. 31, 185-197.

Casey, R., 1982. Genetics of pea seed storage proteins. Qualitas Plantarum, in press.

Casey, R. and Short, M., 1981. Variation in amino acid composition of legumin from *Pisum*. Phytochemistry 20, 21-23.

Casey, R., Sharman, J.E., Wright, D.J., Bacon, J.R. and Guldager, P., 1982. Quantitative variability in *Pisum* seed globulins; its assessment and significance. Qualitas Plantarum, in press.

Davies, D.R., 1976. Variation in the storage proteins of peas in relation to sulphur amino-acid content. Euphytica 25, 717-724.

Davies, D.R., 1980. The r_a locus and legumin synthesis in *Pisum sativum*. Biochem. Genet. 18, 1207-1219.

Derbyshire, E., Wright, D.J. and Boulter, D., 1976. Legumin and vicilin, storage proteins of legume seeds. Phytochemistry 15, 3-24.

Domoney, C., Davies, D.R. and Casey, R., 1980. The initiation of legumin synthesis in immature embryos of *Pisum sativum* L. grown *in vivo* and *in vitro*. Planta 149, 454-460.

Gatehouse, J.A., Croy, R.R.D. and Boulter, D., 1980. Isoelectric-focusing properties and carbohydrate content of pea (*Pisum sativum*) legumin. Biochem. J. 185, 497-503.

Guldager, P., 1978. Immunoelectrophoretic analysis of seed proteins from *Pisum sativum* L. Theor. Appl. Genet. 53, 241-250.

Kooistra, E., 1962. On the differences between smooth and three types of wrinkled peas. Euphytica 11, 357-373.

Matthews, P., Dow, P. and Rumary, C., 1975. Protein quantity and available methionine in *Pisum*. John Innes Annual Report, 26-31.

Millerd, A., Thomson, J.A. and Schroeder, H.E., 1978. Cotyledonary storage proteins in *Pisum sativum*. III. Patterns of accumulation during development. Aust. J. Plant Physiol. 5, 519-534.

Nelson, O.E., 1981. Genetic control of polysaccharide and storage protein synthesis in the endosperms of barley, maize and sorghum. Adv. in Cereal Sci. and Techn. III, 41-71.

Randall, P.J., Thomson, J.A. and Schroeder, H.E., 1979. Cotyledonary storage proteins in *Pisum sativum*. IV. Effects of sulfur, phosphorus, potassium and magnesium deficiencies. Aust. J. Plant Physiol. 6, 11-24.

Richter, E., 1975. Sugar, starch and amylose content of hybrids between smooth and wrinkled peas. Pisum News. 7, 51.
Shia, G. and Slinkard, A.E., 1977. Relationship of seed shape and cotyledon color to percent protein of peas. Crop Sci. 17, 183-184.
Slinkard, A.E., 1981. Production, utilization, and marketing of field peas. Pisum News Letter 13, 49.
Stafford, A. and Davies, D.R., 1979. The culture of immature pea embryos. Ann. Bot. 44, 315-321.
Sussex, I.M. and Dale, R.M.K., 1979. Hormonal control of storage protein synthesis in *Phaseolus vulgaris*; in the plant seed. Academic Press.
Sussex, I.M., Dale, R.M.K. and Crouch, M.L., 1980. Developmental regulation of storage protein synthesis in seeds; in Genome organization and expression in plants. ed. C.J. Leaver. Plenum Press. pp. 283-289.
Thomson, J.A., Schroeder, H.E. and Dudman, W.F., 1978. Cotyledonary storage proteins in *Pisum sativum*. I. Molecular heterogeneity. Aust. J. Plant Physiol. 5, 263-279.
Thomson, J.A., Millerd, A. and Schroeder, H.E., 1979. Genotype-dependent patterns of accumulation of seed storage proteins in *Pisum*; in Seed protein improvement in cereal and grain legumes. Vol. I. STI/PUB/496 IAEA, Vienna pp. 231-240.
Tsai, C.Y., Larkins, B.A. and Glover, D.V., 1978. Interaction of the Opaque-2 gene with starch-forming mutant genes on the synthesis of zein in maize endosperm. Biochem. Genet. 16, 883-896.
Weeke, B., 1973. Rocket immunoelectrophoresis Scand. J. Immunol. 2, suppl. 1, 37-46.

DISCUSSION

C. Calet *(France)* I am pleased to hear you stress the importance of starch not only for plant but also for animal nutrition. Perhaps more attention should also be given to the anti-nutritional factors and other elements which are critical in determining the inclusion rate of legumes in animal feeds. There is no problem in introducing peas at a level of about 20% in a pig diet for instance, but levels of 40% cause problems because of anti-nutritional factors. Is it better to select peas for high protein content and settle for a 20% inclusion rate in the diet, or to select for a lower level of protein but without anti-nutritional factors so that it can be included in the diet at a level of 40%?

D. Boulter *(UK)* It depends partly on which are the important anti-nutritional factors. An increase in seed protein content may allow a reduced inclusion rate in the diet, thus reducing the amount of anti-nutritional factor, but this would not always apply. For example, lectin or trypsin inhibitor levels tend to be positively correlated with protein content. The answer will depend on factors as yet unknown, that is, a clear demonstration of precisely what causes the problem when higher levels of peas or beans are fed in certain animal diets.

G. Kapsiotis *(Italy)* The selection of either round or wrinkled peas will depend on their final use. The anti-nutritional factors in peas, as Spadoni will show, are minimal. If the peas are for human consumption the anti-nutritional factors such as lectin are heat labile and so are eliminated.

Some years ago there was a PAG symposium on nutritional guidelines for plant breeders. If I remember correctly, it was suggested then that neither sulphur amino acids nor protein content should be considered, but only increased production.

V. Haahr *(Denmark)* In both maize and barley, yields are lower for smooth grain types compared with shrivelled grain types. Is this also the situation with the pea, is it necessary to breed for round grains to obtain the highest yields?

D.R. Davies *(UK)* There is good evidence that the yield of round peas is higher than that of wrinkled. But if protein yield is the objective, equivalent yields can be obtained from wrinkled peas and from round peas.

D. Boulter What do you think is the basis of the difference between the Australian results with whole peas and your *in vitro* results? The increase in quality recorded by the Australians is only found at very low sulphur levels, which would not normally be of interest in agricultural practice.

D.R. Davies Sulphur levels used were comparable apart from some of their lower values. We used 50 ppm as our control with a minimum of 4 ppm, whilst their lowest were 0.2 and 0.02 ppm. They obtained an effect at 4 ppm whereas we did not. Why the difference? One possibility is that at the time we excised the embryos, even though they were very small, they contained a sufficient amount of sulphur amino acids. The second point is that under these conditions we require for our purposes a lower level of sulphur deprivation. The third point is that the maximum difference in the amount of legumin will only be obtained by culturing the embryos almost to maturity, because only at this time is most of the legumin synthesised. In our case, they would have been 500 or 600 mg. We did not culture them to that size (500 - 600 mg) and so we may not have seen the effect on legumin synthesis. So, I would accept the Australian results implicitly; our results do not disprove them in any sense.

D. Boulter It is important to bear in mind this agricultural point; that really this result is of interest in the experimental situation, but it is not of significance agriculturally.

THE GENES FOR *PISUM* LEGUMIN

R. Casey

John Innes Institute, Colney Lane,
Norwich, NR4 7UH, UK.

ABSTRACT

Legumin, one of the major storage globulins of Pisum (pea) seeds, is an oligomeric protein composed of acidic and basic subunits, which form disulphide-linked acidic-based pairs in the native molecule.

Extensive genotypic variability has been observed for the acidic subunits, which are present as both major and minor species. Classical breeding studies have shown that codominance seems to be the rule in the inheritance of the major acidic subunits of legumin and that the observed variation in subunit structure is based on allelic alternatives at single loci. Biosynthetic studies have shown legumin to be synthesised as a precursor subunit of mol. wt. 60 x 10^3 and raise the possibility that the acidic (40 x 10^3 mol. wt.) and basic (20 x 10^3 mol. wt.) subunits of legumin may be closely linked within the genome.

Structural and genetic analyses of legumin subunits indicate the existence of a small gene family of homologous polypeptides which has probably arisen through gene duplication and subsequent diversification, while the examination of linkage relationships has shown the structural locus for the major acidic subunits to be sited on chromosome 7, close to the r and tl loci.

INTRODUCTION

The protein of pea (*Pisum*) seeds, in common with that of many legume grains, is limited in its nutritional suitability for man and other monogastric animals by its low content of the sulphur-containing amino acids methionine and cysteine, a deficiency which is due to the fact that the major seed proteins - the storage globulins - contain little methionine and cysteine. Mature pea seeds contain two major storage globulins, vicilin and legumin (Osborne and Campbell, 1898) which may comprise 80% or more of the total seed protein. Vicilin, the 7S protein which is usually the major of the two proteins in most *Pisum* genotypes (Casey et al., 1982), is totally lacking in methionine and cysteine (Croy et al., 1980a), whereas legumin (a 12S protein) contains between 1.0 and 1.8 g% methionine and cysteine, depending on genotype (Casey and Short, 1981). It has thus been considered that an increase in the proportion of legumin relative to vicilin would be a desirable objective in the breeding of *Pisum* seeds with protein of increased nutritional quality. There is evidence that some subunits of

Pisum legumin have higher methionine contents than others (Domoney, 1980); a slightly different way of approaching the matter, therefore, is to identify those types of a polymorphic series having the greater methionine and cysteine contents and to select for genotypes which produce maximal amounts of these proteins; that is, to select for more legumin which itself has a higher proportion of 'better' subunits, in terms of methionine and cysteine content, than the average.

Such an approach requires an understanding of the genetic behaviour of legumin subunits; it also would be desirable to analyse the inheritance of legumin content and to study the basis of such inheritance patterns, but the latter is poorly understood and outside the scope of this article. The present paper, therefore, describes the genetics of legumin in relation to its structure, synthesis and evolution and discusses the practical implications of such information.

THE STRUCTURE OF LEGUMIN

Legumin has a sedimentation coefficient of approximately 12S and a molecular weight of 350 000 - 400 000; it is an oligomeric protein, composed of two types of subunit of molecular weight around 40 000 and 20 000 (Derbyshire et al., 1976). The larger subunits are acidic in nature, the smaller are basic and the majority of a given legumin preparation will conform to the model proposed by Wright and Boulter (1974) for *Vicia faba* legumin; that is, molecular species composed of six acidic and six basic subunits, disulphide-bonded as acidic-basic pairs. Legumin purified from a single *Pisum* genotype exhibits whole protein heterogeneity, which is most clearly seen as multiple species on cellulose acetate electrophoresis (Thomson et al., 1978), or occasionally as multiple arcs during crossed immunoelectrophoresis of certain genotypes (Guldager, 1978).

Electron microscopy of purified legumin suggests a trigonal bipyramid structure (Casey et al., 1980) which differs in detail from the model proposed by Badley et al. (1975) for glycinin, the 12S protein from soybean; each morphological entity visible in the electron microscope is assumed to be a 60 000 molecular weight acidic-basic disulphide-linked dimer. As will be seen later, the acidic subunits apparently show considerably more genotypic variability in their structure than do the basic subunits, which may reflect the relative positions of the subunits within the native molecule. It is possible that the basic subunits are less exposed, with the acidic subunits occupying more of a surface position and that there may be a greater tendency towards change in the amino acid residues at the surface of the molecule than in those occupying internal positions which latter could be more critical in the maintenance of the native protein structure.

The SDS-denatured acidic and basic subunits show no immunological cross-reactivity and antibodies to the denatured acidic and basic subunits both react with native legumin (Casey et al., 1980).

Both types of subunit exhibit heterogeneity on polyacrylamide gels containing sodium dodecyl sulphate (SDS gels) (Thomson et al., 1978; Casey, 1979a; Matta et al., 1981), on polyacrylamide gels containing urea (Thomson et al., 1978) and on two-dimensional isoelectric-focusing/SDS gels (Casey, 1979; Krishna et al., 1979). Although the basic subunits from a given genotype appear as multiple species on such gels, they share a single amino-terminal glycine residue (Casey, 1979) and N-terminal sequence analysis of the mixture of small subunits (Casey et al., 1981) indicates the existence of several homologous subunits, differing at only a few points along their N-terminal sequence. Similar results have been observed for the small subunit of legumin from *Vicia faba* and of the homologous protein, glycinin, from *Glycine max* (Gilroy et al., 1979; Moreira et al., 1979). The sequence data for the basic subunits suggest the existence of a small gene family, the sequence differences between the members of the family being responsible for most, if not all, of the observed multiplicity on gels; some heterogeneity may also result from post-translational modification (Spencer et al., 1980) or from deamidation on storage (Matta and Gatehouse, 1981).

The acidic subunits also exhibit multiplicity on SDS gels (Casey, 1979a; Thomson et al., 1978) and on two-dimensional gels (Casey, 1979a; Krishna et al., 1979); on the latter there is evidence for both major and minor forms of acidic subunit. The

TABLE 1

THE N-TERMINAL AMINO ACID SEQUENCE OF ACIDIC AND BASIC SUBUNITS OF *Pisum* LEGUMIN AND GLYCINE GLYCININ. (ILLUSTRATION OF THE HOMOLOGY BETWEEN GLYCININ AND LEGUMIN)

Subunit	Position															
	1	2	3	4	5	6	7	8	9	10	11	12	13	14	15	16
Pea legumin[a] Acidic (a^M)	Leu	Arg	Glu	Gln	Pro	Gln	Gln	Asn	Glu	Cys	Gln	Leu	Glu	Arg	Leu	Asn
Glycinin[c] Acidic (A2)	Leu	Arg	Glu	Gln	Ala	Gln	Gln	Asn	Glu	Cys	Gln	Ile	Gln	Lys	Leu	Asn
Pea Legumin[a] Basic (mixture)	Gly	Leu	Glu	Glu	Thr	Val^2 Ile^1	Cys	Thr	Ala	Lys	Leu	Arg	Leu^2 Glu^1	Asn	Ile	Ala^2 Gly^1
Glycinin[b] Basic (mixture)	Gly	Ile^2 Val^1	Asp	Glu	Thr Asn	Ile	Cys	Thr	Met	Lys	Leu	Arg	Glu	Asn	Ile	Gly Ala

Subunit	Position													
	17	18	19	20	21	22	23	24	25	26	27	28	29	30
Pea legumin[a] Acidic (a^M)	Ala	Leu	Glu	Pro	Asp	Asn	Arg	Ile	Gln					
Glycinin[c] Acidic (A2)	Ala	Leu	Lys	Pro	Asp	Asn	Arg	Ile						
Pea Legumin[a] Basic (mixture)	Pro	(Ser)	(Ser)	(Ala)	Pro	Asp	Leu	Tyr	Asn	Pro	Glu	Ala	Gly	
Glycinin[b] Basic (mixture)	Gln	Pro	(Ser)	Arg	Pro^2 Ala^1	Asp	Leu	Tyr	Asn	Pro	(Gln)	Ala	Gly	

Taken from: [a] Casey et al., 1981; [b] Gilroy et al., 1979; [c] Moreira et al., 1979. Note that considerably more sequence heterogeneity in the basic subunits of glycinin was reported by the latter authors than is indicated in this Table; see Moreira et al., 1979 for details.

structural relationship between these major and minor subunits is not known; neither is the relationship between individual major subunits. N-terminal sequence analysis of an acidic subunit of *Pisum* legumin (Casey et al., 1981a) has shown striking homology to the acidic subunits of glycinin and N-terminal group analysis of legumins from a wide range of *Pisum* genotypes has shown conservation of the characteristic leucine (major) and threonine (minor) end groups (Casey, 1979a). It is tempting to speculate, by an analogy with the situation in glycinin (Moreira et al., 1979; Nielsen et al., 1981) that a gene family also exists for the acidic subunits of *Pisum* legumin and that in a genotype where there are several major acidic subunits, these will have very similar primary sequences; sequence information on individual acidic subunits of *Pisum* legumin is required to substantiate this.

Although it had been suggested (Basha and Beevers, 1976; Davey and Dudman, 1979), that legumin contains covalently-linked carbohydrate, it now seems unlikely (Casey, 1979; Gatehouse et al., 1980; Hurkman and Beevers, 1980) that legumin is a glycoprotein and thus the molecular basis of legumin subunit heterogeneity will not include differences in subunit glycosylation patterns.

VARIABILITY IN LEGUMIN SUBUNIT STRUCTURE

Although variability in the properties of the whole legumin molecule is well documented (Millerd et al., 1978; Thomson and Schroeder, 1978; Thomson et al., 1978; Casey, 1979a), such variation is related to variation in subunit structure only in an indirect fashion which precludes detailed genetic analysis. Thus, only variability in subunit structure - which is extensive - will be considered here.

There have been several reports of genotypic variability in the acidic and basic subunits of legumin (Thomson and Schroeder 1978; Casey, 1979a; Croy et al., 1979; Przybylska et al., 1979; Davies, 1980; Matta, 1981; Przybylska et al., 1981); charge and size (apparent molecular weight) variants exist for both and a number of different systems of nomenclature have been used to describe the subunits of legumin, their variant forms and the genes which code for them.

Thomson and Schroeder (1978) have defined five variant systems, designated LA - LE. LA is used to describe charge variability in the major acidic subunits (on urea - polyacrylamide gel electrophoresis); LB describes the same phenomenon but for a group of minor acidic subunits; LC is the variant system used to describe differences in apparent molecular weight (on SDS gels) between the acidic subunits; LD is for variability in charge, on urea-polyacrylamide gels, in the small, basic subunits; and LE refers to a minor group of polypeptides of mol. wt. around 26 x 10^3 which are now thought to be contaminants (Gatehouse et al., 1980).

Casey (1979a) noted variability in the apparent molecular weight (on SDS gels) of the acidic and the basic subunits, whilst variability in the charge and apparent mol. wt. of both the major

(α^M) and minor (α^m) acidic subunits of legumin was observed on isoelectric-focusing/SDS gels.

Matta (1981) has defined three types of legumin as 'small' (LgS), 'big' (LgB) and 'conventional' (LgC), as a function of the properties of the whole protein and of the disulphide-bonded acidic-basic subunit pairs. 'Conventional' legumin would represent the major legumin species from a given genotype, with 'small' and 'big' legumin being minor populations. Matta (1981) observed variability in the acidic subunits of all three legumin types, but noted variability in the basic subunits of only 'small' legumin.

INHERITANCE PATTERNS

It is important to understand how the different nomenclature systems relate to each other and this is outlined in Table 2. The LA/LC systems of Thomson and Schroeder (1978), the α^M-subunits of Casey (1979a) and the Lg-1C gene of Matta (1981) all refer to the acidic subunits of the major conventional legumin species. The LB system (Thomson and Schroeder, 1978) may correspond to α^m-subunits (Casey, 1979b); it is not clear how these relate to the Lg1-B and Lg1-S subunits. The LD system (Thomson and Schroeder, 1978) refers to the so-called β-subunits (Casey, 1979a) or basic subunits of conventional legumin (Lg-2C) (Matta, 1981); again the structural and genetic relationships between these and the basic subunits of 'big' and 'small' legumin is not clear.

TABLE 2

THE GENETIC SYSTEMS FOR *Pisum* LEGUMIN

Subunits (Molecular weights x 10^{-3})	Genetic systems and subunit nomenclatures	Linkage group location
c.40	(a) Lg-1c; Lg-1b; Lg-1s	Group 7
	(b) α^M: α^m	near r/tl
	(c) LA/LC; LB	
c.20	(a) Lg-2c; Lg-2b; Lg-2s	
	(b) β	
	(c) LD	

The genes for the acidic major subunits of legumin have been observed to behave as single, codominant alleles (Thomson and Schroeder, 1978; Casey, 1979a; Matta, 1981). In a cross between a variety with a legumin which had only one major acidic subunit and one which had three, the one-subunit and three-subunit patterns behaved as allellic alternatives (Casey, 1979a); there were no recombinant phenotypes in the small number of F2 seeds examined and subsequent studies (R. Casey and J.E. Sharman, unpublished) have failed to detect recombinants in 130 F2 seeds. It has been proposed that the multiple acidic subunit loci may be closely linked and may be the products of gene duplication (Casey, 1979a; Thomson and Doll, 1979). The few instances of legumin variants with single major acidic subunits (Casey, 1979; Casey et al., 1981a) may result from gene duplications without subsequent diversification, or may be the products of unduplicated loci.

The inheritance patterns of the subunits of minor legumin species are not as clear: the exact chemical status of such subunits has not yet been determined and there is a need for structural studies to determine the relationships between major and minor acidic subunits. Thomson and Schroeder (1978) found the F2 phenotypes in their LB variant system too complex to score reliably; similar problems were encountered with the α^m-subunits (Casey, 1979a). Matta (1981) observed that the acidic subunits of 'big' legumin (Lg-1B) did not behave as a single gene, one subunit sometimes being independently inherited from the acidic subunits of 'main' legumin, whilst the remaining subunits were totally linked to 'main' legumin; the acidic subunits of 'small' legumin (Lg-1S) were also inherited independently from those of 'main' legumin. These latter results suggest that the genes for legumin acidic subunits may not all be confined to one linkage group.

An investigation of the LD system (basic subunit charge variants) (Thomson and Schroeder, 1978) leads to the conclusion that its genetic basis is complex, whilst studies of the segregation of variants of the basic subunits of 'small' legumin (Lg-2S) (Matta, 1981) showed them to be inherited independently of the acidic subunits (Lg-1S) to which they are normally disulphide-bonded in the native legumin molecule (Wright and Boulter, 1974).

THE CHROMOSOMAL LOCATION OF THE GENES FOR LEGUMIN SUBUNITS

Although it would be of considerable interest to know whether the genes for the acidic and basic subunits of legumin are located close to each other, such linkage studies have been hampered by the lack of basic subunit variants in suitable genotypes; where variants have occurred, they have been in genotypes with chromosome abnormalities or in *Pisum fulvum*. Studies of the linkage relationships between storage protein subunit loci and other known *Pisum* genes, however, has proved profitable and using such an approach the locus for the major acidic subunits of legumin has been shown to reside on linkage group 7, close to the r_a locus (Davies, 1980; Matta, 1981); Davies (1980) places the locus within 7 crossover units of r_a. Biosynthetic studies (Croy et al., 1980; Evans et al., 1980; Spencer and Higgins, 1980) suggest that the legumin Lg-2c (the basic subunit) locus may also be near r_a on linkage group 7 (see below).

THE BIOSYNTHESIS OF LEGUMIN

The acidic and basic subunits of legumin are synthesised as 60 000 mol. wt. precursors in which an acidic (40 000 mol. wt.) and basic (20 000 mol. wt.) subunit are covalently linked by a bond other than an -S-S- bridge (Croy et al., 1980; Spencer and Higgins, 1980); this bond is then cleaved during the processing of the precursor, three such precursor molecules unite to form a 9S intermediate (Spencer and Higgins, 1980) and two such trimers, or half-molecules, come together to form the final 12S species. Just as there exists heterogeneity for the individual 40 and 20 x 10^3 subunits so there exists heterogeneity in the nature of the 60 x 10^3 mol. wt. precursor (Spencer and Higgins, 1980); that is, there are a number of precursors. The mRNAs for such precursor molecules would be expected to contain around 2 000 nucleotides and legumin has been shown to be synthesised on a mRNA of around 18 - 20S (which would be of sufficient size) (Evans et al. 1980). Thus it seems likely that both of the legumin subunits of a given subunit pair are coded by the same mRNA; this, however, does not provide absolute evidence for the genomic contiguity of acidic and basic subunit genes, since the mRNA could be the product of genomically distant acidic and basic genes whose transcripts have been spliced. The matter should be settled by studies of the structure of genomic DNA coding for the legumin mRNA, from which it should be unambiguously shown whether the acidic and basic loci lie close together within the genome. Should this prove to be the case, then further investigations will be required to determine why the basic subunits of 'small' legumin are inherited independently of the acidic subunits of 'small' legumin (Matta, 1981).

THE EVOLUTION OF LEGUMIN

It is likely that the multiple, heterogeneous subunits of legumin are the products of a small gene family; the basic subunits are quite clearly members of an homologous series (Casey et al., 1981) and although only a single acidic subunit has been examined at the level of protein sequence (Casey et al., 1981a), its close similarity to the acidic subunits of glycinin (Moreira et al., 1979) suggests that the same situation will exist for *Pisum* legumin as for *Glycine* glycinin (Nielsen et al., 1981) - that is, a small family of genes coding for multiple basic and acidic subunits. It is also likely that the structural gene will be for a 60 x 10^3 precursor molecule, rather than individual acidic and basic subunits. The observed heterogeneity of legumin basic and acidic subunits on gels can probably be largely accounted for by minor mutations within duplicated common sequences, since only small changes are required to effect fairly large alterations in mobility on SDS gels; further understanding of the way in which heterogeneity of legumin subunits can arise must, however, await sequence analysis of the proteins or genes.

Some pea genotypes have a legumin which apparently contains only one type of major acidic subunit (Casey, 1979a; Casey et al., 1981a); such genotypes are of fundamental interest in terms of the evolution of legumin heterogeneity. Two of these genotypes (JI 224 and 849) (Casey, 1979a) are *Pisum fulvum* types - the small-seeded, wild, possible progenitors of *Pisum sativum* - and it is tempting to speculate that they represent the 'primitive' situation before gene duplication and/or subsequent diversification took place. The third (JI 497) (Casey et al., 1981a) is a modern cultivar which may have evolved without undergoing the duplication/mutation of legumin acidic subunit genes.

There is no evidence that the major and minor acidic legumin subunits share common sequences, although it is highly likely that they will be structurally very similar for them both to participate in the formation of the native molecule. One feature of the minor subunits is that in any given genotype most of them form a group with a slightly lower p_I_ than those of the major acidic subunits (Casey, 1979a); one could postulate that after an initial duplication, early mutational event(s) gave rise to the difference in p_I_ between the two groups of acidic subunits, which then evolved independently into groups of subunits within the p_I_ classes. At some stage a regulatory mutation must also have occurred to effect the different levels of expression of the major and minor loci. The arrangement of the major and minor acidic (and thus, presumably, major and minor 60 x 10^3 mol. wt. precursor) loci within the genome will be of considerable interest in relation to the control of expression.

PRACTICAL IMPLICATIONS OF LEGUMIN GENETICS

Any attempt at improvement of the quality - that is, the cysteine/methionine content - of pea legumin will require a clear understanding of the distances between individual legumin genes, since it appears that in both glycinin and *Pisum* legumin some members of the 60 x 10^3 mol. wt. precursor series have better methionine contents than others (Domoney, 1980; Nielsen et al., 1981) and thus it will become desirable to effect segregation of these individual genes to take advantage of the phenomenon; present genetic evidence suggests that the legumin genes are closely linked, with minor exceptions (Casey, 1979a; Matta, 1981).

Thomson and Doll (1979) have pointed out that the existence of multiple genes for legumin subunits will have a dilution effect on the benefit accrued from desirable mutations in specific legumin subunit genes; if an approach of breeding for 'better' legumin subunits - or of insertion of legumin genes which have been 'engineered' to improve the quality of their product - is to be adopted, some way will have to be found to overcome this problem.

ACKNOWLEDGEMENTS

I am very grateful to Professor D. Boulter and Dr. N.K. Matta for providing me with their results in advance of publication.

REFERENCES

Badley, R.A., Atkinson, D., Hauser, H., Green, J.P. and Stubbs, J.M., 1975. The structure, physical and chemical properties of the soy bean protein glycinin. Biochem. Biophys. Acta (412), 214-228.

Basha, S.M.M. and Beevers, L., 1976. Glycoprotein metabolism in the cotyledons of *Pisum sativum* during development and germination. Plant Physiol. (57), 93-97.

Casey, R., 1979. Immunoaffinity chromatography as a means of purifying legumin from *Pisum* (pea) seeds. Biochem. J. (177), 509-520.

Casey, R., 1979a. Genetic variability in the structure of the α-subunits of legumin from *Pisum* - a two-dimensional gel electrophoresis study. Heredity (43), 265-272.

Casey, R. and Short, M.N., 1981. Variation in amino acid composition of legumin from *Pisum sativum*. Phytochemistry (20), 21-23.

Casey, R., Hills, G.J. and Shaw, P., 1980. The 3-dimensional structure of legumin from *Pisum sativum*. John Innes Annual Report, 20-21.

Casey, R., March, J.F. and Sanger, E., 1981. N-terminal amino acid sequence of β-subunits of legumin from *Pisum sativum*. Phytochemistry (20), 161-163.

Casey, R., March, J.F., Sharman, J.E. and Short, M.N., 1981a. The purification N-terminal amino acid sequence and some other properties of an α^M-subunit of legumin from the pea (*Pisum sativum* L.). Biochem. Biophys. Acta. (670), 428-432.

Casey, R., Sharman, J.E., Wright, D.J., Bacon, J.R. and Guldager, P., 1982. Quantitative variability in *Pisum* seed globulins; its assessment and significance. Qualitas Plantarum Pl. Foods Hum. Nutr., in press.

Croy, R.R.D., Derbyshire, E., Krishna, T.G. and Boulter, D., 1979. Legumin of *Pisum sativum* and *Vicia faba*. New Phytologist (83), 29-35.

Croy, R.R.D., Gatehouse, J.A., Evans, I.M. and Boulter, D., 1980. Characterisation of the storage protein subunits synthesised *in vitro* by polyribosomes and RNA from developing pea (*Pisum sativum* L.) I. Legumin. Planta (148), 49-56.

Croy, R.R.D., Gatehouse, J.A., Tyler, M. and Boulter, D., 1980a. The purification and characterisation of a third storage protein (convicilin) from the seeds of pea (*Pisum sativum* L.). Biochem. J. (191), 509-516.

Davey, R.A. and Dudman, W.F., 1979. The carbohydrate of storage glycoproteins from seeds of *Pisum sativum*: characterization and distribution on component polypeptides. Aust. J. Plant Physiol. (6), 435-447.

Davies, D.R., 1980. The r_a locus and legumin synthesis in *Pisum sativum*. Biochemical Genetics (18), 1207-1219.

Derbyshire, E., Wright, D.J. and Boulter, D., 1976. Legumin and vicilin, storage proteins of legume seeds. Phytochemistry (15), 3-24.

Domoney, C., 1980. The synthesis of legumin *in vitro*. John Innes Annual Report, 17-19.

Evans, I.M., Croy, R.R.D., Brown, P. and Boulter, D., 1980. Synthesis of complementary DNAs to partially purified mRNAs coding for the storage proteins of *Pisum sativum* (L). Biochem. Biophys. Acta (610), 81-95.

Gatehouse, J.A., Croy, R.R.D. and Boulter, D., 1980. Isoelectric-focusing properties and carbohydrate content of pea (*Pisum sativum*) legumin. Biochem. J. (185), 497-503.

Gilroy, J., Wright, D.J. and Boulter, D., 1979. Homology of basic subunits of legumin from *Glycine max* (L.) merr. and *Vicia faba* (L.). Phytochemistry (18), 315-317.

Guldager, P., 1978. Immunoelectrophoretic analysis of seed proteins from *Pisum sativum* L. Theor. Appl. Genet. (53), 241-250.

Hurkman, W.J. and Beevers, L., 1980. What is pea legumin - is it glycosylated? Planta (150), 82-88.

Krishna, T.G., Croy, R.R.D. and Boulter, D., 1979. Heterogeneity in subunit composition of the legumin of *Pisum sativum*. Phytochemistry (18), 1879-1880.

Matta, N.K., 1981. Structural and genetical studies on legumin of *Pisum sativum* L. and *Vicia faba* L.. PhD Thesis University of Durham.

Matta, N.K. and Gatehouse, J.A., 1981. Modification of the basic subunits of pea legumin on storage. Phytochemistry (20), in press.

Matta, N.K., Gatehouse, J.A. and Boulter, D., 1981. Subunit heterogeneity and structure of legumin of *Pisum sativum* L. - a multi-dimensional gel electrophoresis study. J. Exp. Bot. (32), 183-197.

Millerd, A., Thomson, J.A. and Schroeder, H.E., 1978. Cotyledonary storage proteins in *Pisum sativum* III. Patterns of accumulation during development. Aust. J. Plant Physiol. (5), 519-534.

Moreira, M.A., Hermodson, M.A., Larkins, B.A. and Nielsen, N.C., 1979. Partial characterization of the acidic and basic polypeptides of glycinin. J. Biol. Chem. (254), 9921-9926.

Nielsen, N.C., Moreira, M.A., Staswick, P., Hermodson, M.A., Tumer, N. and Thanh, V.H., 1981. The structure of glycinin from soybeans. Abhdlg. Akad. Wiss. DDR, Abt. Math. Naturwiss. Techn., in press.

Osborne, T.B. and Campbell, G.F., 1898. Proteids of the pea. J. Am. Chem. Soc. (20), 348-362, 362-375, 393-405.

Przybylska, J., Hurich, J. and Zimniak-Przybylska, Z., 1979. Comparative study of seed proteins in the genus *Pisum*. IV. Electrophoretic patterns of legumin and vicilin components. Genetica Polonica (20), 517-528.

Przybylska, J., Hurich, J. and Blixt, S., 1981. Variation of basic legumin components in the genus *Pisum* revealed by isoelectrofocusing. *Pisum* Newsletter (13), 44-45.

Spencer, D. and Higgins, T.J.V., 1980. The biosynthesis of legumin in maturing pea seeds. Biochem. Internat. (1), 501-509.

Spencer, D., Higgins, T.J.V., Button, S.C. and Davey, R.A., 1980. Pulse-labelling studies on protein synthesis in developing pea seeds and evidence of a precursor form of legumin small subunit. Plant Physiol. (66), 510-515.

Thomson, J.A. and Doll, H., 1979. Genetics and evolution of seed storage proteins. In: Seed protein improvement in cereals and grain legumes. IAEA-SM-230/60 Vienna, 1-16.

Thomson, J.A. and Schroeder, H.E., 1978. Cotyledonary storage proteins in *Pisum sativum*. II. Hereditary variation in components of the legumin and vicilin fractions. Aust. J. Plant Physiol. (5), 281-294.

Thomson, J.A., Schroeder, H.E. and Dudman, W.F., 1978. Cotyledonary storage proteins in *Pisum sativum*. I. Molecular heterogeneity. Aust. J. Plant Physiol. (5), 263-249.

Wright, D.J. and Boulter, D., 1974. Purification and subunit structure of legumin of *Vicia faba* L. (broad bean). Biochem. J. (141), 413-418.

DISCUSSION

B. Snoad *(UK)* I noticed that many of the genotypes in your list contained chromosomal rearrangements. Do you think there is a positional effect to be taken into account here in determining, for example, different vicilin/legumin ratios?

R. Casey *(UK)* I do not know, but I should explain that the genotypes were specifically selected for variability in legumin composition, and were morphologically very varied. Many contained chromosomal rearrangements. The objective was to examine the extent to which legumin varied and to determine the methionine/cystine content of our germplasm collection.

M. Poulsen *(Denmark)* Around 1974 - 75 the Australian work on legumin/vicilin ratios led to some optimism in relation to improvement in protein quality through breeding. This has proved to be not possible, and it would be useful to know whether the aim should be to breed for increased yield, rather than improved protein levels.

R. Casey Simple economics dictate that yield must be a primary objective, but any increase in protein content would clearly be valuable. It is unlikely, however, that breeding for increased legumin will effect a great increase in the quality of the protein. There is a suggestion that increases in legumin are associated with a reduction in albumin which is sulphur rich, and positive selection for legumin may result in decreased methionine and cystine content.

W. Williams *(UK)* Are methionine and cystine significant limiting factors in animal production in European agriculture. It is important that we should have a fairly definitive answer in relation to our perspectives on breeding for protein in *Vicia* and *Pisum*.

C. Calet *(France)* There are seldom situations where either methionine and cystine limit efficiency of feed utilisation because synthetic methionine is added to most diets. An amino acid, however, which is important for certain purposes and which may be limiting in pea and lupin based diets is tryptophan. Improved levels of this amino acid through breeding would be useful.

ENDOMITOSIS AND PROTEIN SYNTHESIS IN *PISUM SATIVUM* SHOOTS: THE EFFECT OF TEMPERATURE, CULTIVAR AND GROWTH REGULATORS

A. Callebaut

Centrum voor de studie van de Biologische Produktieprocessen,
Fakulteit Landbouw, Coupure Links 533,
B-9000 Gent, Belgium.

ABSTRACT

In pea shoots, exogenously applied GA_3 and AMO 1618 show their most pronounced effect on mitotic cell populations. Protein synthesis is strongly stimulated by GA_3 at low temperature, very clearly visible when the amount of protein is expressed on a per-cell basis.

The lack of any effect of the growth regulators on endomitotic DNA synthesis might be explained by the high endogenous GA concentrations in intact growing plants.

INTRODUCTION

In aetiolated shoots of dwarf peas the amount of DNA per nucleus of cortical cells, attains a C level of 8. During pea seed formation, parenchyma nuclei reach a 64 C level. Scharpé and Van Parijs (1973) and Broekaert and Van Parijs (1978) suggested a relationship between the high degree of endoreduplication in the parenchyma cells of developing cotyledons and the high rate of RNA and protein synthesis in these cells. This phenomenon was called a gene dosage effect.

In pea shoots light inhibits endomitotic DNA synthesis and the cortical cells reach only a 4 C level. Application of GA3, in the light, induces cell division and cell elongation, without changing the C level of the nuclei.

In this work, the process of endomitosis and its relation to protein synthesis was further studied in our model system, the pea epicotyl. The endopolyploidisation process was studied by chemical estimation of the DNA amount per epicotyl and by cytofluorometric determination of the DNA amount per nucleus.

First the endomitotic DNA synthesis in the epicotyl was followed in two closely related dwarf pea cultivars, Finale and Rondo, in relation to growth in length and protein content. Whereas cv. Rondo was the most important variety among the blue peas up to 1970, as deduced from area of cultivation, nowadays this cultivar is hardly grown at all in the Netherlands. The most important cultivar now is Finale, which was introduced for its improved resistance to disease and for its improved yield under unfavourable soil conditions.

RESULTS AND DISCUSSION

In *Pisum sativum* cv. Rondo, aetiolated shoots grow only by cell elongation, accompanied by endopolyploidisation of the cortical nuclei, within the range 10 to 25°C.

In *Pisum sativum* cv. Finale, growth of aetiolated shoots at 15°C is the result of cell division and cell elongation accompanied by endoreduplication.

In both cultivars the maximum degree of endoploidy in aetiolated epicotyls is 8 C. In the cv. Rondo, the proportion of 8 C versus 4 C nuclei is close to 80 : 20; in Finale epicotyls, close to 60 : 40.

Light inhibits endomitotic DNA synthesis in the cv. Finale, as also in the cv. Rondo. In these cases the cortical cells reach the 4 C level. However, in light-grown Finale epicotyls cell division also occurs.

The main difference between the cultivars Finale and Rondo seems to be the fact that the number of cells in the epicotyl of the cv. Rondo is fixed during seed ripening. Conversely, in the variety Finale, a large proportion of the cortical cells of the epicotyl retain their capacity to divide.

The presence and the activity of both the endomitotic and the mitotic populations seems to be genetically controlled. Seemingly the maximum C level attained by the endomitotic cells in a given tissue is genetically controlled in the varieties Finale and Rondo. On the contrary, Nougarède and Rembur (1980) showed that in the cv. d'Annonay, also a dwarf variety, cortical nuclei reach the 16 C level in the dark and the 8 C level in the light.

Although total protein content of the full grown Finale epicotyl is 1.7 times higher than that of the Rondo epicotyl, the protein/DNA ratio, at any given stage, during growth of the Rondo epicotyl is always higher than the ratio in the Finale epicotyl. This result can be explained by the lack of a mitotic cell population and by a faster endoreduplication in the cv. Rondo.

In the second part of this work, the growth of pea seedlings (cv. Finale) at different temperatures (5°, 15° and 25°C) was compared. We found that the distribution of the C classes was not significantly changed by temperature.

At 5°C, the length and dry matter content of the aetiolated epicotyls was decreased by 20% compared to the growth at 15°C. The DNA amount per epicotyl was approximately the same; the protein content however was increased by 60%.

At 25°C, growth in length and dry matter content were both reduced by 40% compared with those at 15°C. The cell number, as estimated from total chemical DNA determinations per epicotyl and per nucleus from histophotometric DNA determinations, was decreased by 50%, the result of an inhibition of cell division.

The nuclei immediately proceeded to the endoreduplication cycle, whereby after 4 days 65% of the cortical cells had reached the 8 C level. Nevertheless, the protein content was only decreased by 13% compared to that at 15°C, which reflects the high protein synthesising capacity of the endomitotic cells. Relative maximum rates of protein synthesis, deduced from chemical determinations of the protein content, are 228 at 25°C, 117 at 15°C and 77 at 5°C. The ratio of protein to DNA amount varies between 0.40 and 0.43 at 25°C, compared to 0.25 at 15°C and 0.38 to 0.43 at 5°C (during endoploidisation).

At 25°C, the efficiency of protein synthesis is high. At 5°C, the period of net protein synthesis is considerably extended, resulting in higher absolute protein contents of the full-grown epicotyls. Investigations of the proteins of epicotyls grown at 5°C and 25°C, using polyacrylamide gel electrophoresis, revealed mainly quantitative differences in the banding pattern. In the banding pattern of the chromosomal nonhistone proteins, some major bands, clearly present in epicotyls grown at 25°C, were present in a much lower concentration in epicotyls grown at 5°C. In light-grown Finale epicotyls at 25°C, up to 30% of the cortical nuclei reached the 8 C level, in contrast with Finale epicotyls growing at 15°C.

These results show that the switch from the mitotic to the endomitotic cycle is temperature sensitive. Low temperature favoured cell division in pea shoots, while high temperature stimulated endomitotic DNA synthesis.

Finally, the effect of exogenously applied GA_3 and AMO 1618 (an inhibitor of gibberellin biosynthesis) was examined. Neither GA_3 nor AMO 1618 changed the C level of the cortical nuclei of aetiolated epicotyls of either variety in the temperature ranged from 5°C to 25°C, neither was the ratio of 8 C: 4 C nuclei influenced by the growth regulators in fully differentiated epicotyls. Neither growth regulator had any effect on the DNA content or the protein content of aetiolated Rondo epicotyls grown at 15°C, in agreement with the lack of a mitotic cell population in this cultivar.

In Finale epicotyls at 15°C, GA_3 stimulated growth in length most markedly in the youngest seedlings, which contain a high proportion of mitotic cells. GA_3 had a very pronounced effect on aetiolated Finale epicotyls at 5°C; the cell number was increased by 15%, length and dry matter content were enhanced by 40%, while protein content was doubled compared to those of control epicotyls grown at 15°C. AMO 1618 had the opposite effect. At 25°C, GA_3 application resulted in a higher level of cell division and in a slight increase of the protein content. AMO 1618 had no effect, except for a reduction of the length of the epicotyl.

In these experiments the lack of any effect of exogenous GA_3 on endomitotic DNA synthesis might be explained by the presence of high concentrations of endogenous gibberellins, even after AMO 1618 application. During epicotyl growth of γ-irradiated seeds, where all mitotic activity and probably all endogenous gibberellin biosynthesis is blocked, exogenously applied GA_3 induced endomitotic DNA synthesis.

Mohamed and Bopps (1980) showed that in the shoots of the variety 'Kleine Rheinländerin', endomitotic DNA synthesis was induced by GA_3.

REFERENCES

Broeckaert, D. and Van Parijs, R., 1978. The relationship between the endomitotic cell cycle and the enhanced capacity for protein synthesis in Leguminosae embryogeny. Zeitschr. Pflanzenphysiol. 86, 165-175.

Mohamed, Y. and Bopp, M., 1980. Distribution of polyploidy in elongating and non-elongating shoot axes of *Pisum sativum*. Zeitsch. f. Pflanzen 98 (1), 25-33.

Nougarède, A. and Rembur, J., 1980. Extension cellulaire, division cellulaire et niveaux de ploidie des cellules corticales de l'épicotyle du pois en cours d'élongation. Can. J. Bot. 58 (4), 486-501.
Scharpé, A. and Van Parijs, R., 1973. The formation of polyploid cells in ripening cotyledons of *Pisum sativum* L. in relation to ribosome and protein synthesis. J. Exp. Bot. 24, 216-222.

SESSION IV

NUTRITIONAL AND TECHNOLOGICAL ASPECTS

Chairmen: D. Boulter and C. Calet

NUTRITIONAL CHARACTERISTICS OF A PEA PROTEIN CONCENTRATE FOR HUMAN UTILISATION

E. Carnovale, P. Corcos Benedetti and M.A. Spadoni

Istituto Nazionale della Nutrizione,
Via Ardeatina 546, 00179 Rome, Italy.

ABSTRACT

Compared to the other legumes commonly used in human nutrition, field peas contain a higher amount of lysine and an amount of sulphur amino acids comparable to that of soyabean.

By air-classification it is possible to obtain from field peas a protein concentrate containing 50% protein, 5% lipids and a P/S ratio of 3.28. The Fe and Zn content are, respectively, 16.7 and 6.0 mg/100 g protein concentrate. The individual scores of essential amino acids calculated versus the FAO/WHO pattern indicate, as expected, that cys + met are the first, and threonine the second, limiting amino acids, while there is a relative abundance of lysine and of aromatic amino acids. The pea protein concentrate also contains smaller amounts of trypsin inhibitor compared to other legumes.

The biological value measured as RPV shows that pea protein concentrate, compared to other plant protein concentrates, ranks very near to soyabean and it is lower only than that of the chick pea. Using lactalbumin as reference, the pea protein concentrate brings about a higher N deposition in the liver, partly due to a higher arginase activity.

Data are also presented relative to formulations made with non-traditional ingredients, including pea protein concentrate. These products, in the form of artificial rice, present a high biological value.

The capacity of the pea protein concentrate in lowering the cholesterol level in the serum of hypercholesterolemic rats was also considered.

Ample experimental evidence indicates now that, through very little processing and in suitable combinations, legumes are a good and economical source of proteins, quite as nutritive as animal proteins, in human feeding. Among the various legumes grown in Italy, field peas are receiving increasing attention as a high protein crop with good agronomic characteristics (Anon, 1979).

As shown in Table 1, field peas are limiting in sulphur amino acids, although to a lower extent than the majority of the other legumes, and have the highest lysine content. Moreover, their ratio of essential amino acids (EAA) to total amino acids (TAA) is around 0.4, as for the other legumes, and the chemical score is similar to that of the best crops such as soyabean and chick pea.

TABLE 1

LYSINE, CYSTINE + METHIONINE, THREONINE AND TRYPTOPHANE CONTENT OF THE MOST COMMON LEGUMES (g/16g N)

	EAA/AA tot*	Lys	Cys+Met	Thr	Tryp	CS**
Cicer arietinum	0.39	6.8	2.3***	3.8	0.9	0.66
Phaseolus vulgaris	0.34	7.1	1.8	3.9	0.9	0.51
Lens esculenta	0.41	7.0	1.7	4.0	0.8	0.48
Vicia faba	0.38	6.7	1.5	3.7	0.9	0.43
Glycine max.	0.43	6.5	2.6	3.8	1.2	0.74
Pisum sativum	0.39	7.1	2.5	0.8	1.1	0.71
FAO/WHO pattern		5.5	3.5	4.0	1.0	

* EAA = essential amino acids

** Chemical score against FAO/WHO pattern

*** The first limiting amino acid is underlined

In recent years the scope for using peas has been increased by the development of fine-grinding and air-classification processes, yielding a starch-rich fraction and a pea protein concentrate (Anon, 1974). Our interest in this process and in the nutritional characteristics of the products thus obtained was stimulated by participation in a Programme on "New protein sources and new food formulations" sponsored by the CNR of Italy.

TABLE 2

GROSS COMPOSITION (% DRY BASIS) AND MINERAL CONTENT OF THE PEA PROTEIN CONCENTRATE (PPC)

g%		mg%	
Protein (N x 6.25)	52.30	Na	4.5
Lipid	4.73	Ca	214
Starch	23.88	Fe	16.7
Sugar	7.73	Zn	6.0
Fibre	5.81	Cu	1.7
Ash	5.92	P	1219
P/S*	3.28		

* Ratio polyunsaturated fatty acids/saturated fatty acids

The pea protein concentrate (PPC) we obtained with the facilities at our Institute has the composition given in Table 2. The protein content of the PPC is around 50%, the other 50% being represented mainly by carbohydrates, 15 - 16% of which is fibre. The polyunsaturated : saturated fatty acids ratio is quite high due to the levels of linoleic and linolenic acid (1.177 g and 0.258 g/100 g dry weight, respectively). The levels of iron and zinc are also worth mentioning.

The individual scores of essential amino acids (Table 3) calculated versus the FAO/WHO (1973) pattern indicate, as expected, that cys + met is the first, and threonine the second, limiting amino acid, while there is a relative abundance of lysine and aromatic amino acids.

TABLE 3

ESSENTIAL AMINO ACIDS CONTENT (g/16g N) AND CHEMICAL SCORES OF PPC

Threonine	3.4	0.85*
Valine	5.0	1.00
Cys + Met	2.4**	0.68
Isoleucine	4.3	1.07
Leucine	7.0	1.00
Tyr + Phe	7.3	1.22
Lysine	7.2	1.31
Tryptophan	1.1	1.10

* Chemical scores against FAO/WHO pattern (1973)
** The first limiting amino acid is underlined

The nutritional quality of a legume product, however, is determined not only by the quality of the amino acids making up its protein but also by their availability. In this respect, the most important, and also the best known, factors are the inhibitors of trypsin and chymotrypsin, enzymes which play a key role in the digestion of proteins in animals; such inhibitors are widespread among legumes (Liener and Kakake, 1980).

Without entering the subject of the mode of action of these inhibitors, it is important to note that the growth inhibition they evoke in animals is not due simply to an inhibition of the digestion of dietary protein by proteolytic enzymes present in the intestinal tract. It has been known for a long time, in fact, that trypsin inhibitors can also cause hypertrophy of the pancreas, an effect that is accompanied by an increase in the secretory activity of the pancreas. This hyperactivity of the pancreas leads to an endogenous loss of essential amino acids, particularly sulphur amino acids, due to the high level of these amino acids in the enzymes trypsin and chymotrypsin. This loss of sulphur containing amino acids enhances the already critical situation of legume proteins (Lyman and Lepkovsky, 1957). To inactivate protease inhibitors, the heat treatment must be calculated in such a way as to avoid the breakdown of amino acids and the formation of artefacts.

TABLE 4

TRYPSIN INHIBITING ACTIVITY OF SOME LEGUMES (TIU x 10^{-3}/g)

	Flour	Protein conc. (air classified)
Field pea	2.5	4.5
Faba bean	0.9	1.4
Common bean	30.0	-
Soy bean	59.8*	-

* Defatted

As shown in Table 4, when compared to common beans and soyabeans, field peas have a very low trypsin inhibiting activity. Only faba bean shows an activity lower than that of field pea, while soyabean has the highest. Even after air-classification, in which trypsin inhibitors are concentrated with proteins, the level of the inhibitors remains substantially lower than that of common bean and soyabean.

In Figure 1 the biological quality of the protein of PPC is compared with other vegetable protein concentrates produced in the framework of the Programme "New protein sources". Values are expressed as the relative protein value (RPV) standardised against lactalbumin; that is, by a dose : response procedure

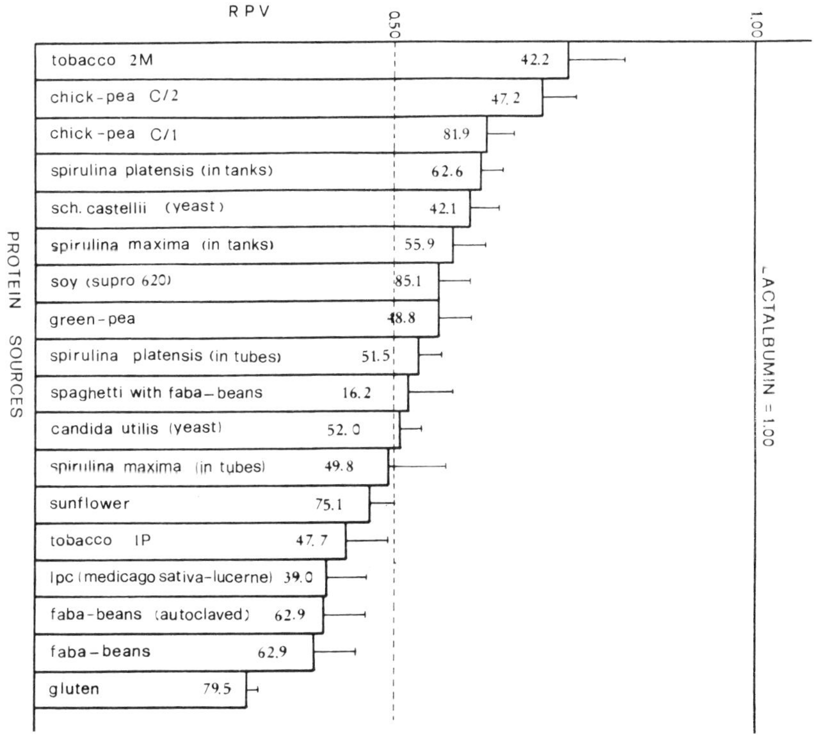

Fig. 1. Relative protein value (RPV) of vegetable protein concentrates. Response parameter : body nitrogen. Figures on the bars are the percentage protein content of the different protein sources.

(Samonds and Hegsted, 1977). Such a procedure has the advantage over the traditional PER of being independent from the amount of food consumed, the amount of protein in the diet and the nutritive quality of the protein in the diet. Moreover, unlike the PER, the RPV is a direct function of the nutritive value of the protein.

Body nitrogen deposition was chosen as a response parameter after we found that with some vegetable protein sources there is a significant discrepancy between values based on body nitrogen and values based on body weight changes. This was due not only to the gut content, when living weight was measured, but also to the modification of body composition consequent to the variation of the total energy intake. The total energy intake is strongly dependent upon the palatability of the diet (Corcos Benedetti et al., 1981). The Figure shows that where the protein concentrates are ranked in decreasing RPV order, PPC belongs to the group of protein concentrates with a protein quality over 0.5. Its RPV is similar to that of soyabean and, among legumes, lower only than that of chick pea concentrate.

The protein quality of PPC was also studied in relation to the capacity for promoting nitrogen deposition in the liver. Originally this was done in a search for a simplified methodology for the evaluation of protein quality (Bises et al., 1981).

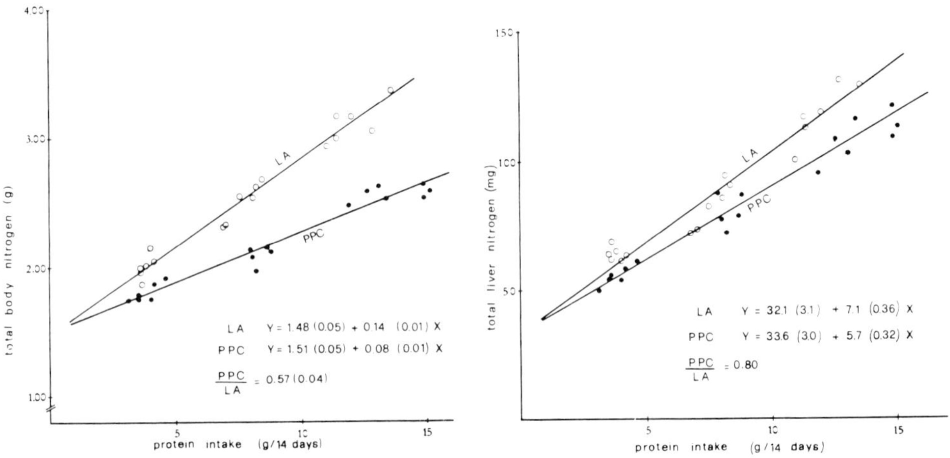

Fig. 2. Dose : response curve, calculated according to the regression equation y = a + bx, where a is the intercept on the y axis and b is the slope. PPC/LA is the ratio between the two slopes. In parenthesis the SE. In (a) total body nitrogen, in (b) total liver nitrogen is plotted against increasing quantities of protein intake.

The effect of protein pea concentrate is compared to that of lactalbumin (LA) in Figure 2, where body and liver nitrogen are plotted against dietary protein. The slope ratio between PPC and LA was higher for the liver than for the whole body. This means that, when compared to lactalbumin, PPC feeding causes a preferential nitrogen deposition in the liver.

The regression equation (Figure 3) that links total body with liver nitrogen deposition of rats fed increasing amounts of PPC shows that about 7% of the nitrogen deposited in the body is taken up by the liver. We found that for lactalbumin, casein and gluten this value was near 5%, while for faba beans it was 11%.

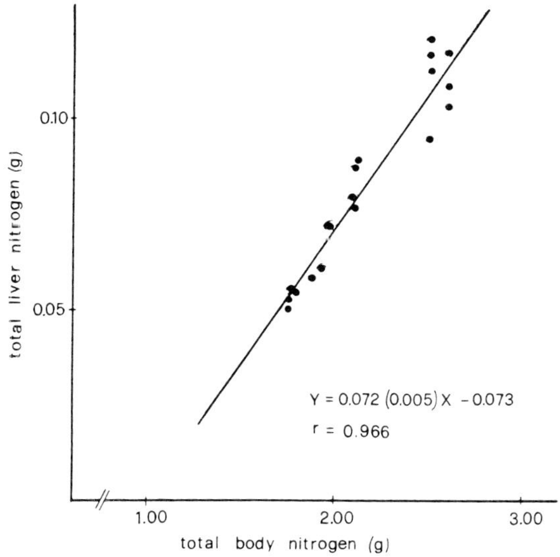

Fig. 3. Correlation between total liver and total body nitrogen of rats fed increasing amounts of PPC.

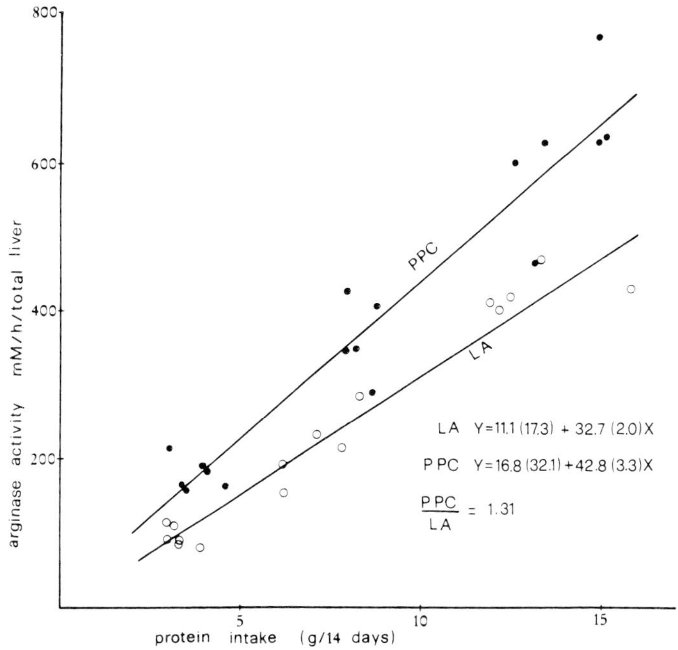

Fig. 4. Dose : response curve, calculated as in Figure 2. The arginase activity is plotted against increasing quantities of protein intake. PPC/LA is the ratio between the two slopes.

Legume protein is characterised by a low level of sulphur amino acids and a relatively high level of arginine. The behaviour of arginase, a protein enzyme whose level adapts to the level of the substrate arginine, may in part explain the preferential protein deposition in the liver of rats fed legumes (Figure 4). In such a case, the slope ratio between PPC and lactalbumin rose to 1.31, while that for liver nitrogen deposition was 0.80 and for body nitrogen deposition was 0.57.

The data presented above clearly show that the pea protein concentrate *per se* is a source of good quality protein, competitive with soyabean, at least with regard to the capacity of promoting protein deposition in the body and the liver.

As is known, the first nutritional function of dietary proteins is to furnish the nitrogen and the essential amino acids needed for the synthesis of tissue protein; in this context, what must be achieved is a final mixture of amino acids matching at its best body needs. The amino acid pattern of the PPC, characterised by an optimal level of lysine, makes this product very suitable for complementation with cereal proteins.

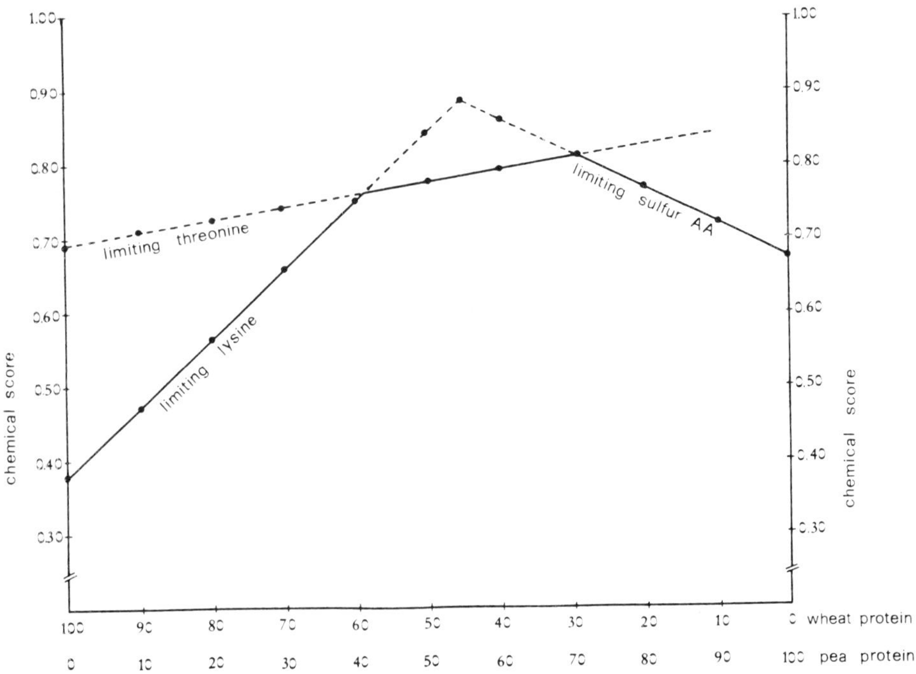

Fig. 5.

Figure 5 shows that for a blend between PPC and wheat to reach the highest score against the FAO pattern, protein must come 70% from PPC and 30% from wheat. In this case, threonine becomes the limiting amino acid. At the optimal combination, that needed to overcome the limitation in lysine and sulphur amino acids, of PPC and wheat, the ratio of the proteins of these two ingredients is 52% and 48% respectively.

Starting from non-conventional sources of starch and protein, including PPC, Resmini and colleagues at the Istituto di Industrie Agrarie dell'Università di Milano prepared new types of pasta, both in traditional shapes as well as in the shape of rice grains (Pagani et al., 1981).

Some formulations utilised for the preparation of artificial rices are reported in Table 5. These formulations have a PPC content ranging from 6% to 45%. A sample with 10% pea meal was also included.

TABLE 5

FORMULATIONS UTILISED FOR THE ARTIFICIAL RICE*

Ingredients	Sample number (1)	(2)	(3)	(4)	(5)
Rice (flour)	80				
Rice (grits)		80		45	45
PPC	10	10	6	45	
Pea flour					10
Dry skim milk	10	10		10	10
Wheat flour			45		
Starch (corn)			45		
Whey protein content (UF)			4		5
Barley flakes					30

* From: Pagani et al. (1981).

Dry skim milk and/or whey protein concentrate were added for their functional properties rather than for any nutritional reason. Table 6 shows the gross composition of these products. Ash, protein and lipid content increased with the PPC percentage; obviously, the higher protein content (36.34%) was obtained when

TABLE 6

GROSS COMPOSITION OF THE ARTIFICIAL RICE

Sample	Water	Ash	Protein (N x 5.7)	Starch	Lipid	Non nitrogenous compounds (by diff.
1	10.24	1.72	13.23	68.00	2.00	83.05
2	9.91	1.64	13.24	66.30	1.60	83.52
3	9.98	0.41	12.00	68.00	1.40	86.19
4	9.72	3.45	36.34	46.00	2.74	57.47
5	10.25	2.03	15.58	57.54	1.98	80.19

Pagani et al. (1981)

the PPC reached the 45% level. The introduction of PPC into these formulations gave pasta products which were acceptable in terms of their technological characteristics but which exhibited a marked pea taste when the percentage of this legume was higher than 10%.

TABLE 7

ORGANOLEPTIC CHARACTERISTICS OF ARTIFICIAL RICE

Sample number	Cooking time (min)	Residue (g/100 g)	Colour	Taste	Stickiness	Global evaluation
1	8	23.65	light hazel	pea	non sticky	sufficient
2	7	20.79	light hazel	pea	twisted	poor
3	10	5.43	white	pea	sticky	barley sufficient
4	10	15.12	light hazel	pea strong	non sticky	sufficient
5	9	11.71	light hazel	pea (light)	non sticky	good

Pagani et al. (1981)

Some organoleptic characteristics of artificial rice are reported in Table 7. The final evaluation of the various products, shown in the last column, takes into consideration the pea taste as a negative factor. Consideration of the pea taste as a negative factor is obviously a subjective matter.

TABLE 8

ESSENTIAL AMINO ACID (EAA) CONTENT OF THE ARTIFICIAL RICE*

EAA	Sample number				
	(1)	(2)	(3)	(4)	(5)
Threonine	3.5**	3.5	3.8	3.5	4.0
Valine	5.7	5.7	5.1	5.3	5.6
Cys + Meth	3.2	3.3	4.7	2.9	4.6
Isoleucine	4.4	4.4	5.7	4.4	4.6
Leucine	8.2	8.2	6.8	7.6	7.6
Tyr + Phe	8.3	8.3	7.0	7.9	7.7
Lysine	5.7	5.6	5.5	6.7	6.0
Tryptophane	1.2	1.3	1.3	1.1	1.2
CS***	0.87	0.87	0.95	0.83	1.00

* Calculated from figures on Table 5
** Underlined the first, double underlined the second limiting amino acid
*** Chemical score against FAO/WHO pattern

The pattern of the essential amino acids, shown in Table 8, looks fairly well balanced. The chemical score, calculated against the FAO/WHO pattern, ranges from 0.83 to 1.00.

For the formulations 1, 2 and 3, threonine becomes the first limiting amino acid, while for formulation 4 with a PPC concentration of 45%, it remains cys + met. In this last case, the second limiting amino acid is threonine, while for formulations 1 and 2 it is cys + met and for formulation 3 it is leucine. Formulation 5 appears to be completely balanced.

These data indicate that, when present in proper combination, PPC may give pasta products with high protein value and acceptable organoleptic qualities. Very recently, Patel et al. (1981) in Canada prepared a PPC-cheese-whey blend using the conventional equipment for processing non-fat dry milk. Such a blend exhibited high functional properties when tried as an ingredient in bread making. The authors suggest that this blend may have additional food uses in products such as pasta, meat patties, egg white substitutes, frankfurters and as extenders for milk supplies.

Another aspect is worthy of mention. In recent years great attention has been paid to legumes as potential hypocholesterolemic foods. Research has been conducted mainly on soyabean and its derivatives (Carroll, 1981) but some work has also been done on faba bean concentrate (Jaya et al., 1981) and other dry beans such as brown beans, chick peas and blackgram (Saraswathi Devi and Kurup, 1973). Results so far indicate that these vegetable protein sources succeed in lowering the plasma cholesterol level of hypercholesterolemic subjects and that this happens as a consequence of a significant decrease in the low density lipoprotein/cholesterol fraction. Recently, a comprehensive study on the effect of the type of dietary protein on serum lipid constituents in adult rats suggested that even pea meal exhibits a marked effect on low density lipoproteins/cholesterol (Eklund and Syöblom, 1980).

The legume protein concentrates obtained by air-classification are quite heterogeneous. Beside the protein moiety, they contain fibre and some minor constituents such as saponins and isoflavones, these last constituents having been indicated as possible hypocholesterolemic factors (Oakenfull et al., 1979; Sharma, 1979). However, from a practical point of view, no clear discrimination can be made among these compounds for their hypocholesterolemic potency. What matters is the well established hypocholesterolemic effect of legume concentrates in hypercholesterolemic subjects, both human and rat.

CONCLUSION

Pea protein concentrate prepared by air-classification represents an ingredient of potentially wide usage. The advantages PPC has over other legume products are derived essentially from the fact that it is obtained through a very simple technique which does not require chemical treatments of any kind. It is consequently free from the formation of hazardous artefacts. Moreover, the content of antitryptic factors is so low that, even following the increase consequent to the protein concentration process, their levels are such as not to require further treatments. The digestibility coefficient of the raw product, determined as apparent digestibility in the rat, is close to

85% and it remains unchanged after thermal treatment in an autoclave.

The widespread usage of PPC in human nutrition is mainly linked to the possibility of obtaining a product that is economically competitive, which depends in the first place on the ability to select, through breeding, cultivars that combine a high yield with good nutritional qualities.

ACKNOWLEDGEMENTS

Results reported in this paper were obtained in collaboration with G. Bises, M. Cappelloni and V. Gentili and supported by a grant from the NRC of Italy, Programme 'New protein sources and new food formulation'.

REFERENCES

Anon, 1979. Prospettive delle proteaginose in Italia: Atti del Convegno di Perugia, 30th May - 2nd June 1979. CNR.

Anon, 1974. Pea flour and pea protein concentrates PEPS; Bull. No. 1. Prairie Regional Laboratory. Nat. Res. Council and College of Home Economics, Un. of Saskatchewan, Saskatoon, Canada.

Bises, G., Corcos Benedetti, P., Crocchioni, G., Gentili, V. and Spadoni, M.A., 1981. Effect of some vegetable protein concentrates on protein deposition in body and liver of the rat. XII Intern. Congress of Nutrition, Aug. 16 - 21, 1981. S. Diego, California, USA, Abstr. No. 382.

Carroll, K.K., 1981. Soya protein and Atherosclerosis. JAOCS 58, 416.

Corcos Benedetti, P., Tagliamonte, B., Bises, G., Gentili, V. and Spadoni, M.A., 1981. Ann. Nutr. Metab., 25, 48.

Eklund, A. and Syöblom, L., 1980. Effects of the source of dietary protein on serum lower density lipoprotein (VLDL - LDL) and tocopherol levels in female rats. J. Nutr. 110, 2321.

Jaya, T.Y., Mengheri, E., Scarino, M.L., Vignolini. F. and Spadoni, M.A., 1981. Evaluation of hypocholesterolemic effect of faba bean protein concentrate on rats fed a high-fat-high cholesterol diet. Nutr. Rep. Intern., 23, 55.

Liener, I.E. and Kakake, M.L., 1980. In: 'Toxic constituents of plant food stuffs', ed. I.E. Liener, Acad. Press. N.Y., pp. 7-71.

Lyman, R.L. and Lepkovsky, S., 1957. The effect of raw soybean meal and trypsin inhibitor diets on pancreatic enzyme secretion in the rat. J. Nutr. 62, 269.

Oakenfull, D.G., Fenwick, D.E. and Hood, R.L., 1979. Effect of saponins on bile acids and plasma lipids in the rat. Brit. J. Nutr. 42, 209.

Pagani, A., Resmini, P. and Dalbon, G., 1981. Formulazione e produzione di paste alimentari a partire da materie prime non convenzionali. Tecnica Molitoria 32, 1.

Patel, P.R., Youngs, C.G. and Grant, D.R., 1981. Preparation and properties of spray-dried pea protein concentrate cheese whey blends. Cereal Chem. 58, 249.

Samonds, K.W. and Hegsted, D.M., 1977. Animal bioassay: a critical evaluation with specific reference to assessing nutritive value for the human. In: 'Evaluation of proteins for humans' ed. Bodwell, p. 68-80, AVI Press, Westport.

Saraswathi Devi, K. and Kurup, P.A., 1973. Hypolipidaemic activity of the protein and polysaccharide fraction from *Phaseolus mungo* (blackgram) in rats fed a high-fat-high-cholesterol diet. Atherosclerosis 18, 389.

Sharma, R.D., 1979. Isoflavones and hypercholesterolemia in rats. Lipids 14, 535.

DISCUSSION

D.W. Griffiths *(UK)* How much of the phosphorus in the pea concentrate was inorganic and how much was phytate phosphorus. There are fairly high levels of phytate phosphorus in field bean protein concentrates which can affect mineral availability.

M.A. Spadoni *(Italy)* Phytate may reduce the availability of calcium and zinc, which are very important in the metabolism of vitamin A, and is common to all legumes.

G. Dardenne *(Belgium)* As Dr. Calet said, tryptophan is one of the most important amino acids. I think it is also very difficult to analyse. What technique do you use here?

M.A. Spadoni It is the technique devised by Louvain. There is a basic hydrolysis with barium hydroxide and then we determine.

G. Dardenne It is important that the same method of analysis is used in all EC laboratories so that results from different laboratories are directly comparable.

R. Davies *(UK)* I should just like to ask for clarification. You showed levels of tryptophan in pea meal which seemed to be adequate and up to FAO standards, and yet I understood Dr. Calet to say earlier that one of our objectives should be to increase tryptophan levels. Perhaps you could clarify this for those of us who are not nutritionists.

M.A. Spandoni My personal opinion is that tryptophan levels in legumes are satisfactory. Moreover, there is no problem in relation to sulphur amino acids, because, nutritionally, it is the final combination of amino acids that is important. Human diets contain a mixture of foods and even a simple mixture of legumes, grain and a little milk provide a complete meal. Improvement of yield is however important but this should not be accompanied by a decrease in protein content. Selection against the flatulence factor as well as consideration of fixed phytate levels are both important in relation to the generalised use of legumes for babies. For adults there is no problem at all.

D. Boulter *(UK)* Your remarks refer to human nutrition in normal diet. There may sometimes be problems relating to tryptophan in animal nutrition and an increase in the tryptophan levels in some animal diets may be necessary.

G. Dardenne Tryptophan is important, but how does threonine rank in relative importance?

M.A. Spadoni Threonine is second in importance when lysine and sulphur amino acids are combined so that tryptophan becomes the limiting amino acid.

(It was generally agreed that an attempt should be made to standardise analytical procedures for protein and amino acid analysis amongst the various laboratories within the EC. This would help to ensure valid comparison of results from different laboratories. No one volunteered to undertake the task of co-ordination. Editor)

IN VITRO NUTRITIONAL PROPERTIES AND THE EFFECT OF BOUND CARBOHYDRATE AND OF CONFORMATION ON THE DIGESTIBILITY OF GLOBULINS FROM LUPIN SEEDS

P. Cerletti, P. Restani, M. Duranti, A. Mariani and A. Semino

Department of General Biochemistry of the University of Milano, Via Celoria, 2 - I-20133 Milano, Italy.

ABSTRACT

The amino acid composition of the total flour, isolated albumins and globulins and of each purified globulin from Lupinus albus has been determined. Chemical nutritional indices are quite poor for some proteins but reasonably good or excellent for others. The overall composition of the flour, of globulins and of albumins is reasonably good.

In vitro digestibility by mammalian endopeptidases has also been studied. The attack by pepsin, and even more by trypsin, of the native globulins is reduced with a specific effect for the different proteins which appears to parallel their bound sugar content. Removal of the carbohydrate by Jack bean glycosidase restores enzyme efficiency. Enzymic breakdown of the proteins is also enhanced when they become heat-denatured. The reduced effect of endopeptidases is attributed to hindrance by bound carbohydrate at the sites of action of the enzyme in the native protein.

INTRODUCTION

For several reasons lupin appears to be a promising source of alternative protein for food and feed uses. Its seed protein content is one of the highest among legumes, from 30 to 45%, according to variety; it grows under climates and soil conditions that soy cannot tolerate; it has been shown that selection can improve its seed oil content and in *L. mutabilis* values of 14.3% have been reached (Cerletti and Duranti, 1979), which is economically very important if the crop is to be competitive with soy. Finally, lupin contains only alkaloids as relevant antinutritional compounds. Antitryptic, allergenic and flatulence factors, haemagglutinins and antivitamins, all of which are present in sizeable amounts in other legumes, are absent or have not been reported in lupin. Moreover, sweet lupin varieties can be obtained where alkaloids are practically absent or very low.

When nutritional aspects are considered, in addition to total amount of protein produced the problem of its nutritional quality arises, both in terms of amino acids composition and metabolic availability, (i.e. digestibility in particular).

We approached this problem on a lupin variety of Mediterranean origin, *Lupinus albus*, which is among those giving high protein yields. The amino acid composition was studied in the flour, in separated albumins and globulins and in each purified globulin. As we shall see later, this information may serve as a guide to genetic selection and agronomic improvement of the plant. Moreover it also supplies information relevant for understanding some molecular properties; in this case, however, the amino acid sequence in the polypeptides is far more useful.

We also studied *in vitro* proteolysis by mammalian digestive enzymes. Reports in the literature speak of limited availability of plant, as compared to animal, proteins and clearcut information on this point is necessary to complement amino acid data. This approach also indicates how far technological treatments, in particular heat denaturation, may influence the digestibility.

RESULTS AND DISCUSSION

Amino acid composition

In a previous communication at this meeting (Duranti and Cerletti, 1982) the different types of proteins existing in lupin seed, with specific reference to *Lupinus albus*, have been described. The data we present here cover all the spectrum of these proteins, either as groups or as separate components. Globulins represent 88% of the total soluble protein and albumins the residual 12%. Unextracted nitrogen is 3%, if alkaloids are excluded. Among globulins, vicilins represent 44% and legumins 33% of the total.

As shown in Table 1, proteins in the groups of vicilins and of legumins are relatively homogeneous as far as amino acid composition is concerned. Differences in globulins not belonging to these groups are, on the other hand, significant. The other globulins all have a distinct amino acid composition. The two major vicilins, proteins 4 and 6, differ significantly only in aspartate and tyrosine content. Proteins 5 and 7, which on the basis of subunit composition, molecular weight and ion exchange behaviour (Duranti and Cerletti, 1982; Duranti et al., 1981) appear strictly related to 4 and 6, respectively, are also

TABLE 1

AMINO ACID COMPOSITION OF SEED GLOBULINS IN *Lupinus albus*.

			vicilins				legumins				Essential a.a. pattern, FAO (1973)
	1	3	4	5	6	7	8	9a	9b	11	
His	5.5	3.4	1.2	1.8	1.4	1.3	1.9	2.1	1.8	2.6	-
Lys	6.3	5.9	4.1	3.8	3.5	3.1	4.0	3.6	1.0	7.9	5.4
Arg	4.0	12.4	12.8	12.4	13.8	13.4	11.9	10.0	11.1	9.9	-
Asp	11.8	9.3	13.4	12.2	11.7	12.9	10.1	10.4	9.8	7.1	-
Thr	7.4	4.7	3.0	2.9	2.6	2.7	3.9	4.8	1.4	3.6	4.0
Ser	8.8	7.1	5.7	6.6	5.5	4.9	4.9	4.1	5.1	7.0	-
Glu	10.1	21.2	23.3	25.3	24.7	27.3	24.8	23.6	41.1	12.1	-
Pro	6.0	5.1	4.0	4.6	4.1	3.7	4.4	4.3	2.0	5.1	-
Gly	4.0	4.9	3.3	4.1	3.9	3.2	3.9	4.3	2.0	7.0	-
Ala	4.2	3.2	2.9	2.7	2.7	2.8	2.6	2.8	0.8	3.9	-
Cys	2.6	0.8	tr	tr	0.0	0.0	1.3	1.1	6.6	0.0	3.5 (Cys + Met)
Met	0.8	0.3	tr	tr	0.0	0.0	0.2	0.5	0.5	3.0	
Val	6.3	4.0	3.1	2.4	3.0	3.1	3.8	3.9	1.7	4.0	5.0
Ile	4.9	3.4	4.9	3.4	4.1	4.5	5.8	5.4	3.3	5.7	4.0
Leu	9.4	6.0	7.3	6.5	7.0	6.8	7.5	8.6	8.8	11.7	7.0
Tyr	3.0	3.6	4.5	5.3	6.7	6.3	5.3	4.8	1.3	4.9	5.0 (Tyr + Phe)
Phe	4.8	3.6	4.9	4.4	4.4	4.1	4.0	3.8	1.1	5.8	
Trp	1.1	1.7	tr	0.0	0.0	0.0	1.0	1.2	tr	n.d.	1.0
MEAA	99.5	80.8	33.2	25.1	25.9	26.4	82.8	84.0	35.1	91.5	100
Index for potential bitter taste	1202	1088	1132	1090	1159	1135	1197	1183	856	1322	-

tr = trace; n.d. = not determined.

alike as far as amino acids are concerned. The two legumins differ in their arginine, threonine and leucine contents. Such minor differences in amino acid composition associated with distinct subunit patterns, support the proposed genesis of the subunits by different fragmentation of a common larger parent polypeptide (Duranti and Cerletti, 1982).

Nutritionally speaking vicilins are extremely poor because of their lack of sulphur-containing amino acids and trytophan, and their low valine, lysine and threonine contents. This produces a low MEAA index. Legumins are better, with only a minor deficiency in lysine, sulphur-containing amino acids and valine; their MEAA index is above 80%. Globulin 9b is peculiar in its absence of tryptophan, low lysine, threonine, methionine, valine and aromatic amino acids but very high cysteine and glutamate. This unbalanced composition yields a low MEAA value. Significant glutamic acid is found also in vicilins and legumins as is frequently the case with plant proteins. Globulin 1, which is also an important seed constituent, has a very balanced amino acid composition which results in an MEAA value of 99%. The index for potential bitter taste calculated according to Ney (1974) is always below 1 300, which is the critical level for beginning of bitter taste, except for globulin 11. This protein, which represents only 0.4 of total globulins, is still in the range of uncertain bitterness.

The total globulin extract is low in methionine and tryptophan but still has an MEAA index of 74 (Table 2). The amino acid pattern of lupin flour is better than for total globulins, which gives rise to a higher MEAA value. This is due to the presence of albumins, which display a more balanced amino acid composition. It is interesting that if the amino acid composition is calculated for the flour, either from the composition of albumins and of total globulins and their relative contributions to the total seed protein, or for total globulins, from the amino acid content of each separate globulin fraction and the percentage of protein in the fraction (adjusting this item to 100% total recovery), the result is very similar to the determined values; this then confirms the observed amino acid values and the recoveries.

If a comparison of the essential amino acids is made with other lupin varieties - our values are compared with literature data in Table 3 - all appear to behave similarly. The best pattern is shown by *L. luteus*, which also has the highest protein content. Favourable results are given also by *L. mutabilis*. Comparison with soy is also good, both for essential amino acids and protein yield; the soybean however appears richer in tryptophan and lysine.

TABLE 2

DETERMINED AND CALCULATED AMINO ACID COMPOSITION OF SEED FRACTIONS IN *Lupinus albus*, %.

	Defatted flour		Total globulin extract		Albumins
	deter-mined	calcu-lated	deter-mined	calcu-lated	
His	2.13	1.90	1.76	1.85	2.06
Lys	4.36	3.26	3.90	3.46	5.33
Arg	11.06	11.07	11.81	11.42	12.40
Asp	10.76	10.47	10.55	10.81	11.79
Thr	3.86	3.23	3.08	3.26	4.89
Ser	5.22	5.05	5.35	5.24	5.23
Glu	25.07	25.68	27.40	24.84	26.24
Pro	4.06	4.04	4.38	3.92	4.01
Gly	3.96	3.46	3.68	3.55	1.97
Ala	3.04	2.69	2.69	2.49	3.57
Val	3.35	3.35	3.16	3.22	3.57
Cys	1.48	1.46	1.48	1.40	1.68
Met	0.51	0.33	0.27	0.23	1.01
Ile	4.24	4.55	4.50	4.52	3.75
Leu	7.31	7.69	7.36	7.54	7.51
Tyr	4.57	4.57	5.58	4.80	1.71
Phe	3.86	3.77	3.58	3.75	3.11
Trp	0.44	0.42	0.49	0.43	0.35
MEAA	78	-	74	-	81
Potential bitter taste index	1102	-	1147	-	1041

In summary, the various protein groups of lupin seeds have quite different chemical nutritional values and for the whole seed this value depends on the ratio between the various proteins. This ratio can be modified by selection and, as we will briefly mention below, by the field treatment of the plant; the data reported therefore become one of the leading criteria for the improvement of this crop. Selection must aim to

TABLE 3

ESSENTIAL AMINO ACID CONTENT OF DIFFERENT SPECIES OF LUPIN SEEDS AND SOYBEAN, %.

	Albus	*Angustifolius**	*Luteus**	*Mutabilis**	Soybean*
Lys	4.4	5.2	6.1	6.0	6.4
Thr	3.9	3.5	4.0	4.0	3.9
Cys + Met	2.0	2.3	3.1	2.7	2.6
Val	3.4	4.0	4.3	4.0	4.8
Ile	4.2	3.8	4.6	4.7	4.5
Leu	7.3	6.6	9.0	7.4	7.7
Tyr + Phe	8.3	7.1	6.9	7.9	8.1
Trp	0.4	0.9	0.9	0.8	1.2
MEAA	78	87	95	92	96

* Hudson et al., 1976.

favourably modify the ratio of legumins to vicilins. These globulins represent by far the major proteins in the seed and their relative proportions can be altered relatively easily. The harvesting time is also important, since legumins seem to be deposited after vicilins, as shown by Danielsson (1952) on *Pisum sativum*. Field treatment may be effective in improving the ratio between the two groups and globulin 1. It has been observed that when soil nutrients are poor in sulphur, legumins and globulin 1 decrease in the seed and nearly disappear, and are substituted by vicilins (Blagrove et al., 1976), thus completely changing the nutritional value of seed proteins for man.

In vitro digestibility

An alternative approach to evaluating the nutritional performance of lupin proteins was to study their breakdown by mammalian digestive proteolytic enzymes. This was done by determining the amount of large and small TCA soluble peptides liberated by endopeptidases.

As seen from Table 4 consistent amounts of peptide become TCA soluble independently of enzyme action; this result from the

TABLE 4

TCA SOLUBLE PEPTIDES FROM UNDIGESTED LUPIN SEED GLOBULINS. LUPIN GLOBULINS AND CASEIN WERE DISSOLVED AT THE pH AND KEPT AT 37°C FOR THE TIME INDICATED. AFTER ADDITION OF TCA TO A FINAL CONCENTRATION OF 10% AND CENTRIFUGATION, MICROBIURET REACTIVE COMPOUNDS WERE ASSAYED IN THE SUPERNATANT. THEY ARE GIVEN AS % OF TOTAL PROTEINS.

Hours at 37°C	Total globulins		Globulin 1		Globulin 4		Globulin 6		Globulin 8		Globulin 9a+9b		Casein	
pH	2.0	*7.9*	2.0	*7.9*	2.0	*7.9*	2.0	*7.9*	2.0	*7.9*	2.0	*7.9*	2.0	*7.9*
0	15.2	*13.0*	17.8	*9.3*	17.3	*10.7*	15.7	*9.7*	9.5	*7.2*	9.1	*13.1*	0	*0*
8		*15.4*		*9.6*		*10.7*		*12.5*		*11.2*		*16.5*		*0*
8	18.8		20.8		18.9		16.1		13.8		12.3		0	

lupin globulins containing 'nicked' subunits (Duranti and Cerletti, 1982). A further effect is observed at pH 2 and this is likely to be a consequence of the dissociation of the oligomeric structure of the proteins at the low pH of incubation. These spontaneously formed peptides were deduced from results obtained in samples where enzymes had acted, and the results are shown in what follows.

As shown in Table 5, pepsin is less effective on lupin globulins than on casein and the decrease differs for the

TABLE 5

PROTEOLYSIS BY PEPSIN. THE PROTEINS WERE INCUBATED WITH PEPSIN AT 37°C IN 0.01 N HCl pH 2.0 IN A 1/60 WEIGHT RATIO. TCA SOLUBLE PEPTIDES PRODUCED, DETERMINED BY MICROBIURET, ARE GIVEN, NET OF BLANK, AS % OF INITIAL PROTEIN.

Digestion time, hours at 37°C	Casein	Total globulin extract	Globulin 1	Globulin 4	Globulin 6	Globulin 8	Globulin 9a+9b
1	35	34	32	34	34	36	30
3	44	38	33	36	36	44	37
6	47	40	39	37	37	44	37
8	48	41	40	37	38	44	46
24	50	41	40	41	38	44	48
Bound sugar, mg/100 mg peptide	0.3	2.3	4.3	2.4	2.1	1.5	0.8

various proteins; vicilins and globulin 1 are less digested than legumins and globulin 9b and the loss in activity apparently parallels the bound sugar content of the proteins. Trypsin is affected to a much larger extent than pepsin and the pattern differs for the various proteins (Table 6). Decreased proteolysis is evident from the very start of incubation for globulin 1 and the vicilins, but legumins and globulin 9b appear less digested than casein only after 3 h incubation. Longer times make the difference more pronounced, and the resultant soluble peptides become smaller in size and react with ninhydrin, whereas biuret reactivity is lost. This is not the case for casein. The above effects are most evident for vicilins and for globulin 1, but less for legumins. Again the decrease parallels bound sugar.

TABLE 6

PROTEOLYSIS BY TRYPSIN. THE PROTEINS WERE INCUBATED AT 37°C IN 0.1 M PHOSPHATE BUFFER pH 7.9 WITH TRYPSIN IN A 1/60 WEIGHT RATIO. TCA SOLUBLE PEPTIDES PRODUCED BY TRYPSIN ARE GIVEN, NET OF BLANKS, AS % OF THE INITIAL PROTEIN. ASSAYS ARE BY MICROBIURET (µbi.) AND NINHYDRIN (nin.) METHODS.

			Vicilins			Legumins			Casein	
Hours	1		4		6	8	9a+9b			
at 37°C	µbi.	nin.	µbi.	nin.	µbi.	µbi.	µbi.	nin.	µbi.	nin.
0.5	4.8		4.6		9.8	19.7	15.6		11.5	
1	6.8	*6.0*	5.6	*5.7*	10.7	22.5	18.0	*5.5*	21.8	*2.7*
3	8.8	*7.1*	11.9	*6.7*	12.7	24.9	20.4	*6.5*	36.4	*3.0*
6	8.6	*8.0*	5.6	*9.9*	16.1	26.0	22.0	*8.5*	40.6	*3.5*
24	4.1	*9.4*	5.6	*26.0*	8.0	16.6	15.8	*11.4*	48.9	*3.5*
Neutral sugar mg/100 mg peptide	4.3		2.4		2.1	1.5	0.8		0.3	

The effect of bound sugar was further investigated by comparing the breakdown by trypsin of lupin proteins in which sugar was present with that of the same globulins from which sugar had been removed. The protein was first incubated with glycosidase (purified in our laboratory from the Jack bean according to Li and Li (1970), to the ammonium sulphate fractionation stage).

As shown in Table 7, glycosidase splits a consistent amount of bound sugar from the globulins. Conditions for glycosidase action are rather mild. We therefore believe, that the protein itself is unchanged, except for its sugar content. Blanks were treated under the same conditions, except for absence of enzyme. Glycosidase treated globulins were then digested with trypsin in parallel with the blanks and the results are shown in Figure 1. Glycosidase treatment strongly increases the effect of trypsin, especially in liberating large peptides, as determined by the biuret reaction. The effect on ninhydrin-reacting compounds is less marked.

TABLE 7

EFFECT OF GLYCOSIDASE ON BOUND CARBOHYDRATES OF TOTAL GLOBULIN EXTRACT AND GLOBULIN 4. THE GLOBULINS WERE INCUBATED WITH OR WITHOUT GLYCOSIDASE IN A 1/70 WEIGHT RATIO IN 0.05 M CITRATE BUFFER pH 4.5. THE pH WAS THEN BROUGHT TO 9.8 WITH 0.2 M BORATE BUFFER. THE DIALYSED REACTION MIXTURE WAS FREEZE-DRIED AND REDISSOLVED IN 0.1 N NaOH AND PROTEINS (Itzaki and Gill, 1964) AND SUGARS (McKelvy and Lee, 1969) WERE ASSAYED.

Digestion time, min. at 25°C	Bound carbohydrates mgx/100 mg protein				Sugar solubilised, % of initial	
	Total globulin extract		Globulin 4		Total globulin extract	Globulin 4
	Blank	+glycosidase	Blank	+glycosidase		
0	2.7	2.7	6.4	6.4	-	-
30	2.7	1.9	n.d.	n.d.	29.1	n.d.
45	n.d.	n.d.	6.0	3.7	n.d.	42.2
60	2.4	1.7	6.0	2.6	33.3	59.4
75	2.3	1.6	n.d.	n.d.	40.7	n.d.

Upon longer incubation, the reactivity to trypsin increases also for the protein which was not treated with glycosidase. This was attributed to the effect of heating during incubation and the following experiments were performed to check this hypothesis.

Control proteins containing little casein (casein, 0.3%) or no (BSA) bound sugar, and one of the globulins were digested

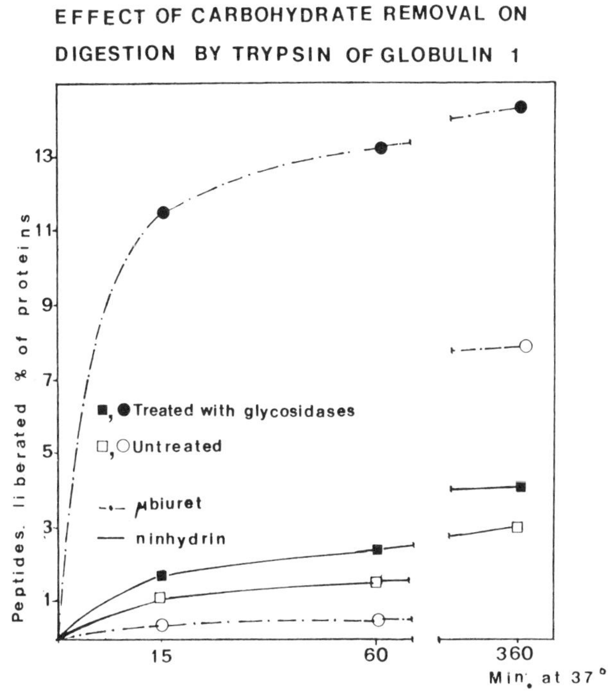

Fig. 1. Globulin 1 was incubated 60 minutes at 25°C with or without glycosidase in a 1/70 weight ratio in 0.05 M citrate buffer pH 4.5. The pH was then brought to 9.8 with 0.2 M borate buffer. The dialysed reaction mixture was freeze-dried and redissolved in 0.1 M phosphate buffer pH 7.9 and was digested at 37°C with trypsin in a 1/60 weight ratio to the globulin. TCA soluble compounds formed determined by microbiuret and ninhydrin are given as % of the initial protein.

with trypsin with and without heating for 30 minutes at 100°C. In all cases heating increased proteolysis. The percentage increase, as measured by the biuret and ninhydrin reactions, is shown in Table 8. The effect of heating is pronounced with lupin proteins, which if untreated are less attacked by trypsin, whereas it is much less for casein and for BSA. For short periods of incubation the relative increase is more relevant for vicilins, which in their native state are less digested, than for legumins. When the incubation is prolonged the effect on globulins and on casein is more in the formation of larger peptides, than on ninhydrin reacting compounds (i.e., small peptides and amino acids). The relative increase is less

TABLE 8

INCREASED DIGESTIBILITY BY TRYPSIN, %, ON HEATING 30 MINUTES, AT 100°C. DIGESTION BY TRYPSIN WAS AS DETAILED IN FIGURE 6. VALUES GIVE % INCREASE, AS COMPARED TO THE NON-HEATED PROTEIN.

Digestion time	Globulin 4		Globulin 8		BSA		Casein	
	µbi.	*nin.*	µbi.	*nin.*	µbi.	*nin.*	µbi.	*nin.*
15 min.	261	*216*	n.d.	*n.d.*	n.d.	*n.d.*	n.d.	*n.d.*
30 min.	n.d.	*n.d.*	159	*169*	51	*25*	27	*71*
1 h	166	*119*	136	*123*	50	*66*	37	*15*
3 h	n.d.	*n.d.*	162	*83*	21	*30*	20	*0*
6 h	66	*18*	n.d.	*n.d.*	n.d.	*n.d.*	17	*6*

marked upon long incubation; this because proteins not treated at 100°C also become more digested when incubated at 37°C for long periods.

CONCLUSIONS

These results indicate that the native conformation of the proteins is the cause of the reduced effect of trypsin, since the diminished enzyme efficiency largely disappears after heat denaturation. In addition to this, reduced proteolysis of native globulins depends on the presence of bound sugar. Therefore, denaturation or removal of bound sugar from the native protein achieve the same effect, namely they expose the site of attack for trypsin on the peptide. When the peptide is protected from enzymic attack by bound carbohydrate, the exposed fragments split are repetitively broken down to yield peptides of very small size or even free amino acids, whereas when enzyme action is not hindered the normal effect of endopeptidases, to produce larger fragments, prevails. This explains the different reactivity with biuret and with ninhydrin.

The results show that the poorer nutritional performance of lupin proteins does not only depend on the amino acid composition but also on reduced digestibility by proteolytic enzymes. By indicating possible causes of this behaviour at the molecular level and elucidating the molecular characteristics of the various storage proteins of the seed, these studies clarify the chemical basis of the nutritional performance and permit a logical approach to improvement of this performance. This can be done by genetical means; i.e., to select varieties with more favourable protein composition; or the relative proportions of nutritionally desirable proteins can be enhanced through agronomic management. Concerning carbohydrate, Basha and Beevers (1976) have shown that the carbohydrate component in glycoproteins from pea changes during seed ripening and thus the period of harvesting may be important to the composition of the proteins. A further approach may be technological treatments which properly denature the proteins and/or remove the carbohydrate, making them more suitable to the attack of mammalian digestive proteases.

ACKNOWLEDGEMENT

The work reported in the present paper has been supported by contracts from the Italian National Research Council (CNR) programme: 'New Protein Foods'.

REFERENCES

Basha, S.M.M. and Beevers, L., 1976. Plant Physiol. 57, 93.
Blagrove, R.J., Gillespie, J.M. and Randall, P.J., 1976. Aust. J. Plant Physiol. 3, 173.
Cerletti, P. and Duranti, M., 1979. J. Am. Oil Chem. Soc. 56, 460-463.
Danielsson, C.E., 1952. Acta Chem. Scand., 6, 149.
Duranti, M. and Cerletti, P. Molecular properties and the composition of the globulins from lupin seeds. This volume.
Duranti, M., Restani, P., Poniatowska, M. and Cerletti, P., 1981. Phytochem. 20, n. 9, 2071-2075.
FAO Report, 1973. Joint FAO/WHO Committee, 63.
Hudson, B.J.F., Fleetwood, J.G. and Zand-Moghaddam, A., 1976. Plant Food for Man, 2, 81.
Itzaki, R.F. and Gill, D.M., 1964. Anal. Biochem., 9, 401.
McKelvy, J.F. and Lee, Y.C., 1969. Arch. Biochem. Bio-phys., 132, 99-110.
Ney, K.H., 1974. In: The Contribution of Chemistry to Food Supplies (Morton, I. and Rhodes, D.N., eds.) p.411, London.
Su Chen Li and Yu Teh Li, 1970. J. Biol. Chem. 246, 5153-5160.

DISCUSSION

R. Casey *(UK)* Glycosidases are often contaminated with proteolytic activity. Did you check to see if your glycosidase treatment released peptides?

P. Cerletti *(Italy)* Electrophoresis revealed no such suggestion. The SDS electrophoresis behaviour of these enzymes, which you saw this morning in Duranti's paper, is identical as far as we can tell.

C. Calet *(France)* You used heat treatment to increase the availability of protein. Is there not a danger of a Maillard reaction? What temperature do you use?

P. Cerletti The temperature used was $100^{\circ}C$. These are purified proteins and I do not know whether a Maillard reaction could be expected to occur.

D. Boulter *(UK)* Were they heated in an oven or in boiling water? If the latter is done carefully, I think there is little Maillard reaction in such a situation, if there is pure carbohydrate present.

I would just like to make one comment in this matter, because I think the literature is very confusing. Very often in the literature, certain information is given to suggest that plant proteins, particularly, say, legume seed proteins are relatively indigestible. This evidence is usually based on *in vitro* study, using just pepsin or trypsin or sometimes mixtures, and often the protein is undenatured. Frequently under these circumstances, a plant protein will give a poor value as demonstrated in this talk. However, the enzymes which are used are unrepresentative of the cascade situation that takes place in normal digestion, and it is possible, if you are prepared to prepare complex mixtures of enzymes, virtually to digest completely these seed proteins from several legumes. The statement that they are undigested or indigestible from *in vitro* studies is usually misleading.

With the *in vivo* situation the information is also misleading, because what is fed to the animal is not seed protein, it is in fact the meal, and the meal may contain certain antiphysiological substances, such as tannins and inhibitors, which may lead to misleading results from rat assays. In those few instances where purified seed proteins have been fed to rats, pigs and chickens, they are completely digested. There are, however, some seed proteins which are not completely digested, particularly the trypsin inhibitors. These are digestible if they are fed alone, but in combination with animal trypsin, they will form a complex and pass through into the faeces. Unfortunately, therefore, this literature is extremely confused, but I would suspect that most basic storage proteins are highly digestible when they are denatured and attacked by a cascade of enzymes, as in the digestive tract.

P. Cerletti I agree that reports in the literature are confused: this is because the various components involved in such complicated systems have not been studied sufficiently at the molecular level. This is necessary in order to have a clear picture of their behaviour and to obtain the necessary information to be able to modify that behaviour if required and this is our aim.

In vitro studies can be significant without the use of a cascade of enzymes, because such studies detect the mode of action of the physiological reactants. For example, our results indicate that the protein is modified upon prolonged incubation at $37^{\circ}C$ with trypsin, and its digestibility increases.

If we now come to the *in vivo* situation, sweet lupin contains very low levels of anti-nutritional compounds and it is possible to use it directly. It, therefore, becomes important to study the action of proteases on protein which has not been modified by pretreatments.

D. Boulter I agree, of course, but I am sure you will admit that the levels of *in vitro* digestion you refer to are about 45% and such a system, with less than 50% digestion of protein, obviously cannot mimic the *in vivo* situation.

P. Cerletti I was taking the molecular approach. Our friends the nutritionists will put it on the operative level.

C. Calet I have a comment to follow your proposition, Dr. Boulter. We say that there are many differences between digestion *in vitro* and *in vivo*. I think there are also many differences within the *in vivo* situation. For instance lupin is good for chickens, but digestibility of the protein in cows is poorer. Protein concentrate from peas is not very good in calves. I think the best model for human, both adult and baby, nutrition is not the rat, but the calf.

THE POLYPHENOLIC CONTENT OF FIELD PEAS AND THEIR POSSIBLE SIGNIFICANCE ON NUTRITIVE VALUE

D.W. Griffiths
Welsh Plant Breeding Station, Aberystwyth, UK.

INTRODUCTION

Considerable interest is currently being shown in both the field pea (*Pisum* spp.) and the field bean (*Vicia faba*) as possible alternative crops for increasing the production of indigenous plant protein in Western Europe. However, the results of various feeding trials (Nitzan, 1971; Palmer and Thompson, 1975) have indicated that the nutritive value of field beans is significantly less than that predicted by its amino acid composition and indeed a number of anti-nutritive factors, including tannins, have been identified. Tannins have also been reported to be present in some varieties of peas (Lindgren, 1975) but the results of some comparative feeding trials (Fowler and Livingstone, 1977; Davidson, 1977) would suggest that peas may be fed at far higher levels of inclusion than beans without any deleterious effects on animal productivity. Consequently, the objective of this study was to compare the levels of tannin present in both species and to examine *in vitro* their potential nutritional significance.

TANNIN AND POLYPHENOLIC CONTENT OF PEAS AND BEANS

Grain samples from three varieties, consisting of two coloured and one white flowered types, were selected from each of the two species and manually separated into testa and cotyledon. These, together with samples of milled whole grains, were analysed for both total phenol (Folin-Dennis method) (Burns, 1963) and condensed tannin (vanillin - H_2SO_4 method) (Swain and Hillis, 1959) contents.

The grain weights of the pea varieties were, as expected, significantly lower than those of field beans but the proportion of testa was also less for the peas, averaging 9.64% of the total seed weight as compared with 13.3% for the field bean varieties. The condensed tannin content of the separated testa and cotyledons (Table 1) indicated that in both species these compounds

TABLE 1

DISTRIBUTION OF CONDENSED TANNINS AMONGST THE TESTA AND COTYLEDONS OF PEA AND BEAN VARIETIES

Variety	Flower pigmentation	% condensed tannins		
		Whole	Cotyledons	Testa
Peas				
Marathon	Coloured	0.23	0.08	2.86
Minerva	Coloured	0.35	0.09	3.75
Filby	White	0.06	0.08	0.03
Beans				
Maxime	Coloured	0.59	0.08	4.20
Minden	Coloured	0.70	0.08	4.20
Triple White	White	0.06	0.07	0.04

were concentrated in the seed coat or testa of the coloured flowered varieties, whilst both the white flowered varieties were devoid of tannins. The reduced levels found in the whole milled pea samples as compared with the whole bean samples would appear, therefore, to reflect not only the slightly lower tannin content of the pea testa but also the reduced amount of testa present in the whole peas. The same samples were also analysed for total phenol content and the values obtained reflected those for condensed tannin content, indicating that polyphenolic production was largely restricted in both species to the testa of coloured flowered varieties.

EFFECT OF TESTA TANNIN ON PROTEIN SOLUBILITY

Protein solubility was determined by extracting a known weight of sample in 5% saline. This process was repeated three times and the resulting residue weighed and analysed for nitrogen content by the macro Kjeldahl method.

Preliminary investigations revealed that in contrast to the white flowered varieties from both species, removal of the testa increased protein solubility by 2 - 3 units for the coloured flowered pea varieties and by 7 - 8 units for the coloured flowered bean varieties. However, since this difference between the two species in relation to the protein binding capacity of the seed coats reflected not only differences in testa tannin content, but also differences in the amount of testa present, cotyledons and testa from one coloured and one white flowered variety from each of the two species were mixed in all possible combinations, such that each mixture contained 15% testa. The protein solubilities of the 16 mixtures were then determined (Table 2) and it can be seen that, at the same rate of inclusion, the coloured flowered testa from both bean and pea species had almost an identical inhibitory effect on protein solubility. This would suggest that the coloured flowered testa from both species had, on a weight for weight basis, a similar affinity for both pea and bean cotyledon protein.

TABLE 2

PROTEIN SOLUBILITY (%) OF VARIOUS COTYLEDON-TESTA MIXTURES

Testa	Cotyledons			
	Minerva	Filby	Minden	T. White
Minerva	74.4	74.7	78.8	71.4
Filby	80.5	81.8	85.4	74.7
Minden	73.6	73.7	78.8	67.4
T. White	80.9	80.7	84.8	71.9

EFFECT OF TESTA EXTRACTS ON ENZYME ACTIVITY

Testa extracts were prepared by refluxing 1 g testa with 850 ml distilled water for 5 h and, after filtering, concentrated by rotary evaporation to 50 ml. The inhibitory activities of these preparations were evaluated using commercially available enzyme preparations and suitable synthetic substrates.

The results for the inhibitory activities towards the digestive enzyme, trypsin, of four varieties consisting of one coloured variety and one white flowered variety from each of the two species, are given in Table 3. The enzyme was strongly inhibited by extracts from both coloured flowered varieties, but was only slightly affected by the white flowered testa extracts. The extent of the inhibition by the coloured flowered testa

TABLE 3

THE INHIBITION OF TRYPSIN BY VARIOUS TESTA EXTRACTS

Variety	% inhibition	
	No PVP	+ PVP
Peas		
Minerva	65.8	8.9
Filby	7.2	8.0
Beans		
Minden	51.4	8.3
Triple White	6.5	6.4

extracts was similar for both species and the addition of polyvinyl pyrrolidone (a tannin complexing reagent) almost completely restored trypsin activity. This would again suggest that the observed inhibitory activity was of a polyphenolic origin and the fact that similar results were obtained for both α-amylase and cellulase indicates that the observed inhibition was non-specific.

CONCLUSION

From the results presented here it is clear that pea tannins, which are restricted to the seed coat of coloured flowered varieties, have a very similar protein binding capacity and enzyme inhibitory activity to those present in field beans. Consequently, since field bean testa tannins have previously been shown to have a significant negative effect on nutritive value (Martin-Tanguy et al., 1977; Moseley and Griffiths, 1979), it is likely that high levels of field pea testa tannins would have a similar negative effect. Although whole peas contain less tannin than whole beans, it is likely that in the preparation of iso-nitrogenous diet this difference would be negated due to the higher content of peas required to achieve the same protein level.

In relation to breeding programmes designed to maximise the nutritive value of field peas it would appear that the tannin content of the selections made should be minimised by either selecting for the desired characteristics within white flowered tannin-free types, or alternatively, should coloured flowered types have non-transferable advantageous characteristics, the testa content of these should if possible be kept to a minimum, thereby reducing the total legumin content of the whole grain.

REFERENCES

Burns, R.E., 1963. Methods of tannin analysis for forage crop evaluation. Georgia Expt. Stn. Tech. Bull NS 32.

Davidson, J., 1977. The nutritive value of field beans and field peas. In: Protein Quality from Leguminous Crops. EUR 5686EN, Commission of the European Communities, Luxembourg, p 243-250.

Fowler, V.R., 1980. The value of field beans (*Vicia faba* L.) and other legumes for the feeding of pigs. In: Vicia faba: Feeding Value, Processing and Viruses (Ed. D.A. Bond) Martinus Nijhoff, The Hague, pp 31-41.

Fowler, V.R. and Livingstone, R.M., 1977. Replacement of soyabean as the protein concentrate in pigs diets by field beans (*Vicia faba*) and field peas (*Pisum sativum*). Anim. Prod. 24, 138.

Lindgren, E., 1975. The nutritive value of peas and beans for hens. Swed. J. Agric. Res. 5, 159-161.

Martin-Tanguy, J., Guillaume, J. and Kossa, A., 1977. Condensed tannins in horse bean seeds; chemical structure and apparent effect on poultry. J. Sci. Food Agric. 28, 757-765.

Moseley, G. and Griffiths, D.W., 1979. Varietal variation in the anti-nutritive effects of field beans (*Vicia faba*) when fed to cats. J. Sci. Food Agric. 30, 772-778.

Nitzan, Z., 1971. *Vicia faba* vs. soyabean meal as a source of protein. J. Sci. Food Agric. 22, 252-255.

Palmer, R. and Thompson, R., 1975. A comparison of the protein, nutritive value and composition of four cultivars of faba bean. J. Sci. Food Agric. 26, 1577-1583.

Swain, T. and Hillis, W.E., 1959. The phenolic constituents of *Prunus domestica* J. Sci. Fd. Agric. 10, 63-68.

DISCUSSION

R. Thompson *(UK)* Is it possible to use PVP, or something like it, to bind tannins in feedstuffs before feeding them to monogastrics?

D.W. Griffiths *(UK)* I am not aware of PVP being used, but polyethylene glycol, which has similar effects, has been used to try to show that protein digestibility is increased by binding tannins. I doubt if it could be used economically though. There are a number of bird resistant sorghum types which have high levels of tannin, and alkali treatment has been suggested to increase the molecular size of the tannin and so reduce its effect on digestibility.

M. Poulsen *(Denmark)* We have found that digestibility of the individual amino acids of triple white *Vicia faba* was very different from that of normal field beans. The digestibility of sulphur containing amino acids was almost doubled, and certainly the biological value of the proteins increased. Is there any explanation for this?

D.W. Griffiths Do you mean that there was an increase in methionine availability in the white flowered types? It may be that tannin may bind the trypsin and so stimulate the pancreas to produce more trypsin.

J. Picard *(France)* The use of isogenic material would make this type of finding more easily interpreted.

P. Cerletti *(Italy)* Are the tannins altered at the acid pH of the stomach?

D.W. Griffiths They may split off from the proteins at that pH, but in the small intestine the pH value is about 7, so they would be available to bind with the proteins and with the enzymes in the small intestine.

EMULSIFYING AND GELLING PROPERTIES OF THE PROTEINS FROM LUPIN SEED

Anna Riccardi, P. Cerletti and G. Bertolotti
Department of General Biochemistry, University of Milan,
Via Celoria, 2 - 20133 Milano, Italy.

INTRODUCTION

The use of new protein material for feed and in food products depends on its nutritional and functional properties. These latter have great technological importance because they direct the behaviour of the material considered under the conditions of treatment (e.g. temperature, pH, ionic strength, type and amount of other substances present etc.), and will determine its contribution to the consumer acceptance of the final product.

Lupin appears a promising protein source for temperate areas, both in terms of its soil and climate requirements, and in terms of the amount of protein produced and the nutritional characteristics of the latter, as discussed in previous communications (Cerletti et al., 1982). The type and molecular properties of its proteins have been investigated (Duranti et al., 1982) and part of this information may serve as a basis towards the understanding and possible modifications of the functional behaviour. This indeed was the approach of the work done in our laboratory which will be briefly reported here. The functional properties were studied on the defatted flour which, due to its high protein content, represents the cheapest and most convenient protein mixture for food and feed and also for the globulins (which quantitatively represent the predominant protein component). In this report an account will be given of the emulsifying and rheological properties.

RESULTS AND DISCUSSION

Emulsifying properties were studied by two different procedures.

Emulsion activity was first determined according to Yatsumatsu (1972). This approach allowed comparison with data from the literature. The assay, however, does not satisfactorily represent the emulsion potential when, as in our case, the protein used dissolves completely and has emulsifying properties of some significance, since the prescribed amount of oil is too little to produce the potential amount of emulsion. As an alternative we used the method of Quaglia et al. (1980) and Ricardi et al. (1981); this method avoids the above shortcomings since it determines the inversion point on emulsions using a constant total volume, but a changing ratio of oil to solution. These differences are clear from Figures 1A and 1B.

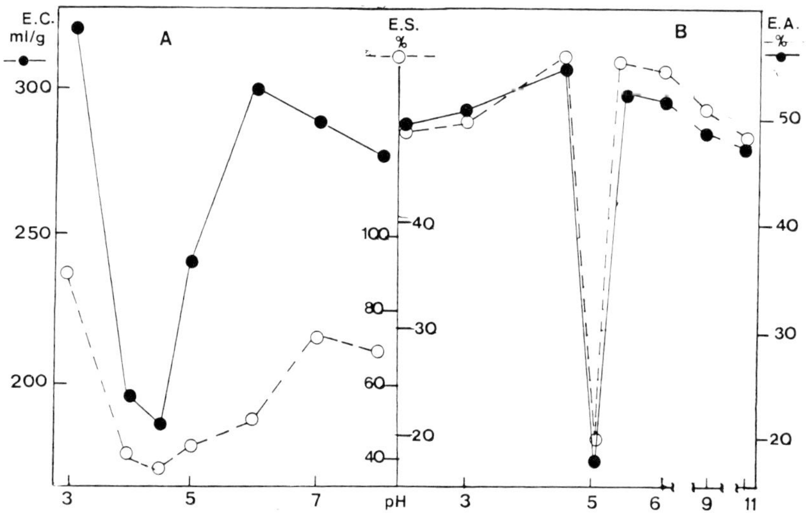

Fig. 1A. For the emulsifying capacity (E.C. ●——●), 1% globulin solutions in citrate and phosphate buffer at the desired pH, I = 0.34 - 0.37, were mixed in different ratios with increasing amounts of soy oil to a final volume of 100 ml, at 4°C. After 1 min shaking at 13 000 revolutions/min, the inversion point was appreciated visually. The ml of oil emulsified per mg of protein in the sample, measure the emulsifying capacity. Emulsion stability (ES, O---O) was measured according to Acton and Saffle (1970).

1B. The emulsifying activity (E.A. ●——●) was measured by mixing the globulin solution, 70 mg/ml, at the desired pH and I = 0.02 with the same volume of soy oil, then shaking 1 min at 710 revolutions/ min and centrifuging for 5 min at 1 200 x g. Emulsifying activity is given by the percent height of the emulsified layer versus total height. For emulsion stability (ES, O---O) the same ratio was measured after heating 30 min at 80°C, 15 min cooling under tap water and 5 min centrifuging at 15 000 x g.

The emulsion capacity is good and changes with pH, with a consistent decrease in slightly acidic medium which can be ascribed to the reduced solubility of the globulins (Riccardi and Cerletti, 1981). At the high salt concentration of the assays (I = 0.35) the pH of minimal solubility shifts towards lower values (Manrique et al., 1974). When measured according to Yatsumatsu (1972) (Figure 1B) modifications become evident only close to pH 5 where the solubility of the proteins at the ionic strength used (I = 0.02) is minimal and therefore not all the available oil is emulsified.

The emulsion capacity of lupin globulins compares well with that of proteins from beef plasma and from favabean. The emulsion activity gives similar values to those reported for soy (Yatsumatsu, 1972).

The method of Yatsumatsu involves a heating step which denatures part of the proteins and the coagulated material increases the emulsion volume artefactually. When we measured, according to Acton and Saffle (1970), the emulsified oil remaining after 24 h standing at room temperature, the emulsions appeared to have degraded to a different extent over all the pH range assayed. Stability appeared to follow emulsion capacity except for reduced stability around pH 6.

Rheological properties

As shown in Figure 2, viscosity in flour dispersions submitted to an increasing shearing gradient behaves differently on either side of the isoelectric point. Stronger resistance to the applied gradient suggests a more consistent aggregation in an acidic environment. This probably correlates with better gelling capacity, as discussed below. The fluids appear non-Newtonian and display a certain degree of hysteresis. Rearrangement of structure under an applied stress is evident also in gels. Viscosity of globulins could not be measured since it was below the sensitivity of the instrument.

Gel formation by the flour and globulins at various pH values and different temperatures is shown in Table 1. Components other than protein in the flour favour the gelling process; indeed it occurs over all the pH range assayed and at the lower temperatures. Globulins require a higher temperature and gels form only on the acidic side of the isoelectric point. Vicilins, legumins and protein 9b, which together represent 90% of the

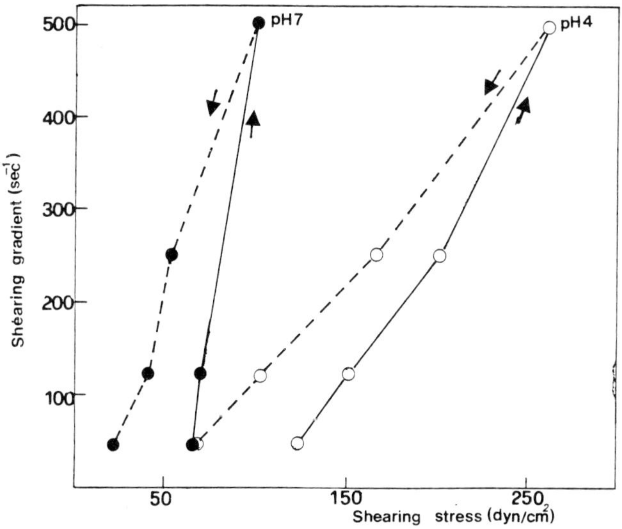

Fig. 2. Flow properties of dispersions of lupin flour. Dispersions of 45 mesh lupin flour (240 mg/ml) was prepared according to Catsinpoolas and Meyer (1970) at the desired pH. Viscosity measurement was made with a RN VEB MLW viscometer at 25°C. The applied gradients were increased and decreased as shown by arrows.

TABLE 1

GEL FORMATION BY LUPIN FLOUR AND GLOBULINS

pH	Flour		Globulins	
	75C	87C	75C	87C
2	+	n.d.	-	-
4	n.d.	+	-	+
5	+	n.d.	n.d.	c
7	+	+	n.d.	c
8	+	n.d.	n.d.	n.d.
10	+	n.d.	-	-
13	n.d.	n.d.	n.d.	-

\+ = gel
\- = no significant modification in the dispersion
c = clotted granules in liquid
n.d. = not determined

globulins, contain high levels of dicarboxylic amino acids (which are amidated only to a minor extent); glutamate is 25%, and aspartate 12%, of the total amino acids in vicilins and legumins, and in protein 9b the two components are 41% and 10% respectively (Duranti et al., 1981). In the pH region mentioned, carboxylic groups became less dissociated and the interaction both between protein molecules and with the solvent improves, with the mentioned effects on gel formation and on viscosity.

At the isoelectric and neutral pH protein coagulates to form granules, and at extreme pH values no visible change is observed. The relatively low content of sulfhydryl groups and of disulfide bridges may be in cause: amino acids are practically absent in vicilins, which represent 44% of the total proteins (Duranti et al., 1981). In the soybean the 11 S protein which corresponds to the legumins, shows better gelling capacities than the 7 S protein, which corresponds to the vicilins (Saio and Watanabe, 1978). Reduced interaction with solvent may play a role in clotting.

The behaviour of gels of flour and of globulins submitted to changing shearing gradient is shown in Figures 3 and 4. The initial globulin gel rearranges to a less resistant structure as soon as submitted to the gradient. A minor effect of this type is observed also with the flour. The overall behaviour is thixotropic with considerable hysteresis and is similar to that observed with soy proteins (Catsimpoolas and Meyer, 1970). It is possible that the tridimensional network of the gel is altered by mechanical stress, and rearranges.

Where indicated, an applied gradient was prolonged until a stable asset was reached at lower consistency values. Those attained with low gradients, however, appeared metastable upon increasing the applied force. The size of change and the time required to attain it depend on the applied gradient.

Final values, apparently stable and reversible, reached by gels of flour correspond to those measured in viscosity assays. In gels of globulins, on the other hand, the consistency decreases to values below the measuring capacity of the instrument. In both cases the product recovered still has gel features.

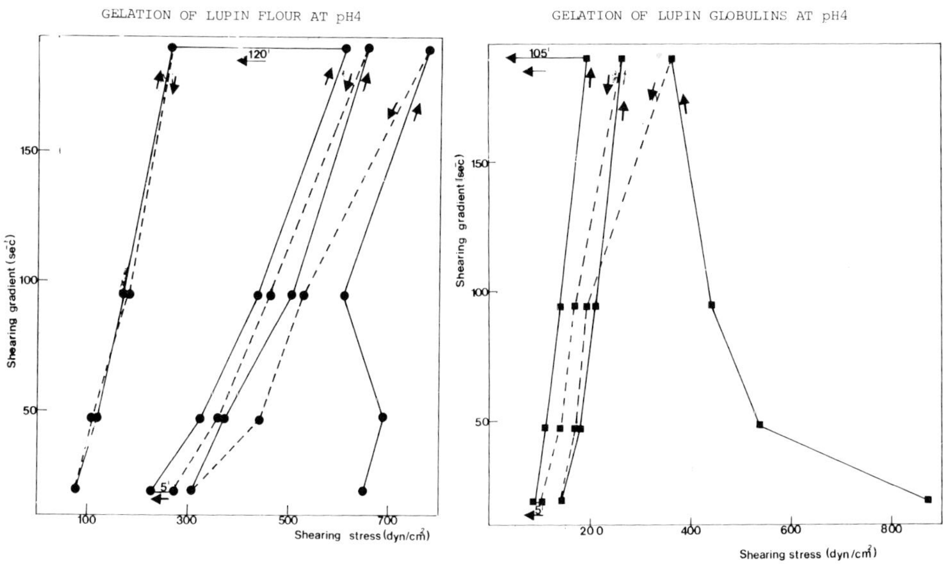

Fig. 3 and 4. Rheological properties of gels. Gels were prepared at 87°C and pH 4 as described in Table 1. The viscometer used was as in Figure 2. The applied gradients were increased and decreased as shown by arrows.

Where indicated by horizontal arrows the gradient was maintained for the necessary time to measure the immediate stress, until stability was reached. The time required and the value attained are indicated. All assays were at 25°C.

Some solvent also separates, corresponding to one third of the initial gel volume.

Gels of lupin flours formed at pH 7 have similar behaviour to those formed at acidic pH, except for a more resistant structure at the start. However, they rapidly rearrange under the applied gradient to reach the same final values as gels at pH 4.

REFERENCES

Acton, J.C. and Saffle, R.C., 1970. J. Food Sci. 35, 852.
Catsimpoolas, N. and Meyer, A.W., 1970. Cereal Chem. 47, 559-570.
Cerletti, P., Restani, P., Duranti, M., Mariani, A. and Semino, A., 1982. *In vitro* nutritional properties and the effect of bound carbohydrate and of conformation on the digestibility of globulins from lupin seeds. In: Perspectives for peas and lupins as protein crops (Eds. R. Thompson and R. Casey). This publication.
Duranti, M., Restani, P. and Cerletti, P., 1982. Molecular properties and composition of the globulins from lupin seed. This publication.

Duranti, M., Restani, P., Poniatowska, M. and Cerletti, P., 1981. Phytochemistry 20, 2077-2083.
Manrique, J., Cole, S.J. and Edwards, R.A., 1974. Proceedings of the 24th Annual Conference, Melbourne. Vic. 7-11 Oct.
Quaglia, G.B., Carnevale, E. and Alessandroni, A., 1980. Congresso SINU, October, Caserta.
Riccardi, A., Alessandroni, A. and Quaglia, G.B., 1981. Proceedings of Symposium: Progress in Food Engineering. Milan, in press.
Riccardi, A. and Cerletti, P., 1981. Proceeding of the European Congress on Vegetable Proteins for Human Food. Nantes, in press
Saio, K. and Watanabe, T., 1978. J. of Texture Studies 9, 135-157.
Yatsumatsu, K. et al., 1972. Agr. Biol. Chem., 36, 719.

DISCUSSION

C. Calet *(France)* In your experiment you compared lupin flour to globulin from flour, but in practice the housewife uses egg yolk and albumin, and in industry casein is used. How do the properties of protein from lupin compare with those of protein from yolk or albumin or casein?

P. Cerletti *(Italy)* For properties such as solubility and foam formation, lupin protein compares very well with albumin. As far as rheological properties are concerned, the flour is good but the globulins are rather poor, which means that poor rheological properties are to be expected from the purified proteins or protein concentrate; the other components of flour also contribute to these properties.

AIR-CLASSIFIED PEA PROTEIN CONCENTRATE: PREPARATION AND CHARACTERISTICS

E. Carnovale and M. Cappelloni

Istituto Nazionale della Nutrizione,
Via Ardeatina, 546-Rome, Italy.

The main ways of utilising legumes, in particular pea, for human consumption as well as the technological processes used, the products obtained and their use are given in Table 1.

TABLE 1

POSSIBLE USES OF PEA SEED

Technology used	Product obtained	Use (actual or potential)
None	Raw seed	Consumption pre or post cooking
Freeze-drying	Raw seed	Consumption after cooking
None	Dry seed	Traditional dishes
Dehulling-grinding	Grits, flour	Bakery products, soups
Pre-cooking, drying grinding	Quick-cooking legumes, instant flours	Soups, beverages
Fractionation:		
Air-classification	Protein concentrate	Pasta, bakery products meat extenders
Precipitation	Protein isolate	Meat analogues, technological aids
Gel-chromatography	Protein	Technological aids

The utilisation of the fresh product is beyond the scope of this contribution. The significant increase however in the consumption of frozen peas, following improved freezing techniques and the introduction of suitable cultivars, should be stressed.

Three main classes of protein extraction products reported

here are protein concentrates, protein isolates and pure proteins. This last class is still somewhat hypothetical, and is represented by proteins isolated in the crystalline state. It is part of a research project conducted by colleagues at the INRA, in France. After developing production on an industrial scale, the utilisation of the proteins will be as technological aids.

The extraction procedure in the preparation of protein isolates is still subject to problems associated with the recovery of proteins and the removal of lipids (Gueguen et al., 1981). The presence in the protein isolate of lipids, 60% of which is represented by polyunsaturated fatty acids, and the treatment at pH 7 - 9 cause serious problems in relation to the storage of the products and hence their nutritional quality.

Protein concentrates from pea, as well as from other non-oleaginous legumes, are best obtained by dry treatment using air classification (Anon., 1974). Such technology is based on physical principles, without the use of chemicals and does not result in protein denaturation, which would lead to loss of availability; it also avoids the formation of artefacts. Moreover, the dry treatment does not involve the production of polluting effluents, which used to be removed. The problem with this method is the retention in the protein concentrate of various anti-nutritional or otherwise undesirable components (Carnovale and Zaza, 1981).

Air classification is quite a flexible technology and by changing few parameters, such as feed rate, air speed and angle of the guide vanes in the classifier chamber, it is possible to obtain a wide range in protein content (Figure 1). It is clear that an increase in protein content results in decreased protein recovery. Moreover, changes in protein content are accompanied by modifications of the functional properties. In particular, we have studied emulsifying properties as a function of the protein content.

Protein concentrates from various legumes including horse bean, pea, chick pea, have been prepared in this way using the air classification equipment at the National Institute of Nutrition. The flow diagram used is given in Figure 2 and

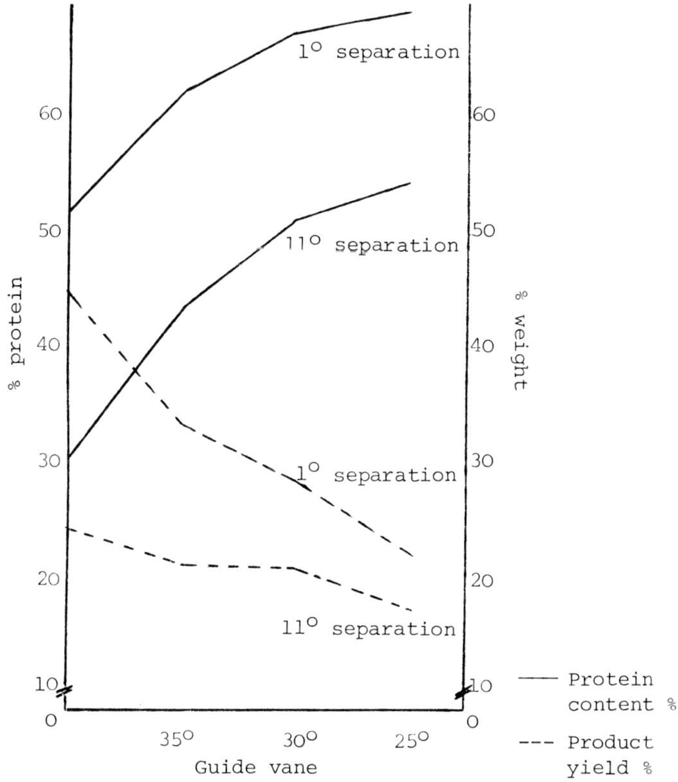

Fig. 1. Relation between protein content and product recovery

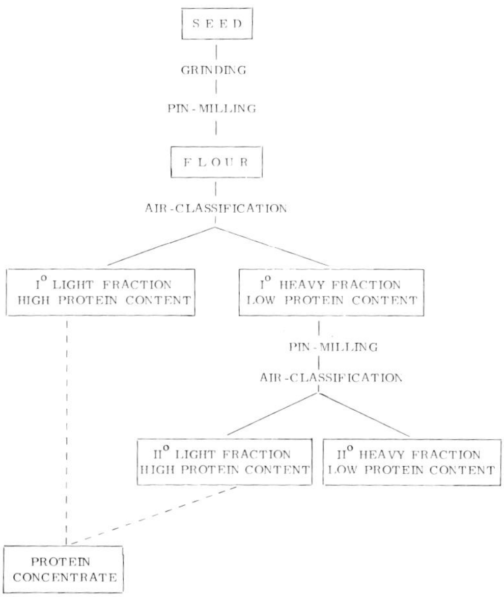

Fig. 2. Flow sheet for dry processing of horse bean and pea

represents the following procedure: an initial grinding, followed by a second milling on a pin-mill and an initial air separation. To increase protein recovery, the heavy fraction was fractionated a second time.

Protein fractionation, protein yield and protein recovery for whole and decorticated horse bean, for smooth and wrinkled pea and for chick pea are reported in Table 2.

TABLE 2

PROTEIN DISTRIBUTION AFTER AIR-CLASSIFICATION PROCESS

	Flour		Concentrate		By product		Protein
	weight (kg)	protein (%)	weight (kg)	protein (%)	weight (kg)	protein (%)	recovery (%)
Horse bean, whole	100	28.0	40	62.3	60	5.2	89
Horse bean, dehulled	82	33.0	36	65.4	46	6.6	87
Smooth pea, whole	100	22.4	40	47.5	60	6.5	85
Wrinkled pea, whole	100	25.0	54	33.2	46	12.9	71
Chick pea, whole	100	23.0	42	47.2	58	6.4	87

In the case of horse bean, the results were satisfactory for both whole and dehulled seeds. For chick pea, the results were not as satisfactory as they may appear. The presence of lipids (6%) interfered with regular feeding of the machine, causing the formation of protein aggregates.

The data for pea relative to imported cultivars are of poorly defined characteristics; wrinkled pea was not suited to the preparation of protein concentrates by air classification. This was mainly due to its structure, especially size and distribution of starch granules (presence of composite starch granules of variable granulometry), which were not satisfactorily separated from the protein bodies.

The protein concentrate from smooth pea was prepared from a commercial blend of yellow pea. Percentage protein, protein recovery and protein distribution obtained following the procedure outlined in the flow diagram are given in Figure 3. From 100 kg of whole pea flour with 25% protein, we obtained 40 kg of protein concentrate with 52% protein, and a protein recovery of 83%. We also obtained 60 kg of a starch-rich fraction, still containing 7% protein, which could be usefully employed in animal feeding. Use for this purpose is still being investigated and would have to be taken into consideration for complete economical evaluation of the process.

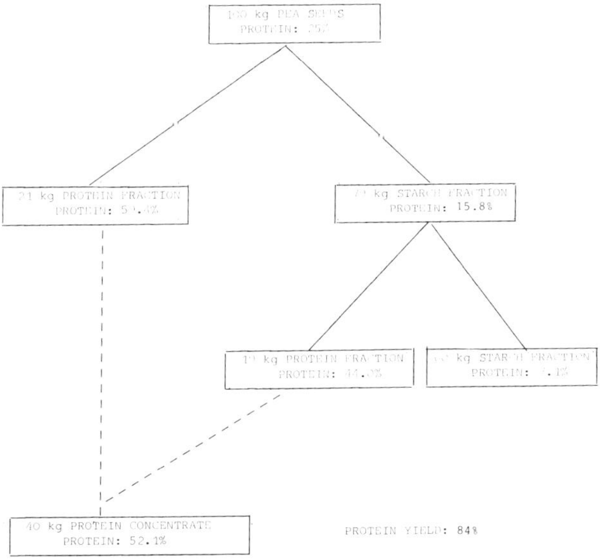

Fig. 3. Product and protein yield during air classification of pea flour

Spadoni has already described the characteristics of the pea protein concentrate including its nutrient composition, its nutritional quality, the presence of antitryptic factors and some possible uses. The results obtained so far suggest the possible commercial application of the technique. There are not any major nutritional or technological problems, except for the 'pea taste' and the dark colour of some of the products obtained.

The main problem, an agronomical one, remains the need of an all year around availability of a product economically competitive with other crops and with good agronomic characteristics associated with high nutritional quality.

In common with other European countries, particularly France, much interest in the pea crop is developing in Italy, mainly for use however, in animal feedstuffs.

An investigation of the chemical and technological characteristics of some pea cultivars has been undertaken. The cultivars used were recommended by the seed producer on the basis of their satisfactory agronomic characteristics, but their nutritional qualities were unknown.

The nutrient composition of the four cultivars examined is given in Table 3. As it can be seen, there are marked differences in their protein and carbohydrate content, both starch and soluble sugars, as well as in their lipid content. The composition of the fibre fraction, representing about 12% by weight of the seed, was also examined (Anon.; McQueen and Nicholson, 1979). There were no significant differences among the components examined (neutral detergent and acid detergent fibre and lignin) and the cultivar Finale had the lowest values (Table 4). Determination of the polyphenol contents showed (Singleton and Rossi, 1965) the cultivar Imposant, with coloured flowers, to have the highest values.

TABLE 3

COMPOSITION OF FOUR PEA CULTIVARS. (% DRY MATTER BASIS)

Cultivar	Protein (N x 6.25)	Lipid	EtOH soluble sugars	Starch	Fibre	Ash
Imposant	24.95	1.84	5.26	50.73	11.75	3.83
Finale	25.00	1.94	4.83	53.17	9.96	3.40
Rondo	28.65	2.49	10.50	40.34	12.71	4.52
Target	28.81	4.02	13.56	37.24	12.20	3.73

TABLE 4

FIBRE COMPOSITION OF PEA (NEUTRAL DETERGENT FIBRE (NDF), ACID DETERGENT FIBRE (ADF) AND LIGNIN) (% DRY MATTER BASIS)

	NDF	ADF	Lignin
Imposant	11.75	9.50	0.97
Finale	9.96	6.03	0.37
Rondo	12.71	9.87	0.69
Target	12.20	9.64	0.85

TABLE 5

COMPOSITION OF PEA PROTEIN CONCENTRATES (% DRY MATTER BASIS)

	Protein (N x 6.25)	Lipid	EtOH soluble sugars	Starch	Fibre	Ash
Commercial blend	52.09	3.93	7.70	23.88	5.79	5.90
Imposant	52.08	4.06	6.62	23.91	5.56	6.67

The cultivars Rondo and Target, although the most promising in relation to protein content, are not suitable for air classification techniques because of their high lipid content, and respond to processing like the wrinkled varieties.

The cultivars Finale and Imposant are smooth seed varieties, and satisfactory for the air classification process for protein production. Using the method described in the flow diagram (Figure 2) the cv. Imposant gave a protein concentrate with 52% protein and 85% recovery. A comparison between the pea protein concentrate from the cv. Imposant and the commercial blend previously described is given in Table 5. Table 6 illustrates the amino acid composition of the now processed flour and of the pea protein concentrate. The pea protein concentrate shows an amino acid composition very close to that of the original flour, indicating the absence of any damage to the protein from the use of such techniques.

To conclude, our data as well as that reported by Spadoni, indicate that pea, in the form of protein concentrate obtained by air classification, could represent a good source of protein in human nutrition, both from a nutritional and technological point of view.

Taking into consideration the variability of the characteristics of the cultivars examined, the need for an accurate nutritional and technological characterisation of the cultivars of interest to a particular region should be underlined. For this purpose closer collaboration among agronomists, nutritionists and technologists should be sought.

TABLE 6

ESSENTIAL AMINO ACID COMPOSITION OF PEA FLOURS AND PROTEIN CONCENTRATES (% DRY MATTER BASIS)

Amino acid	Flour		Concentrate	
	Imposant	Finale	Imposant	Commercial blend
Lysine	7.11	7.01	7.48	7.19
Threonine	3.98	3.73	3.85	3.44
Cystine	1.61	1.60	1.64	1.46
Valine	4.37	5.07	4.57	5.01
Methionine	0.90	0.87	0.93	0.90
Isoleucine	4.30	4.37	4.15	4.27
Leucine	7.30	7.38	7.19	7.05
Tyrosine	3.41	3.23	3.40	2.97
Phenylalanine	4.72	4.97	4.88	4.36
Tryptophane	1.13	0.97	1.04	1.09
Chemical score 1)	0.44	0.44	0.45	0.42
2)	0.70	0.70	0.73	0.67
Limiting AA 1), 2)	Sulphur AA	Sulphur AA	Sulphur AA	Sulphur AA

1) Against FAO/WHO egg

2) Against FAO/WHO pattern

ACKNOWLEDGEMENTS

Thanks are due to C. Bein for her technical collaboration.

This investigation was supported by a grant of the Italian National Research Council, Project 'Research on new protein sources and novel food formulations'.

REFERENCES

Anon., 1974. Pea flour and pea protein concentrates. PFPS Bulletin n. 1 - Prairie Regional Laboratory. Nat. Res. Council and College of Home Economics, University of Saskatchewan, Saskatoon, Canada.

Anon., 1980. Official Methods of Analysis of the AOAC Acid detergent fiber and lignin. 134 (7.069), XIIIth Ed.

Carnovale, E., Corcos, P. and Spadoni, M.A., 1982. Nutritional characteristics of a pea protein concentrate for human utilization. In: Perspectives for peas and lupins as protein crops (Eds. R. Thompson and R. Casey). This publication.

Carnovale, E., and Zaza, G., 1981. Dosaggio chimico della vicina e convicina in campioni diversi di fava e derivati. Atti del Seminario su: Il problema del favismo in vista del potenziamento e del consumo della fava e dei suoi derivati. Consiglio Nazionale delle Ricerche - Roma 39.

Gueguen, J., Berot, S., Briffand, J. and Davin, A., 1981. Pilot plant processing for fababean and pea protein isolates. Proc. European Congress, Plant proteins for human food - Nantes.
McQueen, R. and Nicholson, J.W.G., 1979. Modification of the neutral-detergent fiber procedure for cereals and vegetables by using α - amylase. J. Assoc. Off. Anal. Chem. 676, 62.
Singleton, V.L. and Rossi, J.A., 1965. Colorimetry of total phenolics with phosphomolybdic - phosphotungtic acid reagents. Ann. J. Enol. Viticult. 16, 144.

DISCUSSION

C. Calet *(France)* I think it is important to see that the amount of amino acid in the protein - the composition of the concentrate - is exactly the same for the two flours; but do you use rat or chicken experiments to see if they have the same amino acid availability.

E. Carnovale *(Italy)* There is no difference in digestibility in rat experiments and I agree this is a very important point in this technology.

THE ROLE OF TECHNOLOGY IN LUPIN UTILISATION

C. Pompei
Istituto di Tecnologia Alimentari, Università di Milano,
Via Celoria, 2, 20133 Milan, Italy.

Lupin cultivation and utilisation has gradually declined with the progress of civilisation and the resulting trend towards cultivation of crops which are more easily improved and immediately utilisable. The reasons for this decline are characterised by the modest possibilities of direct utilisation of the seed and the consequent lack of interest of science in this leguminous plant.

These last years, however, have shown a reversal of this tendency and a revival of this leguminous plant endowed with many agro-alimentary benefits. In the 1950 - 1955 period, the increase in lupin growing was slightly greater than the average increase in other grain legumes. Technology will play an important role in this development and may lead to levels of production equal to those of the soyabean. Our research and experiments have led us to outline the possibilities for utilisation shown in Figure 1.

Today we can recognise three kinds of lupin, all of which contain a level of protein which is almost double that of the other commonly used grain plants. The first kind contains, in addition to its high level of protein, a medium lipid content (about 10%) and a modest quantity of alkaloids (less than 0.05%); the second kind, with a medium content of oil, has a level of alkaloids greater than 0.2%; it is, therefore, bitter and useless as it is; the third kind of lupin, which has the same alkaloid content as the second type, is rich in oil having more than 18 - 20%. The potential for utilisation is dictated mainly by economic, rather than technological, reasons.

The lupin of the first kind, owing to its limited content of alkaloids, is the more easily utilised and is the only one able to provide products for consumption with little preparation. The fresh product, harvested at a convenient stage of ripening, can be used for the production of genuine snack products, or for frozen products which can be used to replace green peas or mixed

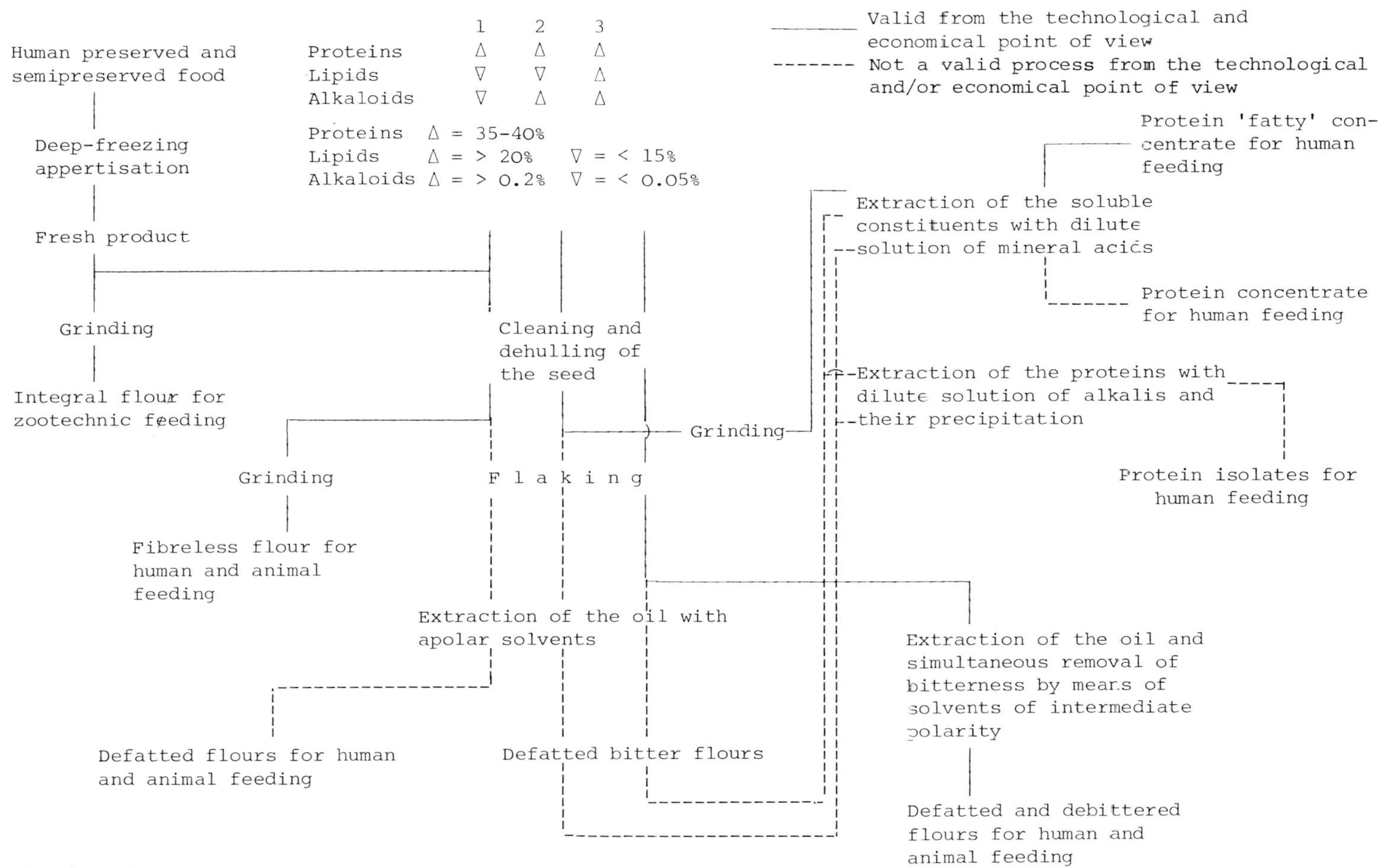

Fig. 1. The role of technology in the utilisation of lupin.

with other vegetables in frozen preparations for soups or various other dishes. The dried grain can be utilised, after grinding, as proteinic-lipidic flour for the feeding of monogastric animals. Owing to its relatively high oil content and to the presence in the oil of polyunsaturated fatty acids, the integral flour has to be stabilised against oxidation. In order to be able to use the flour for human feeding, it is necessary to carry out the dehulling by means of a traditional technological process which will be explained below. The dehulling leads to a reduction in fibre and a proportional increase (of about 10%) in protein content. For a better utilisation of the flour in human feeding, it is desirable to remove the oil; however, as these types of lupin contain about 10% of oil, such extraction is not economically viable. The economy of the process requires a minimum of 18 - 20% oil in Europe and of 16 - 18% in the Latin American countries. Lupin flour of this kind, delipidised in the laboratory, has been used as a protein wheat flour strengthener in bread-making. The replacement of up to 20% of the wheat flour with lupin flour has reduced the bread specific volume (which remained, however, at the same level as that of the various kinds of bread in the market), and has given rise to an important increase in the protein value with a slight improvement of the chemical index and a very slight reduction of the *in vitro* digestibility of the proteins.

Lupins of the second kind are less easily utilised: the only process which seems, at present, to be valid - but with certain reservations - is that of the production of a 'full-fat' protein concentrate. The process results in a flour with a slightly increased protein content compared with that of the fibreless flour; it contains practically all the oil and is not bitter. It is a product which can be used for foods of second transformation, containing oil or fats and stabilised against oxidation. They can be used in meat preserves and semipreserves, in mixtures for stuffed preparations, etc. Naturally, an improvement of the concentrate characteristics could be obtained if it were possible to remove the oil, but this is not viable for the reasons given previously. A concentrate from defatted flour, produced at an experimental level, has been utilised for the production of 'pasta'! The results have been interesting since the replacement of 20% of fine corn with this concentrate allowed the use of fine soft corn instead of fine hard corn, which latter is more expensive, thus obtaining a hyperproteinic 'pasta' having the same cooking characteristics as those of the traditional 'pasta'. The elimination of the oil would also allow the production of protein isolates by means of a process explained below.

The third kind of lupin is certainly more interesting than the second one; it is more or less as bitter as the second type, but its oil content is of such a value that extraction becomes economically viable. A drawback is the residual oilcake which is bitter and has to be submitted to further processing in order to eliminate the bitterness. This oilcake can then be used for the preparation of concentrates and isolates which may replace the similar products derived from the soyabean. In order to utilise this kind of lupin it is therefore necessary to rid it of its bitterness and to do this, two extractions are required: the first by an apolar solvent to extract oil; the second by a polar solvent (acidified or alkaline water, according to the purpose) producing a concentrate or an isolate. This double extraction can reduce the economic feasibility of the process.

We have investigated the possibility of extracting oil and alkaloids with one extraction only. The results we obtained on *Lupinus albus*, taken as a model, indicate that the process is possible and we are now extending it, at a laboratory level, to *Lupinus mutabilis*, for this is the kind of lupin for which the process has been studied. In addition to the advantage of a major saving, the process allows the production of an oilcake which is practically white, thus widening the potential utilisation of the product itself.

Figure 2 illustrates the outline diagram for the cleaning, grinding, dehulling and flaking of the seed on the pilot scale. This is not an original process, since it is used for soya, but some modifications have been made to suit the shape and size of the seed. The flakes obtained are 0.4 mm thick. The only problem at present is the lack of use for hulls and 'fines', which represent 20 - 25% of the clean starting seed.

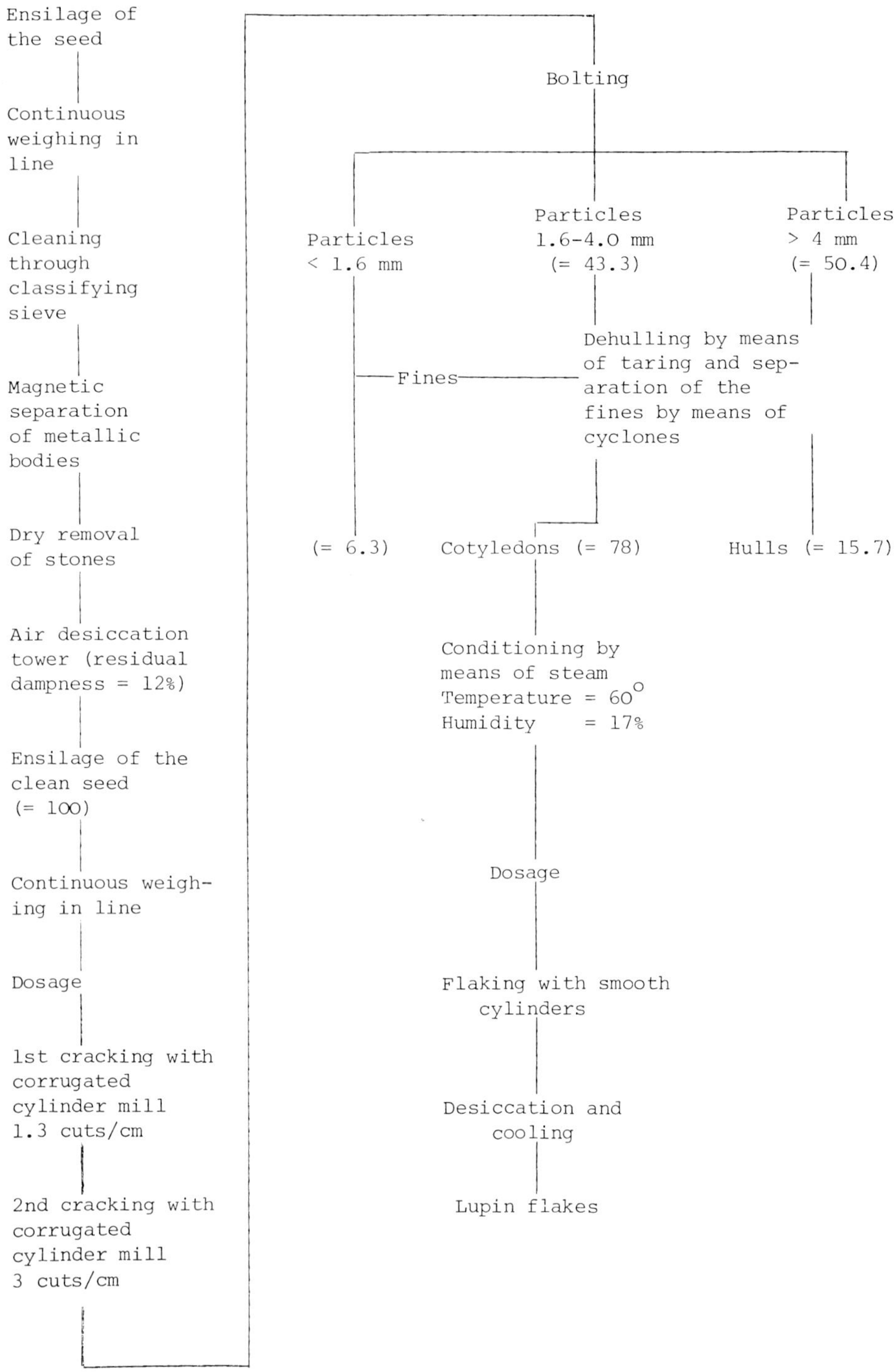

Fig. 2. Outline for cleaning, dehulling and flaking of the seed.

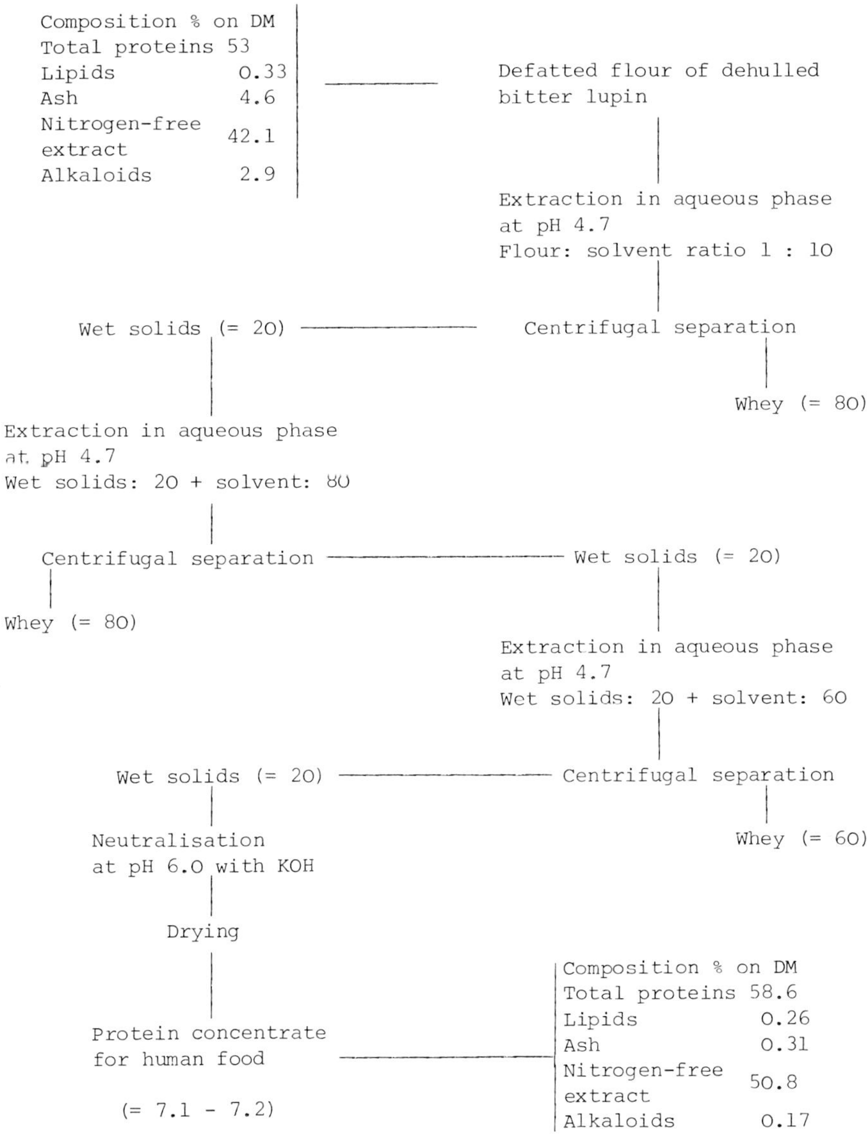

Fig. 3. Outline of production of the protein concentrates.

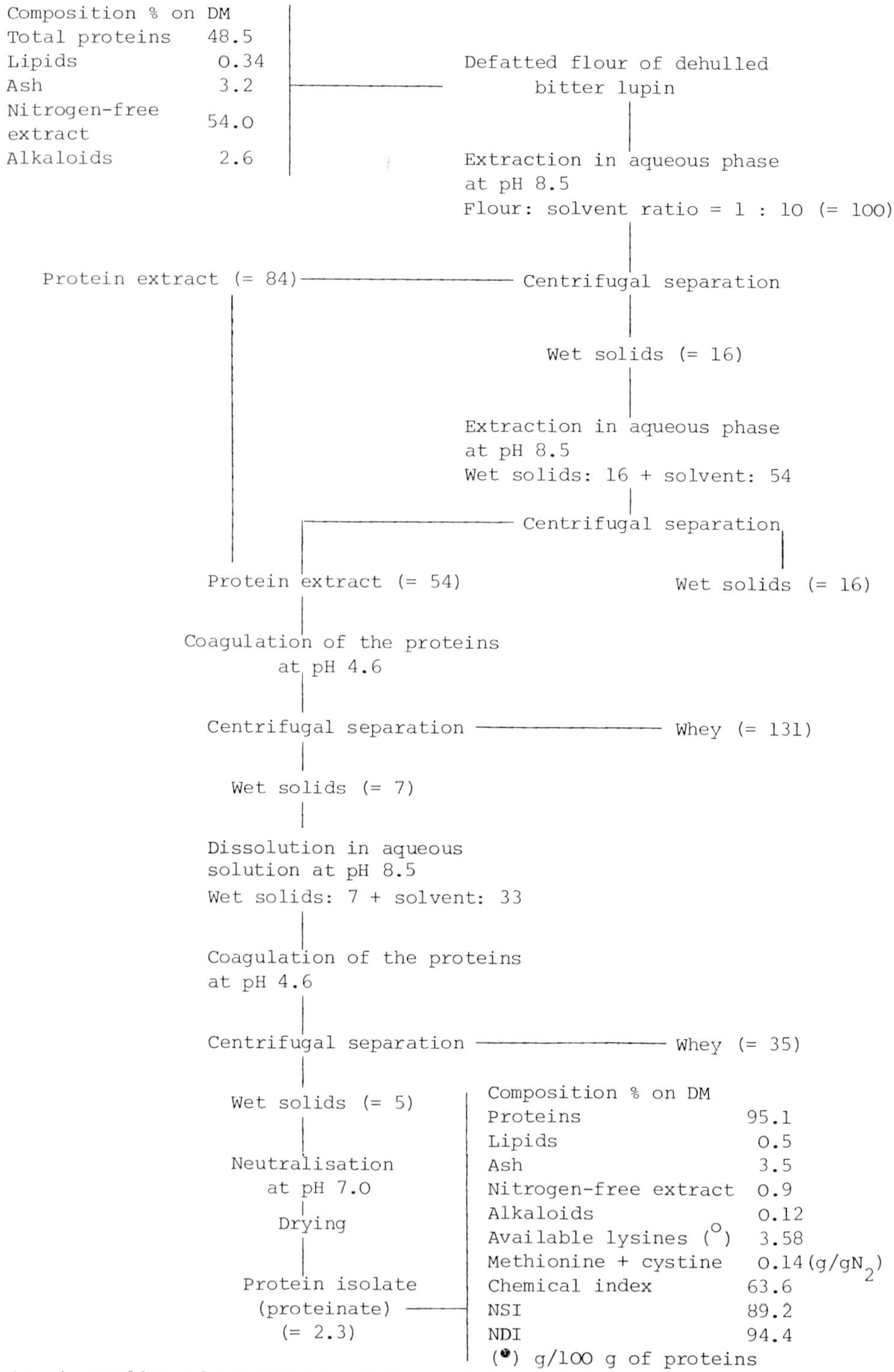

Fig. 4. Outline of production of the protein isolates.

Figure 3 illustrates the production of the protein concentrate. The process is for the case which we have previously defined as not economically viable, in that it begins with a defatted flour of *L. albus* which contained 11.2% oil; the process is, however, identical to that which can be carried out with a defatted flour from *L. mutabilis*. The process is laborious, expensive and results in only a modest increase in protein concentration. The yield is about 70%. An important inconvenience of this process is the great residual volume of effluent (about 30 l/kg of treated flour) which pollutes the environment and the purification of which adds to the cost.

Figure 4 shows the production outline of the protein isolates. Here too we can advance the same economic considerations as expressed for the concentrates concerning the oil contents of the starting seed. The process allows a considerable purification of the proteins, which represent 90 - 95% of the final product. The yield is about 20 - 25%, compared to the raw material. An important mass of a wet by-product is left by the process; this by-product is hardly useful. Moreover 18 - 20 l of polluting effluent is produced per kg of treated flour. The process is very expensive and the finished product is of limited use as a technological flocculent. Nutritional tests carried out on rats show that if methionine is added to the isolates it gives an acceptable response. The production of concentrate and the production of isolates represent the only valid systems for the elimination of the oilcake bitterness when the oil extraction is carried out with polar solvents.

The process set up for the simultaneous extraction of the oil and of the bitter principals is based on the use of a solvent of intermediate polarity. Many solvents have been tested alone as well as in azeotropic mixtures. The best results have been obtained using 95% ethyl alcohol. A good improvement in the extraction efficiency of the oil was obtained by working at the boiling temperature of the solvent while an improvement in the extraction efficiency of the lupanine was obtained using alcohol acidified with strong mineral acids. Figure 5 shows the results we have been able to obtain with a multiple extraction process. The data refer to *L. albus*. We can observe how, with

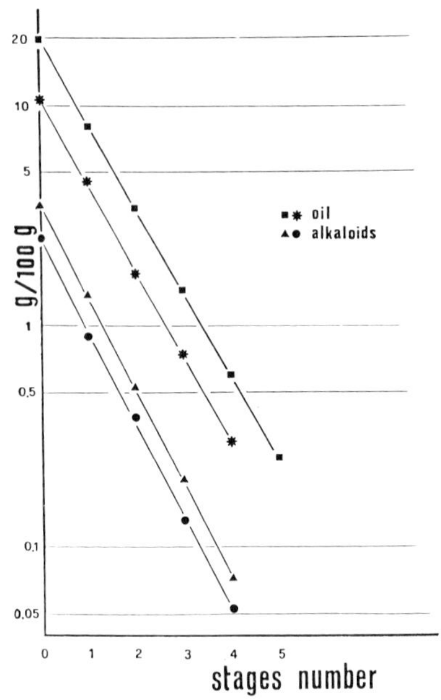

Fig. 5. Process of multiple extractions with ethanol 95% acidified with HCl. *L. albus* ★●; Estimation for *L. mutabilis* ■▲.

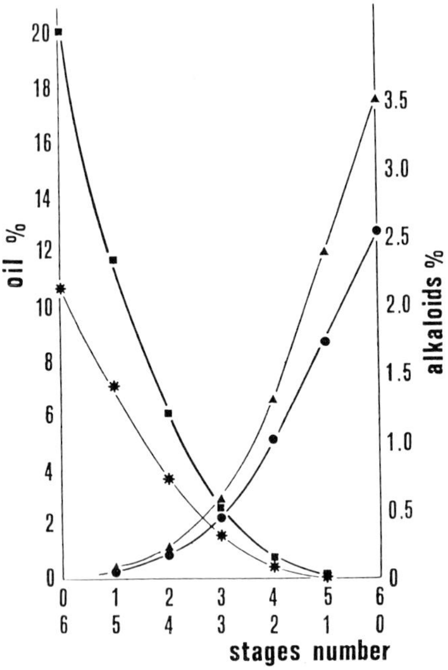

Fig. 6. Process of extraction in counter-current with ethanol 95% acidified with HCl. *L. albus* ★●; Estimation for *L. mutabilis* ■▲.

a four-stage process, it is possible to reduce the oil content from the initial value of 10.7% to 0.3% and the lupanine content from 2.45 to 0.054%. From the semi-log representation of the figure it is possible to estimate the number of stages required to reach satisfactory levels of residual oil and alkaloids when the raw material is *L. mutabilis*, or others, with oil contents of 20%, if we assume that the efficiency of extraction is the same in both species; a five-stage process is then sufficient to produce an extracted flour containing 0.25% of oil and less than 0.05% lupanine. Obviously a process of this kind, multiple extraction, is not feasible at the industrial level since it requires a very high consumption of solvent and consequent high costs to recover it through distillation. A hypothetically viable industrial process is a process of extraction in counter-current with a ratio flour solvent = 1 : 5 (p : v). In our experiments we have realised, in one laboratory, a three-stage perfect system of extraction in counter-current. From our results we have calculated the number of stages required to reach a conveniently low level of lipid and bitterness. As shown in Figure 6 we can obtain from *L. albus* an extracted flour having an oil content less than 0.2% and lupanine contents of about 0.05%, using a four-stage process. With *L. mutabilis* , a five-stage process will be required in order to obtain a defatted flour with virtually zero residual oil and alkaloids.

From these results it is logical to expect that in an industrial process, where 15 - 20 stages are normally carried out, it is possible to reach fully acceptable oil and alkaloid levels.

The cultivation of the lupin, however, cannot be separated from the consideration of energetics. This concept is shown in Table 1. The energetic inputs and outputs relevant to the cultivation of the bean in California are shown, on the basis of which a hypothesis related to the cultivation of the lupin has been formulated. In the energetic inputs for the lupin, the item 'preparation of the soil for sowing' has been taken as being energetically similar to that of the bean because the work required for the lupin, even if it may be less accurate, will be carried out in more difficult soil conditions. The items 'weed-cleaning' and 'irrigation' have been cancelled, since the lupin vegetates in the autumn-winter-spring period and does not require a water supply. The item 'fertilisation' has been reduced by 25% owing to the lupin's modest requirements and to the fact that, possessing a rich rooting system, it is able to explore soil areas which other cultivated plants cannot reach. One million and a half Kcal have been added to the total inputs for lupin for the extraction of the oil. It can be observed that the energetic balance favours lupin cultivation. If the value in Table 1 is plotted on the Malcolm Slesser plot (Figure 7) in which the energetic inputs per pound of produced protein are shown, we see that while a pound of proteins in the bean requires an expense of energy which is higher than that required for the production of one pound of proteins contained in a fish protein concentrate from Peru, the pound of lupin proteins requires an energetic cost equal to that required to produce wheat in Illinois.

TABLE 1

ENERGETIC BALANCE RELEVANT TO THE GROWING BEAN, TO THE GROWING LUPIN AND ITS TRANSFORMATION INTO DEFATTED FLOUR

	Bean (California)		Lupin (Hypothesis)	
	Quintals/ha	Kcal	Quintals/ha	Kcal
Yield	20		20	
Proteins	4.48	1 836 800	8	3 280 000
Lipids	-		4	3 720 000
Nitrogen-free extract	12.6	5 166 000	3.2	1 312 000
Energetic output		7 002 800		8 312 000
Preparation of soil and sowing		1 081 900		1 081 900
Weeding		363 100		-
Irrigation		684 200		-
Fertilisation		1 474 600		737 300
Antiparasitics Anticryptogamics		130 400		130 400
Transport		76 600		76 600
Harvesting		379 100		379 100
Energetic input		4 189 900		2 405 300
Oil extraction		-		1 000 000
				3 405 300
$\frac{\text{Input}}{\text{Output}} \cdot 100$		59.8		41.0
$\frac{\text{Input}}{\text{lb of protein}}$		4 236		1 928

The second group of considerations, based on the requirements of the technologist, are directed towards those who are working on genetic improvement. We have pointed to the possibility of utilising fresh seed for the production of frozen or snack foods. The problems to be solved for this use, which is the more direct, must be looked for in the direction of the gradual ripening of the pods on the plant and the difficulty of

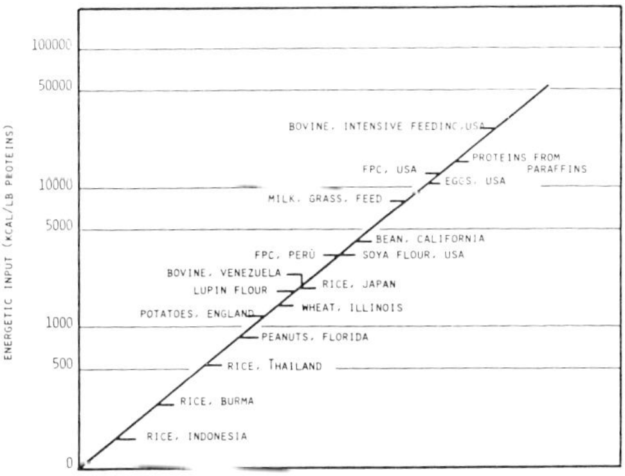

Fig. 7. Energetic input relevant to the production of some foods in the world, by Malcolm Slesser. The values relevant to the bean cultivated in California and to the defatted lupin flour, obtained as indicated in Table 1 are included.

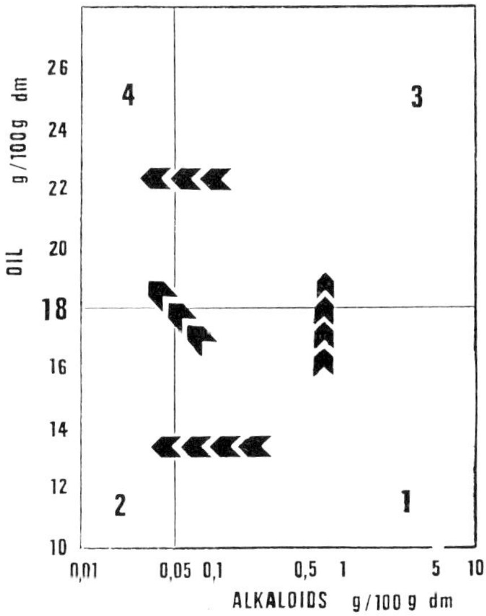

Fig. 8. Indications of genetic selection of the lupin relevant to the oil and bitter alkaloid contents.

the hulling of the seeds. Both the problems could be solved through genetic selection and through plant improvement.

The utilisation of the dry seed would also be even more convenient, from the economic point of view, if it were possible to induce genetically some modifications to the seed composition. The course of genetic selection, indicated in Figure 8, concerns the lipid and alkaloid content of the seed. Area (1) is of little interest, the oil content being too low to justify its extraction and the alkaloid content being too high to allow direct consumption. Lupins which have the composition included in this area can be used exclusively for the production of full-fat concentrates, but we have seen that this is of little interest. Lupins whose composition is included in area (2) can be used for the production of integral flours. Owing to the high fibre contents, they can be useful in animal feedstuffs, that is, after dehulling they can find some use also in human feeding. Lupins whose composition is included in area (3) have oil contents which justify their extraction, but they contain high levels of alkaloids; the utilisation of these lupins is possible, with the extraction of first the oil with an apolar solvent, then the alkaloids with a polar solvent. They can be used for the production of concentrates or isolates, or even better, extracted with the combined system of extraction by means of a solvent of intermediate polarity. It is clear that lupins whose composition is included in area (4) would be the most interesting because the oil could be extracted by a traditional process and also because the residual oilcake from the extraction could be directly utilised in the same way as the defatted flour from soya.

Therefore, and perhaps contrary to what is generally thought, the utilisation of lupin seed does not depend on its protein content, which is always very high, or even on its alkaloid content, but mainly on its oil content. Lupins of low, as well as high, alkaloid contents with a lipid concentration less than 16 - 18% force the technologists to face exclusively economic problems. Therefore, we must seek to reduce the alkaloid content, but also, and above all, we must seek to increase the oil content in order to obtain values at least similar to those of soya; soya itself would never have gained its present importance if it had contained levels of oil of 10 - 12%.

DISCUSSION

M. Cousin *(France)* May we confirm that the bean used for comparison was *Phaseolus* and not *Vicia faba*?

C. Pompei *(Italy)* Yes, it was *Phaseolus vulgaris*.

W. Williams *(UK)* In what way does the level of oil in the seed affect either the technology of processing or the economics of production?

C. Pompei Increased oil content simplifies the technology, because high levels justify extraction. If the amount of oil in seed is between about 7 - 10%, it is not worth extracting and this poses problems. The presence of acid polyunsaturated fat makes the flour very unstable, and it must, therefore, be used as soon as it has been ground or it must be stabilised with antioxidants. Of course, if the flour is used for livestock feedstuffs, which already contain antioxidants there is no problem.

W. Williams Is it possible to stabilise the flour present in cultivars of *Lupinus albus* with about 14% oil?

C. Pompei Yes, it is possible by using antioxidants.

W. Williams Following on from that point, how does alkaloid level affect processing and economics of production?

C. Pompei It is very simple. I know most about *Lupinus albus*, so I shall stick with that. Once the alkaloid content drops below 0.06 to 0.07% the flour does not have a bitter taste and can therefore be used immediately.

W. Williams Yes, I understand that. But suppose that you had used low alkaloid seed to start with, how much cheaper would the technology have been, and what would you have gained? In other words, is it worth breeding for low alkaloids?

C. Pompei I don't know. It is better to have 0.3% alkaloid than to have 3.5%, but not much better.

C. Calet In other words, what sort of saving does that represent?

C. Pompei There would be some saving on extraction costs with low alkaloid seed, because fewer stages in the extraction process would be needed to reduce levels adequately. If existing industrial plant developed for extracting oil from other seeds can be used for lupins, there will have to be 15 to 20 stages in the extraction process, so whether there is 3.5% alkaloid or 0.3% will make very little differecne as the same equipment will be used. The great disadvantage of the presence of alkaloids is the bitter taste.

C. Calet In France field bean flour is added to our bread - largely to whiten it. Are there regulations or acceptability criteria which dictate whether lupin flour could be used in the same way, because this might be a worthwhile development.

C. Pompei I am not well qualified to talk about legislation in Italy. Soya protein has been tested for this purpose and this raised a few problems. Soya protein is allowed in meat, but adding it to bread puts the bread into the category of a speciality bread, which gives greater scope for the baker to include other materials in the bread. Some of our results show that the bread does become a little yellower but it is not unpleasant. There is almost no difference in taste. The crust is a little darker, but not unpleasant and above all the volume of bread is very acceptable, even when the flour is 20% lupin.

J. Picard *(France)* The incorporation of bean flour in bread making flour in France is very low with on average about 0.5% and the maximum allowed is 2%.

M. Cousin In France the incorporation of field bean in bread is not for protein, but for enzyme and only very small amounts are necessary. M. Godon in Nantes also made bread with pea flour from wrinkled seed; the required activity was higher and required only one half of that needed with beans.

CONTRIBUTION TO THE MEASUREMENT OF QUINOLIZIDINIC ALKALOIDS IN LUPIN SEEDS

J.C. Chenieux, P. Levillain, J.M. Merillon and M. Rideau
Faculté de Pharmacie,
F37032 Tours, Cédex, France.

ABSTRACT

Two species of lupin are interesting in France, both for their agronomic characteristics as well as for their chemical composition: Lupinus albus L. and Lupinus mutabilis Sweet. Rapid measurement of the quinolizidinic alkaloids in the seeds, especially the two main alkaloids (sparteine and lupanine) is necessary. Various authors have proposed techniques of photometric assay by the acid-dye method with bromcresol purple (Nowacki et al., 1972; Von Baer, 1978), bromthymol blue (Wiewionowski and Skolik, 1959; Reifer and Niziolek, 1957) or methyl orange (Reifer and Niziolek, 1957). We have studied such an assay based on the fact that sparteine and lupanine form, with methyl orange, ethylene dichloride-extractable ion-pairs. Figure 1 shows the effect of pH on the extractability of the alkaloid/methyl orange complex. At pH 6, both alkaloids can be measured at 425 nm. A linear relationship is observed between the concentration of sparteine and the absorbance of the complex. For lupanine, the relationship is linear between the square root of the concentration and its absorbance (Figure 2).

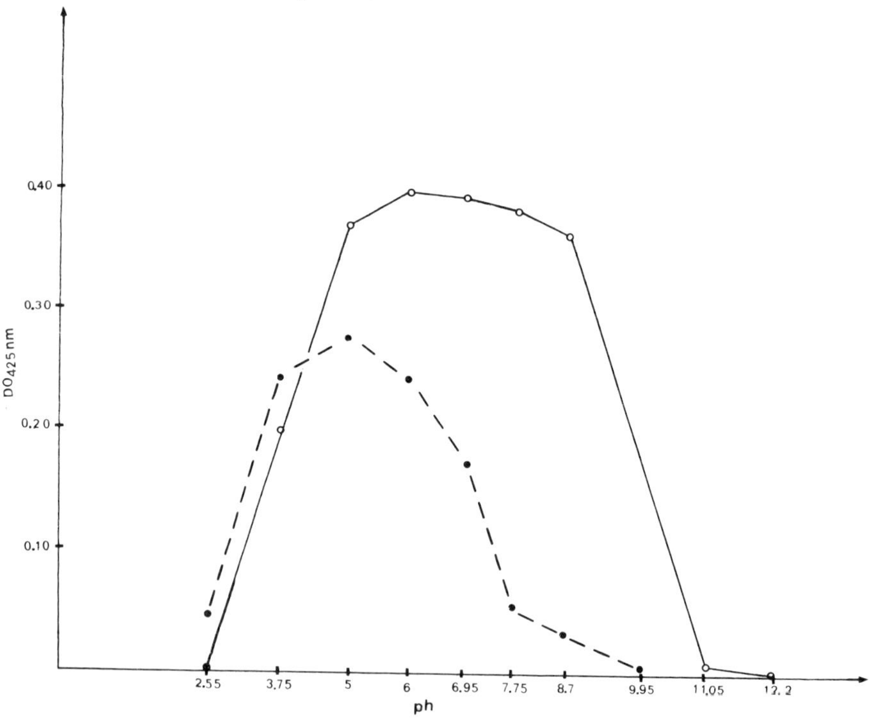

—— sparteine sulphate   --- lupanine perchlorate

Fig. 1. Effect of pH on extraction of the alkaloid/methyl orange complex.

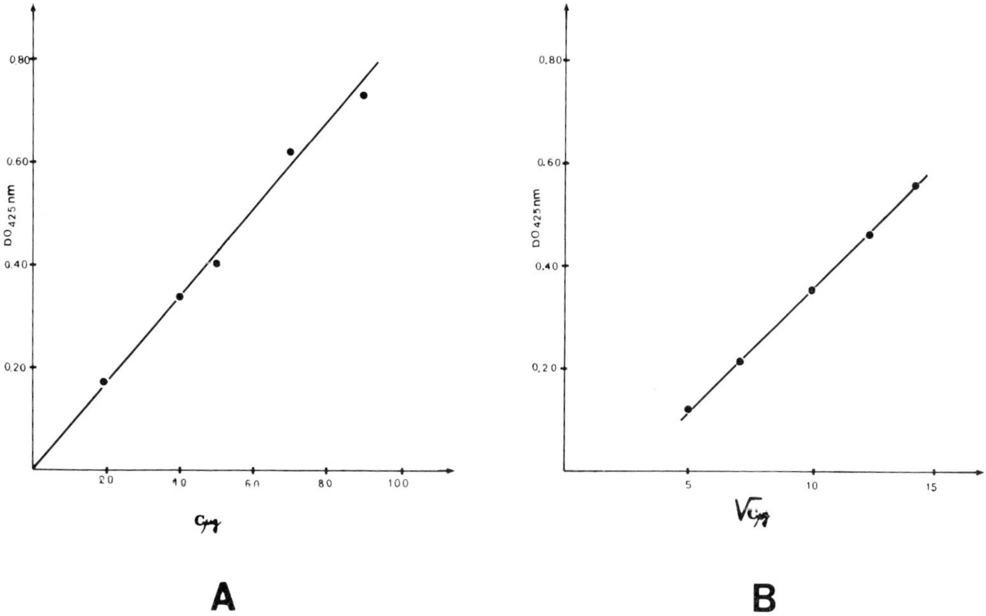

Fig. 2. Calibration curves of sparteine sulphate (A) and lupanine perchlorate (B) using methyl orange at pH 6.

INTRODUCTION

The following points have been studied: extraction procedure of alkaloids from the seeds; separation of the alkaloids on TLC plates, followed by elution; assay by methyl orange.

EXTRACTION OF ALKALOIDS FROM GROUND SEED

<u>High alkaloid lupin seeds</u>

Finely ground lupin seeds (150 mg) were placed in a test tube (12 x 80 mm) and titrated with 0.4 ml 25% NH_4OH. Alkaloids were extracted by 3 x 4 ml ether/chloroform (2/1 v/v) during 12 h, 1/2 h then 1/4 h. Extracts were evaporated ('Brinkmann') at 50°C, then the residue was dissolved in 200 µl MeOH.

Drying the material with Al_2O_3 (Von Baer, 1979), silica gel, Na_2SO_4, or the use of other extraction solvents (ether, $CHCl_3$, MeOH) did not give better results.

<u>Low alkaloid lupin seeds</u> (< 0.5%)

One thousand mg ground seeds were soaked in 3 ml 25% NH_4OH

and extracted by 3 x 20 ml ether/chloroform as above. Alkaloids were purified by successive extractions with 0.1 N HCl (3 x 10 ml) then after being made alkaline (NaOH), with ethylene dichloride (3 x 10 ml). The organic phase was allowed to evaporate to dryness (50°C), and then the residue was dissolved in 50 µl MeOH.

SEPARATION OF THE ALKALOIDS

Ten to 40 µl of alkaloidic extracts were spotted on silica gel plates (Merck G60) and alkaloids were separated by acetone/toluene/96° ethanol/25% NH_4OH (30/10/6/1 v/v) (8 extracts on each plate). Twenty five and 50 µg sparteine sulphate, and 50 and 200 µg lupanine perchlorate were co-chromatographed on the same plates.

After migration (15 cm), the solvent was dried off. The alkaloids, which cannot be seen under UV, were located by spraying with a methanolic solution of iodine (10 g/l). The alkaloidic spots corresponding to lupanine and sparteine were cut out and placed into test tubes (12 x 80 mm).

Sparteine was eluted by 3 x 4 ml methanol/$CHCl_3$/acetic acid (50/50/1, v/v). Lupanine was eluted by 3 x 4 ml MeOH/$CHCl_3$ (1/1). Percentage recoveries were 99% and 90% for lupanine and sparteine respectively. The reaction with iodine did not interfere in the assay, and elution was identical whether done by scratching off the gel or cutting out the spots.

MEASUREMENT OF THE ALKALOIDS

Alkaloidic extracts were evaporated and the residue was dissolved in 2 ml of a solution of methyl orange buffered at pH 6, then transferred into test tubes. Three ml ethylene dichloride was added. After stirring, the tubes were centrifugate (4 000 rpm, 5 min) to break the emulsion, and the ethylene dichloride phase was drawn off with a Cornwall syringe to which a 19 G2 Terumo needle was fitted. The absorbances were read at 425 nm.

This method allows measurement of 10 µg sparteine and 20 µg lupanine with an error below 5%. We have used it to estimate the content of alkaloid in sweet or bitter seeds, or to evaluate debittering treatments of *Lupinus mutabilis*. Thus, we found that the use of trichlorethylene on powder mixed with ammonia removes practically all the lupanine and 85% of the sparteine.

REFERENCES

Nowacki, E.K., Nowacka, D. and Weznikas, T., 1972. Metody ilosciowego i jakosciowego oznaczania alkaloidow lubinowych. Przystosowane do badan genetycznych. Hodowl. Rosl. Aklim. Nasienn., 16, 59-76.

Reifer, I. and Niziolek, S., 1957. Kolorymetryczna mikrometoda oznaczania alkaloidow w nasionala lubinu, Acta Biochim. Polon., 4, 165-180.

von Baer, D., 1978. Procedimiento rapido para la determinacion de los alcaloides totales en *Lupinus mutabilis* con purpura de bromocresol (BKP), Inf. Projecto Jupino No. 3. 102-108.

von Baer, D., Reimerdes, E.H. and Feldheim, W., 1979. Methoden zur Bestimmung der Chinolizidinalkaloide in *Lupinus mutabilis*, Z. Lebensm. Unters. Forsch., 169, 27-31.

Wicwiorowski, M. and Skolik, J., 1959. Méthode de dosage des alcaloides totaux du Lupin, (traduit du polonais à l'INRA de Lusignan), Roczn. Chem. 33, 461-469.

DISCUSSION

W. Williams *(UK)* Can you foresee any problem in using this method on green tissues? Alkaloids in lupin are inherited entirely maternally. So the F1 seed reflects the genotype of the female parent. The F2 seed represents the genotype of the F1, and you have to go to the F3 seed before you can see the segregation for the F2. It would be quite useful, therefore, if your method could be used on leaves from the F2.

P. Levillain *(France)* The method was originally used to monitor debittering of powder and then we used it for selection of seeds. I think the main problem with all these techniques is that they are two-step techniques, extraction and determination. Only Mr. Dardenne's infra-red technique is one-step. I think determination is good, but I do not know enough about the extraction methods relevant to green leaves.

C. Sator *(FRG)* Have you any experience with gas chromatography or high pressure liquid chromatography, and if so, could you tell us something about it?

P. Levillain Gas chromatography is the more precise technique. It is a reference method for determination of separated alkaloids, but it takes far longer than high pressure liquid chromatography. HPLC is the solution for fast, simple and not very precise determination. The main problem of HPLC is detection of the substance at the end of the column. As these alkaloids do not absorb UV and are not fluorescent, they are very difficult to see. They can be separated with HPLC but it is very difficult to measure the quantity at the end of the column, so a chemical reaction is required before injecting in the column, or at the end of the column, and such reactions at the end of the column are generally quite difficult. So theoretically, I think that gas chromatography is better than HPLC. But maybe we could change if there was good colour development or a suitable rapid chemical reaction.

BEHAVIOUR OF CARBOHYDRATES DURING TECHNOLOGICAL PEA PROCESSING

P. Colonna and Christiane Mercier
Institut National de la Recherche Agronomique,
Laboratoire de Biochimie des Aliments,
Chemin de la Géraudière, 44072 Nantes Cédex, France.

The distribution and the composition of carbohydrates - α-galactosides, starch and cell wall material - have been studied on smooth and wrinkled peas (*Pisum sativum*) before and after dry (air classification) or wet (aqueous protein extraction) processing of dehulled legume seed flours.

The dry process involves successive pin milling, air classification and refining techniques. The wet concentration method uses alkaline solubilisation of proteins and removal of the by-products by centrifugation.

During air classification, proteins are separated from starch by density and particle size, leading to high protein (≃ 60%) and low protein (≃ 6%) content fractions. With smooth pea flours, the high starch content (≃ 80%) of the low protein fraction, with a reduction of both the cell wall polysaccharides and α-galactoside contents, was a way to prepare starch. By laboratory scale purification, starch was isolated in high yield and high purity, with no damage of the granules as viewed by light and electron microscopy. This result shows the efficiency of the dry process in the case of smooth peas, due to the uniform large sized starch granules (15 - 30 μ), compared to the population of small (0.5 μ) and medium (10 - 20 μ) diameter cereal starch granules (Colonna et al., 1980).

With wrinkled peas, the air classification was of a low efficiency, due partly to the higher lipid content (4%) of the initial flour, which prevents a good separation of the particles in the air stream, and to a broad distribution of the granule size (6 - 80 μ). However, from an air classified flour containing only 45.5% starch and by a laboratory scale purification, starch granules have been obtained, contaminated mainly by proteinaceous (0.2%) and hemicellulose (0.17%) material. No general figure can be given for the variation of cell wall polysaccharides for either smooth or wrinkled peas. The hemicellulosic fractions and the ratio arabinose/xylose vary between air classified fractions, indicating that the hemicelluloses still present in the air classified fractions are not chemically identical with the total initial hemicelluloses. α-galactosides are still present in the high starch content air classified flour with no variation for wrinkled peas and a decrease of 25% for smooth peas.

In the wet process, different conditions for the aqueous extraction of protein after neutral (pH = 7) or alkaline (pH = 9) solubilisation led to insoluble polysaccharide residue. Pea flours have been separated on a pilot scale into starchy and cell wall material fractions using different sieving (200 - 60 μ) and washing conditions (Colonna et al., 1981). After centrifugation, starch was completely recovered in the by-product, which contains only traces of α-galactosides, but higher fibre and protein contents, than that from air classification. By operating at pH 9 and using different screens (200 - 60 μ), native starch granules were prepared, contaminated mainly with cell wall polysaccharides (4%), proteins being negligible (0.3%). High cell wall content fractions, obtained as the insoluble residue remaining on the different screens (200 - 60 μ) constitute more than 30% furfural generators (pentosans) and ADF. Their water absorption capacities (about 10 to 20 g water per g dry material) were highly correlated (r = 0.96) with furfural generator contents when proteins were eliminated previously. The removal of the alkaline-soluble proteins by centrifugation before screening did not improve the purity of starch, but has increased the yield of starchy by-product. The washing conditions (ratio water/by-product) were observed to be critical to the recovery and the purity of the starchy fractions. The wet process, applied to wrinkled peas, was inefficient due mainly to prevention of efficient sieving and centrifugation by the high swelling of cell wall polysaccharides.

In conclusion, smooth peas, known as a protein source, can also be considered as a potential source of starch and cell wall material, which can be extracted either by air classification or by a wet process. Simultaneous studies have shown that broad beans behave similarly. In contrast, wrinkled peas cannot easily be fractionated into the major components, proteins and starch, either in the dry or wet process, mainly because of their higher lipid content, broader distribution of starch granule size and higher swelling capacity of the cell wall polysaccharides.

REFERENCES

Colonna, P., Gallant, D. and Mercier, C., 1980. *Pisum sativum* and *Vicia faba* carbohydrates: studies of fractions obtained after dry and wet protein extraction processes. J. Food Sci., 45, no. 6, 1629-1636.

Colonna, P., Gueguen, J. and Mercier, C., 1981. Pilot scale preparation of starch and cell wall material from *Pisum sativum* and *Vicia faba*. Sci. Alim., 1, no. 3, 414-426.

DISCUSSION

R. Casey *(UK)* You mentioned that there was a difference in lipid content between round and wrinkled peas. Have you any idea of the extent of variation in lipid contents within the *Pisum* world collection or within your collection, perhaps? There is between 2% and 4% between round and wrinkled types, but do you know how much lipid content varies between genotypes?

P. Colonna *(France)* In five varieties of smooth pea we found between 1 - 2% and in two varieties of wrinkled pea it was 4% and 4.3%. It is quite common for lipid content to increase in seeds when the starch content is decreased.

C. Calet *(France)* Is there an increase or decrease in the amount of galactoside in the protein fractions after concentrate preparation.

P. Colonna In the case of dry processing the protein concentrate has a similar galactoside content to the initial flour. That is also true of the starch concentrate. In wet processing galactosides are solubilised so that there is no galactoside or sucrose in the starchy by-product if washing is efficient. After precipitation for protein, alpha-galactosides are still in solution so that wet processing completely removes galactoside. So for human food, with wet processing, we have to remove sucrose and alpha-galactoside because we are unable to use this fraction of galactoside.

C. Calet Wet processing is very expensive, and consequently if we could remove more galactoside by dry processing it would be better.

P. Colonna Unfortunately, this is not possible.

THE LIMITS OF LUPIN SEED INCORPORATION IN BROILER DIETS (*LUPINUS ALBUS* VAR. KALINA)

L. Lacassagne

Station de Recherches Avicoles, INRA,
Centre de Recherches de Tours, Nouzilly,
37380 Monnaie, France.

ABSTRACT

Three experiments to determine the maximum incorporation rate of Lupinus albus var. Kalina in simple corn-soya diets without depressing performances were carried out on 1 248 Hubbard male chicks aged from 1 to 28 days (first and second experiments) and from 29 to 49 days (third experiment). In each experiment the diets were isoenergetic and provided the same lysine and sulphur amino acid levels.

A 10% lupin incorporation in 1 - 28 day diets has no depressing effect on the 4 week body weight. On the other hand, 20% lupin lowers the body weight at the same age. This depressing effect occurs during the first week of the bird's life. The group fed with the control diet during the first week and with the 20% lupin diet thereafter perform as well as the control group, judging by the 28 day body weight.

The incorporation of 20% lupin seed in the diet increases the food conversion efficiency during the first four weeks of the chick's life.

Kalina can be included at levels of 25, 30 and even 35% in the 29 - 49 day diets without any deleterious effect on body weight and food conversion efficiency.

INTRODUCTION

In France the only lupin which may be developed in the coming years, depending on progress made in selection, is *Lupinus albus* var. Kalina, of Polish origin. Experiments performed up to now on other white lupins (Cubillos, 1976; Koreleski et al., 1974; Perez Cuesta et al., 1972; Perez Cuesta et al., 1973; Perez Cuesta and Tirado Serrano, 1974; Perez Cuesta and Tirado Serrano, 1978; Perez Cuesta and Perez Hernandez, 1978; Perez Hernandez, 1978) and on Kalina (Lacassagne, 1979) do not permit firm conclusions to be drawn as to the usefulness of this legume under the usual French conditions of broiler production. The aim of the experiments described below, carried out on a large number of broilers, was to determine the nutritional value of different levels of Kalina in corn-soya diets without addition of any synthetic lysine and tryptophan.

MATERIALS AND METHODS

Three experiments were conducted to study the effect on broiler performances of various levels of lupin seed incorporation in 1 - 28 day and 29 - 49 day diets. Each experiment involved 1 248 one day old Hubbard male chicks allocated at random to 24 floor pens of 52 birds each. The first experiment was conducted in the winter, independently of Experiments 2 and 3. The latter two experiments were conducted in the autumn using the same batch of chicks, housed identically, and the same raw ingredients for diet formulation. The diets were given *ad libitum* as a meal during the first five days, then as 2.5 mm pellets. The control diets were corn-soya type, while the experimental diets were corn-soya-lupin type. For each experiment the diets were calculated to be isoenergetic, the energetic value being balanced with tallow in Experiments 2 and 3 and with tallow plus 1% arachid oil in Experiment 1. The lysine and sulphur amino acid (SAA) levels were identical. The SAA were balanced with synthetic dl-methionine and equivalent lysine levels were obtained by increasing the crude protein level in corn-soya-lupin diets.

EXPERIMENT 1

In this experiment the following comparisons were made. Three diets were compared during the first growth period (1 to 28 days of age): the control diet; one experimental diet containing 20% lupin; and a second containing 10% lupin. The latter was obtained by mixing the first two diets in equal proportions. The calculated energy, crude protein, lysine and SAA levels are listed in Table 1. Each diet was tested on eight replicates of 52 birds each. During the second period (29 to 49 days of age) the control group was fed on corn-soya diet and the other two on the same corn-soya-lupin diet with 26% lupin seed. The characteristics of these diets are described in Table 2.

EXPERIMENT 2

This experiment compared four groups of chicks from 1 to 28 days of age: one with a control diet; one with a 20% lupin diet; one with a 10% lupin diet obtained by mixing the first two diets in equal proportions. The fourth experimental group received the control diet during the first five days then the 20%

TABLE 1

COMPOSITION AND CALCULATED ANALYSIS OF 1 - 28 DAY BROILER RATIONS

Ingredient	Experiment 1		Experiments 2 and 3	
	Control	Lupin 20%	Control	Lupin 20%
Yellow corn	59.06	48.24	58.24	47.81
Soybean meal, 50% protein	33.00	22.20	33.30	22.40
Lupin	-	20.00	-	20.00
Tallow	2.90	4.50	4.40	5.70
Arachid oil	1.00	1.00	-	-
Dicalcium phosphate	1.95	2.00	1.95	2.00
Calcium carbonate	0.85	0.80	0.85	0.80
Dl-methionine	0.19	0.21	0.21	0.24
Premix*	1.05	1.05	1.05	1.05
Crude protein, calculated	20.65	21.80	20.50	21.70
Crude protein, observed	21.05	21.80	20.80	22.05
Metabolisable energy	3090	3090	3060	3060
Lysine	1.10	1.10	1.10	1.10
Methionine and cystine	0.90	0.90	0.87	0.87
Tryptophan	0.24	0.22	0.24	0.22

*Premix provided per kg of diet: in IU: Vit. A = 10 000, Vit. D_3 = 1 500
in mg: Vit.E = 15, Niacine = 25, Vit. B_{12} = 0.008, Calcium Panthotenate = 8, Vit. K_3 = 5, Riboflavin = 4, Pyridoxin = 1, Folic acid = 0.2, Biotin = 0.1, BHT = 125, Choline = 250, Fe = 35, Cu = 8.5, Mn = 110, Co = 1, I = 1.28, Se = 0.1, Zn = 100
in g: NaCl = 4, Coccidiostatic = 0.5.

TABLE 2

COMPOSITION AND CALCULATED ANALYSIS OF 29 - 49 DAY BROILER RATIONS

Ingredient	Experiment 1		Experiments 2 and 3			
	Control	Lupin 26%	Control	Lupin 25%	Lupin 30%	Lupin 35%
Yellow corn	62.06	48.17	63.30	49.87	47.17	44.36
Soybean meal, 50% protein	30.00	16.00	29.00	15.70	13.10	10.50
Lupin	-	26.00	-	25.00	30.00	35.00
Tallow	2.90	5.80	3.70	5.40	5.70	6.10
Arachid oil	1.00	-	-	-	-	-
Dicalcium phosphate	1.70	1.80	1.70	1.80	1.80	1.80
Calcium carbonate	1.10	1.00	1.10	1.00	1.00	1.00
Dl-methionine	0.15	0.18	0.15	0.18	0.18	0.19
Premix (as Table 1)	1.05	1.05	1.05	1.05	1.05	1.05
Crude protein, calculated	19.10	20.60	19.15	20.60	20.95	21.30
Crude protein, observed	19.50	21.10	19.45	20.80	21.30	21.35
Metabolisable energy	3080	3080	3070	3070	3070	3070
Lysine	1.00	1.00	1.00	1.00	1.00	1.00
Methionine and cystine	0.77	0.77	0.77	0.77	0.77	0.77
Tryptophan	0.22	0.19	0.22	0.19	0.18	0.18

lupin diet (Table 1). Each diet was tested on six replicates of 52 birds. From 29 to 49 days of age, the control group was fed on a corn-soya diet and the other three groups on the same diet with 25% lupin seed (Table 2).

EXPERIMENT 3

From 1 to 28 days of age all chicks received a corn-soya control diet identical to that used in Experiment 2. At 28 days of age, the birds were allocated on a weight basis to four treatments, with six replicates of 52 birds per treatment. Then, from 29 to 49 days of age, the following diets were compared: a control, and three others containing 25, 30 and 35% lupin seed respectively (Table 2).

RESULTS

The results are summarised in Table 3. Although the lysine, SAA and tryptophan requirements were provided, the 28 day body weight was significantly reduced when 20% lupin was added to the diet. This was found both in Experiment 1 and in Experiment 2 in spite of a lower average body weight in Experiment 2 due to an unusually high level of females remaining after sexing, to a different batch of birds and to a different environment due to the season.

It appears that performances are depressed during the first five days of the chick's life. When a control diet is fed during this period the chicks subsequently fed on a 20% lupin diet do not show a depressed 28 day body weight (Experiment 2, Table 3).

On the other hand, during the 29 - 49 day period, the birds perform as well as controls even at a 35% lupin level where 59% of the crude protein is provided by this legume. In fact, the depressed 49 day body weight of groups fed on a 20% lupin diet until 28 days of age (significant in Experiment 1, almost significant in Experiment 2) seems to be due to the depressed weight gain during the first week of the chick's life.

These results are similar to those found by Vogt et al., (1979) with *Lupinus luteus* obtained under slightly different conditions (isonitrogenous diets); namely, the maximum level of lupin accepted by broilers increases with age.

TABLE 3

BROILER PERFORMANCES AT 28 AND 49 DAYS OF AGE

	1 to 28 day period			29 to 49 day period	
Diet type	Body weight at 28 days of age	Feed conversion ratio	Diet type	Body weight at 49 days of age	Feed conversion ratio*
Experiment 1					
Control	919±4 x**	1.719±0.009	Control	2017±10 x	2.48±0.02
Lupin 10%	925±5 x	1.709±0.009	Lupin 26%	2017±11 x	2.51±0.02
Lupin 20%	904±5 y	1.700±0.007		1966±10 y	2.51±0.01
Experiment 2					
Control	846±5 a	1.673±0.004 a	Control	2060±14	2.37±0.02
Lupin 10%	832±6 a	1.653±0.008 b		2044±14	2.34±0.02
Lupin 20%	812±6 b	1.632±0.004 b	Lupin 25%	2020±15	2.32±0.02
Control then Lupin 20%	846±6 a	1.643±0.009 b		2041±15	2.38±0.02
Experiment 3					
Control	850±6	1.657±0.006	Control	2014±13	2.36±0.02
	845±6	1.647±0.008	Lupin 25%	2012±16	2.35±0.01
	850±6	1.655±0.008	Lupin 30%	2036±14	2.36±0.01
	849±5	1.652±0.004	Lupin 35%	2030±14	2.35±0.01

* Partial food conversion ratio

**means ± standard error. Values followed by a different letter are significantly different (P = 0.5 for a and b, P = 0.01 for x and y: Newman - Keal's test)

With respect to feed conversion efficiency, the incorporation of 20% lupin into 1 - 28 day diets lowers the food conversion ratio (food consumed : weight gain) in spite of the negative effect of the 20% lupin added on the growth rate (almost significant in Experiment 1; significant difference in Experiment 2). These improvements in food efficiency may be due to the fact that the crude protein level of the lupin diet was higher than that of the control diet (+ 0.75% in Experiment 1, + 1.25% in Experiment 2).

Nevertheless, when considering the difference in the results in Experiments 1 and 2, it is difficult to judge if the decrease in the food conversion ratio observed in Experiment 2 was due solely to the increase in the level of crude protein. It is likely that, in Experiment 2, the unsaturated fatty acids contained in lupin, combined with those of the corn, improved the digestibility of the tallow which was included in the diet to achieve balanced energy values (Artman, 1964; Lewis and Payne, 1966). Therefore the apparently lower feed efficiency of the lupin diets in Experiment 1 could be due to the addition of 1% peanut oil to all Experiment 1 diets. Bearing in mind its economic impact, the beneficial action of lupin on the saturated fatty acid efficiency of tallow should be more thoroughly investigated.

REFERENCES

Artman, N.R., 1964. Interaction of fats and fatty acids as energy sources for the chick. Poult. Sci., 43, 994-1004.

Cubillos, A., 1976. Reemplazo de afracho de Maravilla por semilla de *Lupinus albus* var. Astra y *Lupinus luteus* var. Aurea en alimentacion de broilers. Archivos de Zootecnia, 25, 369-377.

Koreleski, J., Rys, R. and Kuchta, M., 1974. Nasiona bobiku, pelszki i lubinu jako podstawowe zrodlo bialka w dawce paszowej v brojlerow. Act. agraria et Silvestria, series Zootechnica, 14, 57-69.

Lacassagne, L., 1979. Valeur alimentaire du lupin blanc doux, variété Kalina, chez le poussin en croissance. Proc. Matières premières et Alimentation Animale. Ed. by INRA, Nouzilly, 43-56.

Lewis, D. and Payne, C.G., 1966. Fats and amino acids in broiler rations. Brit. Poult. Sci., 7, 209-218.

Perez Cuesta, M., Tirado Serrano J., Conrado Martinez, M. and Jodral Gutierrez, A., 1972. Eficienca nutritiva del altramuz dulce (*Lupinus albus* var. Multolupa) comparada con la de harina de soja en pollos para carne. IIème Congrès Mondial d'alimentation animale, Madrid, 641-646.

Perez Cuesta, M., Tirado Serrano, J., Conrado Martinez, M. and Jodral Gutierrez A., 1973. Investigacion de la eficacia nutritiva del altramuz dulce (*Lupinus albus*, var. Multolupa) comparada con la harina de soja, en el racionamiento de pollos para carne. Archivos de Zootecnia, 22, 1-27.

Perez Cuesta, M. and Tirado Serrano, J., 1974. Evaluation of the feed conversion and chain production in broilers fed on sweet lupin. XV World Poultry Congress and Exposition, New Orleans, 604-606.

Perez Cuesta, M. and Tirado Serrano, J., 1978. Valoracion de la harina de semillas de altramuz dulce (*Lupinus albus* var. Baeticus) como fuente proteica en las dietas para pollos de ngorde. 16ème Congrès Mondial d'Aviculture, Rio de Janeiro, 553-559.

Perez Cuesta, M. and Perez Hernandez, M., 1978. Rendimientos comparativos en pollos para carne de la harina de semillas de *L. albus* Neuland y de la torta de soja. 16ème Congrès Mondial d'Aviculture, Rio de Janeiro, 545-551.

Perez Hernandez, M., 1978. Valoracion biologica de la proteina de las semillas de altramuz dulce *L. albus* Neuland en pollos para carne. 16ème Congrès Mondial d'Aviculture, Rio de Janeiro, 521-528.

Von Vogt, H., Harnish, S. and Krieg, R., 1979. Der Einsatz von Süsslupinenschrot im Geflügelfutter. Arch. Geflügelk., 43, 229-238.

DISCUSSION

M.A. Spadoni *(Italy)* Is there any indication that alkaloid levels caused reduced intake by the chicks?

M. Lacassagne *(France)* The alkaloid level in this lupin is about 0.02%, which is very low, but we did not determine whether taste affected intake.

C. Calet *(France)* It is perhaps surprising that lupins have a deleterious effect in the first days of life, but interestingly I found the same results with rapeseed meal. Rapeseed meal at 20% inclusion depresses growth in the first eight days, after which there are no problems.

M. Lacassagne Consumption declined during early growth despite the fact that lysine and tryptophan levels were correct.

J. Picard *(France)* It may be of interest that in France we have 200 or 300 hectares of experiments and farm trials on lupins. The main objective is to determine whether our research findings can readily be translated into commercial practice. There are about 100 plots each of between 1 and 3 hectares throughout the country which will give estimates of seed yield and utilisation by various farm animals. Currently only spring lupins are included but winter lupin trials are planned for 1982/83.

G. Dardenne *(Belgium)* Is silage production from lupin being assessed?

C. Calet I have heard nothing about silage from lupins, but results for pea and *Vicia faba* silage are good, with high palatability.

J. Picard There is one experiment with silage from lupins. The silo has only just been opened and it looks satisfactory, but no feeding results are available yet.

FINAL SESSION

Chairman: C. Calet

FINAL SESSION

C. Calet *(France)* The following conclusions relating to future research needs have been drawn up by the Chairmen of the sessions here, largely in response to findings reported at this meeting

As a generalisation, we consider that peas, faba beans and lupins should attract continued support, with the emphasis placed on each depending on the requirements in the regions or countries best suited to their culture.

Of the three suggested recommendations for consideration by the EC Standing Committee on Agricultural Research, the first proposal is a short term one; to improve yield, but more importantly, to improve yield stability of seed legumes.

The second proposal which has medium term objectives, should be to continue to make the best possible progress through modern techniques in genetics and breeding. This would be facilitated by improved knowledge of the physiology and metabolism of the plant. This is so, particularly with regard to energy conservation, for example in relation to nitrogen and phosphate utilisation. Symbiotic nitrogen fixation through *Rhizobium*, and mycorrhizal association which enable a better utilisation of phosphate are the most important such areas for attention. This work would of course be relevant also to other legumes including those used for forage. Perhaps a future seminar could be devoted to this subject.

The third proposal is long term in nature and is largely concerned with genome transformation through sophisticated techniques relevant to genetic engineering. As emphasised during the seminar, peas are likely to be especially well suited to such studies because of the accumulated knowledge of their genetics, cytogenetics and physiology. Urgent European involvement is needed to maintain the momentum generated so far.

B. Snoad *(UK)* asked whether the inclusion of additional or more clearly identified topics to the list of three general areas was possible, and if not,what the mechanism is for making specific recommendations.

C. Calet informed the meeting that it seemed prudent to restrict the number of topics and present only a general format for consideration by the Directorate-General for Agriculture in Brussels. Specific topics should be notified to a national representative on the Expert Committee - in the case of UK that would be Dr. Miflin or Dr. Watkins.

J. Picard *(France)* wished to ensure that there was general agreement for the three proposals which emerged from the Chairmen's meeting. It was important to decide whether equal priority should be given to each.

M.A. Spadoni and P. Cerletti *(Italy)* agreed with the general conclusions but thought a specific reference should be made to the need for a closer link between food technology and other disciplines.

J. Picard pointed out that food technology was only partly the responsibility of the Protein Expert Group, and that another group dealing with human foodstuffs also had responsibility in this area.

P. Guldager *(Denmark)* observed that with representatives from such a broad spectrum of disciplines there was a good opportunity for co-operation between disciplines, especially with reference to standardisation of procedures e.g. chemical techniques, cultivars, methods of measurement etc.

C. Calet said that this was the sort of comment that provided much needed background information to help in committee in Brussels.

B. Snoad stressed that a major problem in pea experimentation was that of accurately determining grain yield and protein yield. This aspect of technique required additional study if sensible comparisons were to be made between experiments made throughout Europe.

C. Calet thought this was a subject for future consideration. He then presented two proposals from Dr. Hardwick:

"1. Crop physiology: This meeting has shown that between us, we in Europe have enough physiological and biochemical information on *Pisum* for attempting a synthesis. For example, the light interception data of Thompson, the $^{14}CO_2$ data of Cousin and the N uptake data of Haahr might be put together in a quantitative statement (or model) describing yield production in *Pisum*. This would also need the information on biochemical levels of control which is coming from laboratories in Durham, Norwich and elsewhere."

"2. Seminar format: It would be good if meetings such as this could have 'discussion' or 'workshop' sessions. Time could be made available if some paper-reading sessions were replaced by poster sessions."

LIST OF PARTICIPANTS

Dr. V. Abbate
Istituto di Agronomia
Università di Catania
Via Valdisavoia 5
95100 Catania
ITALY

Dr. G. Barbieri
Istituto di Agronomia
e Coltivazioni Erbacee
Università di Napoli
Facoltà di Agraria
80055 Portici
ITALY

Dr. F. Basso
Istituto di Agronomia
e Coltivazioni Erbacee
Università di Napoli
Facoltà di Agraria
80055 Portici
ITALY

Dr. A. Borrelli
Istituto di Agronomia
e Coltivazioni Erbacee
Università di Napoli
Facoltà di Agraria
80055 Portici
ITALY

Prof. D. Boulter
Department of Botany
University of Durham
Science Site
South Road
Durham
UK

Dr. C. Calet
INRA
Rue de Grewelle
75341 Paris
FRANCE

Dr. A. Callebaut
ISO
Museumlaan 5
Tervuren
BELGIUM

Dr. E. Carnovale
Istituto Nationale
della Nutrizione
Via Ardeatina 546
00179 Rome
ITALY

Dr. F. Carone
Istituto di Agronomia
e Coltivazioni Erbacee
Università di Napoli
Facoltà di Agraria
80055 Portici
ITALY

Dr. R. Casey
John Innes Institute
Colney Lane
Norwich NR4 7UH
UK

Dr. D. Ceci
UFF Richerche Agricole
COVALPA
Viale Gramsci 39
41037 Mirandola (Modena)
ITALY

Dr. P. Colonna
INRA Nantes
Chemin de la Geraugière
44072 Nantes Cédex
FRANCE

Dr. M. Cousin
Station de Génétique
et d'Amélioration des Plantes
INRA
Route de St. Cyr
78000 Versailles
FRANCE

Dr. L. Cuoculo
Istituto di Agronomia
e Coltivazioni Erbacee
Università di Napoli
Facoltà di Agraria
80055 Portici
ITALY

Dr. G. Dantuma
Centre for Agrobiological
Research
PO Box 14
6700 AA Wageningen
NETHERLANDS

Dr. G. Dardenne
Lab. Chimie organique
et biologique
Faculté des Sciences Agronomiques
5800 Gembloux
BELGIUM

Dr. R. Davies
John Innes Institute
Colney Lane
Norwich NR4 7UH
UK

Dr. M. Duranti
Istituto Biochemica Generale
Via Celoria 2
Milan
ITALY

Dr. A. Errico
Cattedra di Miglioramento
Genetico
Istituto di Agronomia
e Coltivazioni Erbacee
Università di Napoli
Facoltà di Agraria
80055 Portici
ITALY

Dr. D. Fior
Amon Fitochimica
Via Piave 2
Bolzano
ITALY

Dr. M. Frauen
Institut für Planzenbau
und Pflanzenzüchtung
Universität Göttingen
V. Siebold Strasse 8
D-3400 Göttingen
FEDERAL REPUBLIC OF GERMANY

Dr. L. Frusciante
Cattedra di Miglioramento
Genetico
Istituto di Agronomia
e Coltivazioni Erbacee
Università di Napoli
Facoltà di Agraria
80055 Portici
ITALY

Dr. D.W. Griffiths
Welsh Plant Breeding Station
Plas Gogerddan
Aberystwyth
Dyfed
Wales
UK

Dr. S. Grillo
Cattedra di Miglioramento
Genetico
Istituto di Agronomia
e Coltivazioni Erbacee
Università di Napoli
Facoltà di Agraria
80055 Portici
ITALY

Dr. P. Guldager
Royal Veterinary and Agricultural
University
Department of Genetics
Bülowsvej 13
DK 1870 Copenhagen
DENMARK

Dr. V. Haahr
Agricultural Research Department
RISØ National Laboratory
DK 4000 Roskilde
DENMARK

Dr. R. Hardwick
National Vegetable Research
Station
Wellesbourne
Warwick CV35 9EF
UK

Dr. C.L. Hedley
John Innes Institute
Colney Lane
Norwich NR4 7UH
UK

Dr. R.J. Heringa
Foundation for Agricultural
Plant Breeding
PO Box 117
6700 AA Wageningen
NETHERLANDS

Dr. F.H. Jensen
Pajbjergfonden
Grindsnabevej 25
Dyngby
DK 8300 Odder
DENMARK

Dr. G. Kapsiotis
Consiglio Nazionale
della Richerche
Via Benedetto Musolino 41
00153 Rome
ITALY

Dr. C. Knott
Processors and Growers Research
Organisation
The Research Station
Great North Road
Thormhaugh
Peterborough
UK

Dr. L. Lacassagne
INRA
Station de Recherche Avicoles
Nouzilly
37380 Monnaie
FRANCE

Dr. Lassandro
Amministrazione Provinciale
Sivan SPZ
Milan
ITALY

Dr. M. Lenoble
INRA
Station d'Amelioration
des Plantes
86600 Lusignan
FRANCE

Dr. P. Levillain
Universite de Tours
Faculte de Pharmacie
2bis Boulevard Tournelle
BP 3213
37032 Tours
FRANCE

Dr. D. Lucatini
Capo Divisioni Cereali
Via XV Settembre
00100 Rome
ITALY

Prof. L.M. Monti
Cattedra di Miglioramento
Genetico
Istituto di Agronomia
e Coltivazioni Erbacee
Universita di Napoli
Facolta di Agraria
80055 Portici
ITALY

Dr. F. Mucci
Istituto di Agronomia
e Coltivazioni Erbacee
Università di Napoli
Facoltà di Agraria
80055 Portici
ITALY

Dr. R.F. Murphy
The Agricultural Institute
Kinsealy
Malahide Road
Dublin 5
IRELAND

Dr. J. Picard
INRA
BV 1540
21034 Dijon Cédex
FRANCE

Dr. P. Plancquaert
ITCF
8 avenue du Président Wilson
75116 Paris
FRANCE

Dr. C. Pompei
Istituto Tecnologie Alimentari
Università di Milano
Via Celoria 2
Milan
ITALY

Dr. L. Postiglione
Istituto di Agronomia
e Coltivazioni Erbacee
Università di Napoli
Facoltà di Agraria
80055 Portici
ITALY

Dr. M.H. Poulsen
Department of Crop Husbandry
and Plant Breeding
Research Station Hojbakkegard
Agrovej 10
DK 2630 Tastrup
DENMARK

Dr. A.G. Prendergast
Commission of the European
Communities
200 rue de la Loi
1049 Brussels
BELGIUM

Dr. C. Rubini
CERAC
Via Calzoni 1/3
Bologna
ITALY

Dr. F. Saccordo
Laboratorio Colture Industriali
CNEN
St. Maria di Galeria
Rome
ITALY

Dr. G. Sargiani
SIVAM
Via Scarlatti 30
Milan
ITALY

Dr. Christine Sator
FAL PZ
Bundesallee 50
D-3300 Braunschweig
FEDERAL REPUBLIC OF GERMANY

Dr. G.T. Scarascia-Mugnozza
Facoltà di Agraria della Tuscia
Viterbo
ITALY

Dr. M.G. Schiratti
Commission of the European Communities
200 rue de la Loi
1049 Brussels
BELGIUM

Dr. M.J.A. Simpson
Department of Agricultural Botany
University of Reading
Whiteknights
Reading RG6 2AS
UK

Dr. B. Snoad
John Innes Institute
Colney Lane
Norwich NR4 7UH
UK

Dr. A. Sonnino
CNEN
Laboratorio SVILAGR
CSN Casaccia
SP Anguillarese
Km 1 + 300 Rome
ITALY

Dr. Sotti
Amministrazione Provinciale
Sivan SPZ
Milan
ITALY

Dr. M.A. Spadoni
Istituto della Nutrizione
Via Ardeatina 546
00179 Rome
ITALY

Dr. R. Thompson
Scottish Crop Research Institute
Invergowrie
Dundee
UK

Dr. P. van Oostveldt
Lab. Biochemie
Fak. Landbouw RUG
Coupure Links 533
9000 Ghent
BELGIUM

Dr. E.J. Walsh
University College Dublin
Lyons
Newcastle
Co. Dublin
IRELAND

Prof. W. Williams
Department of Agricultural Botany
University of Reading
Whiteknights
Reading RG6 2AS
UK

MANUSCRIPT PREPARED BY:

JANSSEN SERVICES
33a High Street
Chislehurst
Kent BR7 5AE
UK